PHYSICS—*Once Over—Lightly*

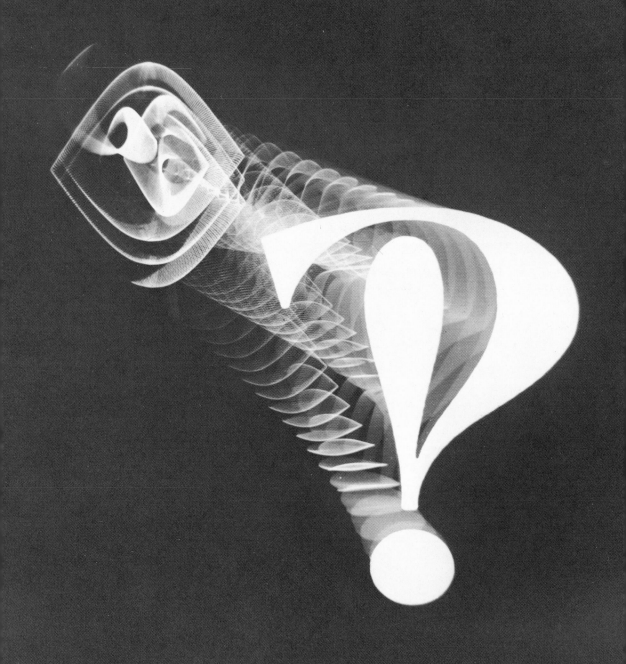

PHYSICS—*Once Over—Lightly*

KENNETH R. ATKINS, *University of Pennsylvania*

JOHN WILEY & SONS, INC. *New York* · *London* · *Sydney* · *Toronto*

Library of Congress Catalogue Card Number: 76-177880

ISBN 0-471-03622-6

Printed in the United States of America.

10 9 8 7 6 5 4 3 2 1

The Interabang used as the basis of the computerized graphics of the frontispiece and the cover was introduced in 1967 by the American Type Founders Co. as the first new punctuation mark to be devised for print since the adoption of quotation marks in the late 17th Century. It is intended to express a simultaneous quality of exclamation and questioning. To the author it seems to symbolize perfectly his own personal philosophy and the attitude to physics embodied in this book. Exclamation is the appropriate reaction to physical theory, with its beauty, elegance, strangeness, and depth of revelation. Nevertheless, since physics is a science, this must always be accompanied by a questioning—questioning about meaning—questioning about significance—questioning about further consequences—and, perhaps the most important of all, a continuous critical questioning of its truth.

COVER & FRONTISPIECE PHOTOGRAPHS: BOB BLANSKY, DOLPHIN COMPUTER IMAGE CORPORATION

Preface

This is a modified version of my book *Physics* shortened and simplified to make it more suitable for a one-semester or one-quarter course. It is much more than just the earlier book operated on with scissors and paste. Many sections have been rewritten. Some new sections have been introduced. The mathematical demands on the reader are even less than they were in the longer version. Trigonometry has been completely eliminated and only the most elementary aspects of algebra and geometry are required. Examples of problem solving have been removed from the main text, although simple problems are still included at the end of each chapter to be used or ignored by the instructor at his own discretion.

Anyone who has taught a short, self-contained course covering the basic concepts of physics knows that it presents an especially difficult challenge. This book meets that challenge in the following way. I have attempted to include no more material than can reasonably be covered in the time available. I have achieved this basically by eliminating the less essential topics. For example, Maxwell's equations are still included, but the electric generator has been omitted. The abbreviation has not been carried to the point of eliminating the essense of the essential. To illustrate what I mean by this, a reader pursuing the claim that Maxwell's equations are included will not find in the text merely an isolated sentence such as *"In 1865 the British physicist James Clark Maxwell summarized electromagnetic theory in a set of elegant mathematical equations."* This is history with little physical content. Instead, turning to Section 7-7, the reader will find a descriptive nonmathematical discussion of Maxwell's equations and an attempt to explain their physical significance. In Section 8-3, he will find a discussion of the relationship between these equations and the phenomenon of electromagnetic waves.

Certain characteristic features of the longer version of the text have been retained. CGS units are used for reasons that the author is prepared to advocate vehemently at great length, but not here. In the diagrams, vectors representing different physical quantities have been distinguished from one another by the use of several different kinds of arrows, and throughout the book a particular vector quantity is always represented by the same kind of arrow. It is not necessary for the student to memorize this vector code, but it may improve the clarity of the diagrams as he subconsciously comes to realize that, for example, a certain kind of broken arrow always represents a velocity. The student is encouraged to use powers of ten instead of special units or prefixes. If the wavelength of an x-ray is expressed as 5×10^{-8} cm and the wavelength of visible light as 5×10^{-5} cm, their relative magnitudes are immediately obvious to a student who has expended the small amount of effort needed to understand the procedure. The alternative practice of expressing the x-ray wavelength as 5 Å and the light wavelength as 500 mμ is much more confusing, particularly for nonscientists whose experience of physics may be almost entirely confined to this one course.

The most obvious application of this book is to a short, terminal course for nonscientists, but it may find other uses in these days of educational innovation. I have used it myself for the first semester of a two-semester course given to business students. It provided an appropriate background for the second semester, when the basic physical concepts were applied to some exciting aspects of astronomy such as cosmology, stellar evolution, quasars, pulsars, and black holes.

Philadelphia
May, 1971

Kenneth R. Atkins

Contents

PHYSICS—*Once Over—Lightly*

1 *The Way Ahead*

1-1 *Describing the World Around Us*

Physics is an attempt to describe, in as fundamental and penetrating a way as possible, the nature and behavior of the world around us. Before dismissing this sentence as an abstract and profitless philosophical statement of the type that authors must use to get their books smoothly under way, the reader should accept the challenge, lift his, or her, eyes from the page, and look at the world around. As I do this myself, I am first conscious of a multitude of shapes and colors; a subtle interplay of patches of color on the wallpaper; an expanse of bright red carpet; through the window, the intricate shape of a green tree against a flat expanse of blue sky. On the wall is a reproduction of Picasso's "Boy with a Horse." On the bookshelf the most conspicuous item is a bulky *Complete Works of Shakespeare*. From the record player I hear the sounds of Schubert's Octet in F Major. Whatever the circumstances surrounding the reader, he will probably not have to look far to find things equally complicated and marvellous, and certainly the mind

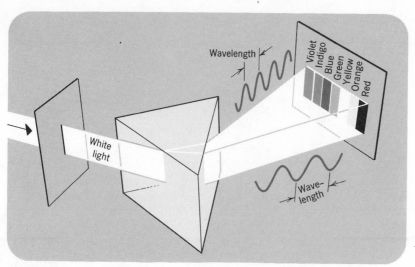

Figure 1-1 *A glass prism resolves white light into a spectrum of basic colors. The different colors correspond to different wavelengths.*

that he is using to consider these matters is an example of one of the most complicated and marvellous structures that can be contemplated.

Obviously, a complete description of the world around us must include the tree, the mind of man, the art of Picasso, the plays of Shakespeare, and the music of Schubert. Even more obviously, these things are not the primary concern of physics. However, if I lift my Picasso off its hook and then let it go, it falls to the ground. The same thing happens with the *Complete Works of Shakespeare* and the record of Schubert's Octet. Moreover, if all three objects are held at the same height and released at the same time, they all strike the ground at the same time (ignoring small, almost imperceptible differences in falling time mainly due to the resistance of the air). This is the sort of thing that is the concern of physics. Whatever is relevant to the aesthetic appeal of the Picasso is clearly irrelevant to the rate at which it falls to the ground, and we are led to make a very general, fundamental statement that: *"All unsupported bodies fall toward the center of the earth at the same rate."* Of course we shall have to be more precise about the meaning of the phrase "at the same rate," and also we might be well advised to specify that the bodies are to fall through an evacuated space, in order to eliminate the effect of the air which, in an extreme case, makes a balloon rise. But when, following in the steps of Newton, we have refined the idea to the point where the statement becomes: *"Any two bodies attract one another with a force that is proportional to the product of their masses and inversely proportional to the square of the distance between them,"* then we have arrived at one of the most important and basic laws of physics, Newton's law of universal gravitation.

The patches of color that are an essential part of our immediate experience can be analyzed by means of an instrument known as a spectroscope, which is essentially a prism of glass. If white light, such as the sunlight now streaming in through my windows, falls upon this glass prism, it is spread out into a spectrum of the colors of the rainbow, as shown in figure 1-1. All the rays of light are bent by the prism, but the red light is bent least and

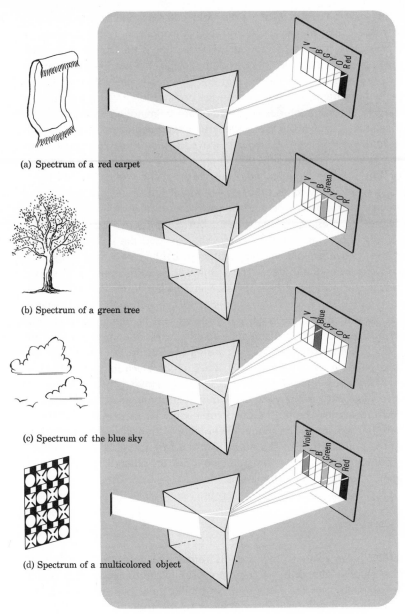

(a) Spectrum of a red carpet

(b) Spectrum of a green tree

(c) Spectrum of the blue sky

(d) Spectrum of a multicolored object

Figure 1-2 *The glass prism analyzes various samples of light into basic color components.*

the violet light most, the other colors coming in between, as shown in the diagram. The colors of this spectrum are the basic components of all visible light. If the spectroscope were arranged to accept only light from my red carpet, then only the red region of the spectrum would be illuminated; if the light came from the green tree, the spectrum would be illuminated mainly in the green region; and if the light came from the blue sky it would be the blue region that would be illuminated (see figure 1-2). However, if the spectroscope were made to accept light from my multicolored wallpaper, then several regions of the spectrum would be illuminated, as shown in figure

1-2d. In this way, any sample of visible light can be analyzed into some or all of the basic colors of the spectrum mixed in appropriate proportions.

We are again making the sort of analysis that is the concern of physics, but physics is not satisfied to stop at this stage. It must penetrate more deeply into the heart of the matter and ask, "What is the nature of light, and what is the difference between various basic colors?" This particular question led to several centuries of speculation, culminating in the early nineteenth century with a burst of ingenious experimental activity that provided the following answer: *"Light is a wave, similar to a wave on the surface of the ocean."* The quantity that characterizes the color is the **wavelength,** or the distance between adjacent crests of the wave. For red light this distance is about twice as great as for violet light (see figure 1-1).

There is much more to be said about the nature of light. An ocean wave requires the presence of the ocean, but a light wave can travel through empty space, as it does between the sun and the earth. The explanation of this, provided later in the nineteenth century, was that *"Light is a wave in an electromagnetic field."* Now the analysis has penetrated so deeply that it can no longer be expressed in familiar terms and the uninitiated reader cannot be expected to understand even vaguely what is meant. The explanation is in fact so lengthy that it will not be attempted until later in this book. Even so, it is a nineteenth century explanation, and today a physicist would rather say that *"Light consists of particles (or quanta or photons) whose behavior is governed by a wave in an electromagnetic field,"* all of which we shall attempt to explain later on. Having thus overreached ourselves to illustrate how physics is always penetrating more and more deeply into the mystery of the behavior of things, let us also emphasize again the other aspect of the matter. Whatever is true in these statements about light is equally true of the light from the red carpet and the light from the painting by Picasso. The statements are an attempt to find something common to all visual experiences.

1-2 Is It True?

It is generally believed that physics deals with indisputable facts and absolute truth. Actually, although the facts are usually indisputable, they are frequently incomplete, and the interpretation of the incomplete facts leads to theories that are only rough approximations to the truth. The situation can be fully appreciated only when one has an understanding of the whole of physics, but we shall attempt to illustrate it by reference to the atomic theory with which, it is safe to assume, the reader already has some familiarity.

In the nineteenth century, when this theory was progressing from triumph to triumph in its clarification of chemistry and the behavior of gases, its proponents might have expounded it in the form of the following postulates.

Postulate A

Matter cannot be divided indefinitely into smaller and smaller pieces: at a certain stage it will be found to consist of very small, submicroscopic entities called atoms, which cannot be divided any further.

Postulate B

Atoms cannot be created or destroyed.

Postulate C

All the atoms of a particular chemical element have identical properties; for example, they all have the same mass.

Postulate D

Atoms of different chemical elements have different properties; for example, they have different masses.

Postulate E

An atom of one chemical element cannot be changed into an atom of a different chemical element.

Now let us examine these postulates in the light of modern knowledge, pointing out the extent to which they are false, but also (to avoid generating too sceptical an attitude) pointing out the extent to which they are good, useful approximations to the true situation.

Postulate A

Atoms can be further subdivided into small fundamental particles known as electrons, protons, and neutrons. It is an open question whether these fundamental particles themselves can be subdivided still further. However, an atom is a stable, compact entity that can remain unchanged for long periods of time, and exceptional measures have to be taken to break it up into its constituent parts.

Postulate B

Modern physicists are aware of processes during which atoms are created out of energy, and reverse processes during which atoms are destroyed and changed into energy. However, these processes are very rare in everyday experience.

Postulate C

1-2 Is It True   The atoms of a particular chemical element can occur in different forms, known as isotopes, with different masses. The element is determined by the number of electrons in the atom (which is the same as the number of protons), but the number of neutrons can then vary to give various isotopes. However, each element has only a small number of stable isotopes and the naturally occurring form of the element usually contains these stable isotopes in fixed proportions.

Postulate D

It is possible for an isotope of one element to have almost the same mass as an isotope of another element. The two atoms are then called isobars. However, isobars do have slightly different masses, and so this postulate is strictly true.

Postulate E

It is well known that nuclear physicists can readily convert an atom of one element into an atom of another element. This process is basic to the performance of a nuclear reactor and an atomic, or hydrogen, bomb. However, the process is not very common in everyday life and is certainly not relevant to ordinary chemical reactions.

In the course of this book we shall discuss several cases of well-established theories that have been found to be not quite true and have had to be modified to bring them a little closer to the truth. An outstanding example is Newtonian mechanics, which is quite adequate to describe the motion of bodies that are not moving extremely fast, and is therefore all that is needed to predict the path of a guided missile, but which breaks down when the bodies begin to move with velocities comparable with the velocity of light and has to be replaced by Einstein's special theory of relativity. It follows inescapably that our best modern physical theories are probably only rough approximations to the truth and will eventually have to be replaced by something better. For example, the special theory of relativity does not cope adequately with the subject of gravitation, and so Einstein was led to formulate the general theory of relativity. However, there are very few experimental tests of the general theory and it has not been accepted with the same degree of confidence as the special theory. Many physicists believe that in this particular field there will be some surprising developments during the next few decades.

Physics must therefore be regarded as an evolving subject approaching closer and closer to the truth but never quite attaining it.

1-3 Classical Physics and Modern Physics

During its evolution, physics has passed through two major phases, which are commonly referred to as classical physics and modern physics. Because these two phases differ so radically, in their attitude toward the nature of the universe and in their philosophical implications, their separation is a profitable one that we shall continually emphasize. In order to give the reader some perspective on the detailed explanations of the succeeding chapters, we shall now present a very brief summary of the evolution of physics.

Classical physics started in earnest in the seventeenth century when Galileo and Newton discovered the laws governing the motion of bodies. It came to regard the universe as a collection of isolated bodies separated by regions of empty, featureless space. The bodies exerted forces on one another in spite of the lack of any obvious direct connection through the intervening space. At first, the bodies discussed were the sun, the planets, the earth's moon, the moons of the other planets, bodies falling to the earth, cannonballs shot from cannons, and so on. The forces were initially gravitational forces, electric and magnetic forces, and certain incompletely understood forces such as the upward force that the top of a table exerts on a plate to prevent it from falling under gravity.

In the nineteenth century the idea that all matter is composed of **atoms** was fully accepted, and the universe then came to be regarded as a collection of isolated atoms moving through empty space and exerting forces on one another. This point of view seemed capable of providing a complete explanation of all the phenomena associated with heat, sound, electricity, magnetism, and the various properties of matter, such as elasticity, viscosity, and surface tension. However, the nature of electricity and the origin of the interatomic forces remained obscure until early in the twentieth century when it was discovered that atoms are themselves complicated structures built up out of **fundamental particles.** The most abundant and most important of these fundamental particles are the **electron,** which is a very light particle and the unit of negative electricity; the **proton,** which is a more massive particle than the electron and the unit of positive electricity; and the **neutron,** which has about the same mass as the proton, but no electric charge. With the discovery of these particles it became necessary to modify the picture of the universe and to think in terms of electrons, protons, and neutrons moving through empty space. This promised to give an even more complete and fundamental description of physical phenomena.

An important feature of this view of the universe was the implication that once the fundamental laws of physics had been formulated, it would be possible to describe completely the motion of the particles and therefore to predict the state of the universe at all future times. This view therefore encouraged the belief that everything that will happen in the future is determined by what has happened in the past and that our lives are governed by Inescapable Destiny. The two capital letters are used to give the phrase an ominous ring and the rest is left to the imagination of the reader.

This brief summary of the situation is an oversimplification, and not all classical physicists would have subscribed to all the above views, particularly the philosophical implications. Moreover, there is an alternative view of the universe that has received considerable attention. This view refuses to regard space as empty and featureless, and prefers to endow all points in space with physical properties. It is sometimes expressed in the extreme form that all of space is filled with a continuous, indivisible fluid frequently called the **ether.** This point of view received considerable support at the beginning of the nineteenth century when Young and Fresnel performed a series of decisive experiments to demonstrate that light is a wave, a wave that can travel through "empty space." At the same time Faraday performed some crucial experiments on the nature of electricity and magnetism. He came to the conclusion that all the space surrounding an electric charge must be visualized as a field of electric force somewhat similar to a continuous all-pervading liquid (the ether) in a state of strain and that, similarly, the region surrounding a bar magnet must be regarded as a field of magnetic force. This implied that every point in space, whether it is occupied by matter or not, is associated with two quantities that specify the nature of the electric and magnetic fields at that point.

This concept of an **electromagnetic field** proved very profitable and the difficult subject of electricity and magnetism yielded to theoretical treatment during the second half of the nineteenth century, when Maxwell expressed the fundamental equations of the subject, not in terms of the behavior of electrically charged particles, but in terms of the behavior of the electric and magnetic fields at all points in space. One consequence of these equations is that they predict the existence of waves that travel through the electromagnetic field. Maxwell was able to show that these waves have all the known properties of light. Light is therefore basically an electromagnetic phenomenon. As the nineteenth century neared its end, Hertz succeeded in generating electromagnetic waves of long wavelength, which we now call radio waves, and he found that their velocity was exactly equal to the velocity of light.

All this emphasis on the behavior of light raised the interesting question "How does the velocity of light depend upon the velocity of the source of the light and the velocity of the observer who sees the light?" The search for an answer to this apparently simple question produced **Einstein's theory of relativity** and drastic modifications in Newtonian mechanics, which had been the basis of classical physics. Einstein presented his theory in 1905. In 1900, Planck had taken the first step in the direction of quantum mechanics. The dawn of the twentieth century thus coincided almost exactly with the dawn of modern physics.

Planck's idea was that light is not emitted continuously but in little bundles of energy called **quanta.** Evidence soon accumulated that light, which had been so firmly established in the nineteenth century as a wave motion, frequently behaves like a stream of particles. To complete the dilemma, electrons, protons, atoms, and molecules, which behaved obviously like particles in many experiments, were found in other experiments to manifest unmistakable properties of wave motion. The answer to this riddle of the duality of particle behavior and wave behavior is contained in the elaborate mathematical formalism of **quantum mechanics.** At its heart is the famous **Heisenberg uncertainty principle,** which tells us that even if an entity like an electron is to be regarded as a particle, it is impossible to know both the exact position and the exact velocity of this particle. The role of the wave associated with the particle is to tell us the probabilities of the particle's being found in various places. Probability and chance then lie at the core of our description of the universe. It becomes impossible to predict precisely any future state of the universe; one can merely state the probabilities of various possible occurrences. Inescapable Destiny is replaced by Fickle Chance!

Modern physics includes both particles and fields. The number of types of fundamental particles has increased rapidly in recent years and it is relevant to ask if they are all fundamental or whether something simpler is hidden beneath. Some elegant relationships are beginning to reveal themselves among these fundamental particles, and the next giant step in the evolution of physics may not be far ahead.

The Way Ahead

1-4 *The Limitations of This Book*

The reader is now beginning to realize what this book is all about. It places the emphasis on "pure" physics rather than "applied" physics. Although applied physics is not part of our main purpose, we do not wish to underestimate the importance of applied physics in furthering the welfare of man and improving his control over his environment. Moreover, pure and applied physics have always been inseparable. To mention only one example, the discovery of the electron was a consequence of the simple desire to understand the inner working of nature, but it led inevitably to the development of electronic devices, radio communication, radar, electronic automation, and large computers. Conversely, most of modern experimental physics, however "pure" its motivation, would be impossible without these electronic devices.

Even within the confines of what would normally be considered pure physics, this book will always place the emphasis on the fundamental principles and will rarely be able to do justice to the way in which these principles have been applied to describe the detailed behavior of matter. For example, we shall consider the structure of the atom and the principles of quantum mechanics which govern the atom's behavior, but it will not be possible to consider in detail all the elements in the periodic table and the way in which the electronic structure of each atom determines the physical and chemical properties of each element. Neither shall we attempt to describe how in recent years the fundamental principles have achieved enormous successes in the explanation of the detailed properties of solids. Again, it is not our intention to underestimate the interest or importance of these matters; but they are complicated, and to discuss them properly would distract us from our main purpose of describing the foundations on which the detailed explanations are built.

Modern physics will be given the attention it deserves, but has not always received, in a comprehensive survey of physics. This does not mean, however, that we shall sweep aside the errors of the past and present only the latest fashion in physical theory. Modern physics is only a halfway house toward the ultimate truth and it should not be presented with an air of finality. Also, since physics is an evolutionary progress towards better and better understanding, modern physics is a structure built on top of classical physics, and it would not be possible to give a simple exposition of modern physics without first having discussed classical physics. We shall therefore adopt a type of historical approach, but not in the form of a narrative arranged in strict chronological order. Even when discussing the physics of the seventeenth century we shall assume a knowledge of the physics of the twentieth century. The policy of King George III toward his American Colonies takes on its full significance in the light of the events of the nineteenth and twentieth centuries. Similarly, Newtonian mechanics is revealed in its full power and elegance when allied with the atomic theory of the nineteenth century, and its limitations are realized in the light of the twentieth century developments embodied in the theories of relativity and quantum mechanics. The reader should therefore be cautioned that this pseudohistorical approach is aimed purely at clarity of exposition and may therefore create some false impressions of how physical theory did in fact evolve. The reader who values the fullness of his education would be well advised to read some of the many admirable books on the history of science.

Above all else is the hope that the reader will be left with an overwhelming feeling of the intriguing mystery underlying the nature of existence and the beauty and elegance of man's attempt to understand this mystery. Physical theory has an aesthetic appeal that bears comparison with the paintings of Picasso, the plays of Shakespeare, or the music of Schubert.

1-4 The Limitations of This Book

Question

You no doubt start out with a preconceived attitude toward physics, based on a previous course you have taken or on what you have heard and read about the subject. Make a careful, introspective analysis of your reasons for this attitude. Can you justify it convincingly to yourself, your classmates, and your instructor?

Problems

Before embarking on the detailed discussion of the succeeding chapters, check that you are familiar with the necessary mathematical techniques. Read the Mathematical Appendix and attempt the problems given there.

*The Way
Ahead*

2 *Motion*

2-1 *The Universe as a Collection of Particles*

Scientists of the seventeenth century were particularly interested in astronomy, and it is therefore easy to understand why some of them regarded the universe as a collection of isolated bodies separated by vast regions of empty space. The appearance of the sky on a clear night illustrates this well (figure 2-1). The stars are points of light whose diameters are clearly very much smaller than the distances between them, and most of the space in the sky appears to be empty. The actual situation is even more impressive than it appears to be. The distance from the sun to the nearest star, Alpha Centauri, is about thirty million times the diameter of the sun! If the sun were represented by a small black dot the size of a printed letter "o" on this page, then Alpha Centauri would have to be represented by a similar dot 20 miles away! Newton was not aware of this fact, since the first measurement of the distance of a star was not made until the year 1838, but he was aware of the distances involved in the solar system.

Figure 2-1 A portion of the night sky. *"Isolated bodies separated by vast regions of empty space."* (*Mount Wilson and Palomar Observatories.*)

Let us consider how the solar system would appear to an observer viewing it from outside. If we shrink the scale until the sun is about the size of a football placed in the center of a football field, the planets can then be represented roughly as follows: Mercury is a grain of sand at a distance of about 10 yards from the sun; Venus is an apple seed at a distance of about 20 yards from the sun; the earth is another apple seed at a distance of about 25 yards from the sun; Mars is a poppy seed at a distance of about 40 yards from the sun, just inside the field; Jupiter is a golf ball somewhere in the stands; Saturn is a table tennis ball in the parking lot outside; Uranus is a pea three blocks away; Neptune is a pea about half a mile away; and Pluto is a grain of sand about three quarters of a mile away. The moon would be represented by a grain of sand about $2\frac{1}{2}$ inches from the apple seed that represents the earth.

For some of the purposes of astronomy, then, the earth may be treated as a small particle in space—but we all know that it is much more complicated than that. What about the air, the rocks, the oceans, and the myriad forms of complicated living organisms that teem over the surface of the earth? Modern physics tells us that any object, however complex, can be visualized as a sufficiently complicated structure built out of atoms, and that the atom itself is a structure composed of the three fundamental particles, electrons, protons, and neutrons. The protons and neutrons are packed tightly together into the nucleus, which occupies a region of diameter about 10^{-12} cm at the center of the atom. The electron has a diameter of about

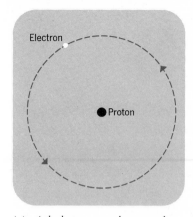

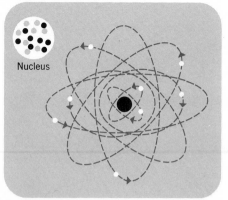

(a) *A hydrogen atom has a nucleus consisting of a single proton, and a single electron is revolving about this nucleus.*

(b) *In the oxygen atom the nucleus contains eight protons and eight neutrons tightly packed together. Eight electrons revolve about this nucleus.*

Figure 2-2 *The structure of the hydrogen and oxygen atoms.*

10^{-13} cm and is about 10^{-8} cm from the nucleus. The atom is therefore about 10,000 times larger than the nucleus and about 100,000 times larger than the electron. If the nucleus were represented by a football placed in the center of a city, the electrons would be tennis balls bouncing around in the suburbs! The atom, like the solar system, is mainly empty space.

Figure 2-2 shows the structure of two atoms; the simplest atom, hydrogen; and a slightly more complicated atom, oxygen. By bringing together two atoms of hydrogen and one atom of oxygen, and by modifying the orbits of some of the electrons, we obtain a molecule of water, as shown in figure 2-3. Using mainly atoms of carbon, hydrogen, oxygen, and occasionally atoms of other elements, we can build up complicated structures representing organic molecules and eventually, when the structures become extremely complex, living organisms. In this way one might hope to describe something as complex as a nerve cell in a human brain.

Consider, then, the following suggestion for giving a complete descrip-

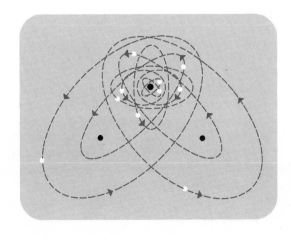

Figure 2-3 *The construction of a water molecule.*

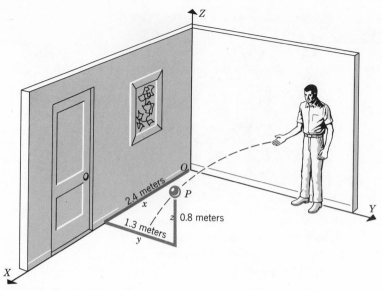

Figure 2-4 *Cartesian coordinates.*

tion of the whole universe. Assume that the universe is composed entirely of small particles such as electrons, protons, and neutrons. At a given instant of time, specify exactly the position of each particle in the universe. This gives a complete description of the instantaneous appearance of any structure in the universe. For example, given the positions of all the particles inside a human brain, we have a complete description of the state of that brain. Now repeat this procedure at all instants of time. Then, according to this viewpoint, everything that can be known about the universe is known. For example, by describing in this way the changing pattern of particles in the human brain, we might hope to have a complete description of all the thoughts passing through that brain.

It is not possible! In the year 1927 it was finally shown not to be possible. For three centuries before that, it might well have been. The developments of the year 1927 (the Heisenberg uncertainty principle) will not be discussed until Chapter 12. Meanwhile it will serve our purpose well, and considerably simplify the exposition, to assume that the universe *is* a collection of particles that *can* be located precisely at points in space.

2-2 Fixing the Positions of the Particles

How shall we set about this program of specifying the exact position of each particle at each instant of time? Suppose that I toss a piece of chalk across the room and ask where it is half a second after leaving my hand. A physicist cannot be satisfied by a vague statement such as: "The chalk is somewhere over there about midway between the painting and the door." He must be more precise, and he achieves precision by making use of numbers.

We can proceed in the following way. We can choose one corner of the room as our reference point and call it the **origin** O (see figure 2-4). We

Motion

14

can then say that, to move from the origin O to the piece of chalk, we must go 2.4 meters along the length of the room, then go 1.3 meters across the width of the room, and finally rise 0.8 meter above the floor.

If you have had sufficient previous mathematical training, you may recognize that this is equivalent to the mathematical procedure of choosing the three edges of the room, OX, OY, and OZ to be **Cartesian axes** (examine figure 2-4 carefully). The point P at which the chalk is located is specified by giving its three **Cartesian coordinates,** $x = 2.4$ meters, $y = 1.3$ meters, and $z = 0.8$ meter.

Notice that we need three numbers to fix the position of the chalk. There is no way of doing it unambiguously with only one or only two numbers. This is what is meant by saying that space has three dimensions.

However, there is a fourth number that is equally important. This is the time t (equal to 0.5 second) that has elapsed since the chalk left my hand. To describe the motion of the chalk completely, we must know the values of the three numbers x, y, and z corresponding to each value of the time t for all instants from the start to the end of the motion. This will then not only tell us about the shape of the curved path represented by a broken line in figure 2-4, but will also tell us how fast the chalk moves along the various parts of its path. Returning to our proposed description of the entire universe, at any instant of time t we must know the three Cartesian coordinates (x, y, z) for every single particle in the universe. Moreover, we must have this information for all instants of time. Then, according to this idea, we know everything there is to be known about the universe. Do you believe that?

2-3 The Units of Mass, Length, and Time

One aspect of the above discussion requires further consideration. We have made use of a unit of length, the meter, and a unit of time, the second. In the next chapter we shall need a unit of mass. How are these units defined?

To measure the length of the line on this page in centimeters, you would take a rule graduated in centimeters, lay it alongside the line, and adjust the zero mark on the scale to be opposite one end of the line. You would then notice that the other end was opposite the mark labeled 9 and you would then say that the length of the line is 9 centimeters. There are two aspects of this statement. One is the *number 9*, and the other is the *unit of length*, the centimeter. The full significance of the statement is that you are provided with an agreed unit of length called the centimeter and nine of these units laid end to end have the same length as the line. The laying end to end is done by the manufacturer of the rule who has taken pains to see that the distance between adjacent marks is as nearly as possible equal to this agreed unit of length. The unit is just as important as the number. If the centimeter had originally been chosen to be just one half as long as it now is, then the length of the line would be 18 centimeters.

The unit of length used to be defined in a very practical way by reference to a certain platinum-iridium bar that is kept under very careful conditions at Sèvres in France. The distance between two scratches on the surface of this bar was *defined* as exactly 1 **meter,** or 100 **centimeters.** The "very careful conditions" are elaborately designed to ensure that the length of the bar does not change; for example, the temperature is kept constant to avoid thermal expansion or contraction of the metal. However, physicists have long been uneasy about the arbitrariness of this standard and the possibility that the distance between the scratches might vary slightly due to unknown factors. In 1960 it was finally decided by international agreement that the unit of length should be defined with reference to the wavelength of an orange line in the spectrum of atoms of a single, pure isotope of krypton (krypton 86). The wavelength is, of course, the distance between adjacent peaks of the wave (see figure 1-1). The meter is now defined to be exactly 1,650,763.73 times the wavelength of this particular krypton line. To the best of our knowledge, this is the present distance between the two scratches on the platinum-iridium bar.

In the same vault at Sèvres where the standard meter is kept there is a cylinder of platinum-iridium which, by international agreement, is taken to represent a mass of exactly 1 **kilogram** or 1000 **grams.** The physical nature of mass will be discussed in Chapter 3.

The unit of time is the **second,** which of course is related to the rotation of the earth on its axis. The earth rotates once in 1 day, which is equivalent to $24 \times 60 \times 60 = 86,400$ seconds. However, small erratic variations in the rate of rotation of the earth make it a very unsuitable basis for the definition of the unit of time, and so in 1967 a new international agreement based the definition of the second on the more fundamental, more reproducible behavior of an "atomic clock." In the type of atomic clock chosen, atoms of cesium 133 are made to vibrate by radio waves similar to those used in radar. The second is now defined as the time interval during which a cesium atom makes 9,192,631,770 such vibrations.

There are two major systems of units used in physics, the MKS system and the CGS system.

MKS System *or Meter-Kilogram-Second System*

The unit of length is the meter (m), and is 1,650,763.73 times the wavelength of the krypton-86 line. The unit of mass is the kilogram (kg), and it is the mass of the platinum-iridium cylinder kept at Sèvres. The unit of time is the second (sec), and it is the time interval during which a cesium atom vibrates 9,192,631,770 times in an atomic clock.

CGS System, *or Centimeter-Gram-Second System*

The unit of length is the centimeter (cm), and it is $\frac{1}{100}$ of a meter or 16,507.6373 times the wavelength of the krypton-86 line. The unit of mass is the gram (gm) and it is $\frac{1}{1000}$ of the mass of the platinum-iridium cylinder. The unit of time is again the second (sec).

Since the fundamental standards are the same in both systems, it might appear that there is no real difference between them. The point is that in the MKS system all lengths must be expressed in meters and all masses in kilograms, whereas in the CGS system all lengths must be in centimeters and all masses in grams. A velocity might be given as 3 meters per second in the

MKS system, but in the CGS system it would be 300 centimeters per second. The density of water is 1 *gram* per cubic *centimeter* in the CGS system, but 1000 *kilograms* per cubic *meter* in the MKS system. The two systems must never be mixed by expressing masses in grams and lengths in meters, for example, as in the highly undesirable statement that the density of water is 1,000,000 grams per cubic meter. It is also inadvisable to depart from either system, as in the expression of a velocity as 3 millimeters per second. This should be either 0.3 centimeters per second or 3×10^{-3} meters per second. Expressing a speed as 36 kilometers per hour is quite acceptable when driving an automobile in France, but when computations in physics are concerned, it should be changed to 10 meters per second (MKS system) or 1000 centimeters per second (CGS system).

The Golden Rule, then, is

> Decide at the very beginning of a calculation whether you are going to use the MKS or the CGS system, and stay with this decision. If it is the MKS system, express all lengths in meters, all masses in kilograms, and all time intervals in seconds. If it is the CGS system, express all lengths in centimeters, all masses in grams, and all time intervals in seconds.

(Throughout the book statements and equations of particular importance will be enclosed in a colored frame.)

The measurement of many physical quantities involves only measurements of mass, length, and time, and this fact can be used to express the units of these quantities. For example, a velocity is a length divided by a time and may therefore be written

MKS system: 7 meters per second, or 7 m/sec.

CGS system: 700 centimeters per second, or 700 cm/sec.

A density is a mass divided by a volume, and a volume is a length multiplied by a length multiplied by a length. The density of water may therefore be written

MKS system: 1000 kilograms per cubic meter, or 1000 kg/m^3.

CGS system: 1 gram per cubic centimeter, or 1 gm/cm^3.

2-3 The Units of Mass, Length, and Time

There is a third system of units, the **English system** or **gravitational system,** in which the unit of length is the **foot,** and the unit of time is the **second.** Instead of defining a unit of *mass* it defines a unit of *force* called the **pound.** This system is extensively used in engineering and in practical affairs, but almost never by physicists. Its use in an introductory text on pure physics would lead only to unnecessary confusion, with no compensating gain, so it will be studiously avoided. In this book preference will be given to the CGS system, mainly because this simplifies the presentation of the fundamental equations of electricity and magnetism. The MKS system will be needed occasionally, particularly in connection with the practical units of electricity, the ampere and the volt.

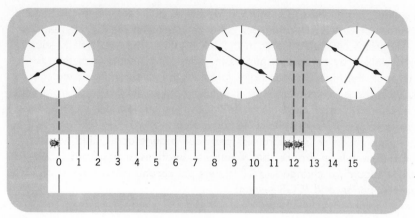

Figure 2-5 *Illustrating the definition of speed.*

2-4 Motion Along a Straight Line

With the help of figure 2-5, imagine a bug crawling along a ruler marked in centimeters. Suppose that you first see him as he crosses the zero mark. You then go away for 10 minutes and return to find him just crossing the 12 cm mark. You might say that he has traveled 12 cm in 600 seconds and so his speed is $\frac{12}{600} = 0.02$ centimeters per second. In actual fact he may have crawled to the 10 cm mark, rested there for five minutes and then resumed his journey with renewed vigor. Under these circumstances 0.02 cm/sec is an *average* speed for the ten-minute interval, but it bears no obvious relationship to the speed with which the bug started out or the speed with which he crossed the 12 cm mark.

   The thing to do is to watch the bug carefully for the next few seconds. Suppose that he moves from the 12.0 cm mark to the 12.5 cm mark in 5 seconds. Then the speed with which he crosses the 12 cm mark is 0.5/5 = 0.1 cm/sec. We are still assuming, of course, that he is not so perverse as to keep changing his speed even during these 5 seconds. The only way to be quite certain would be to time him over a very short time interval, say one hundredth of a second, giving his reflexes insufficient time to enable him to change his speed.

   We conclude that, when a body is moving with variable speed, the way to determine its speed at a particular instant is to measure the very short distance it travels in a very short time interval immediately following this instant. The interval of time should be as short as we can possibly make it and still obtain accurate measurements.

Motion

Instantaneous speed

$$v = \frac{\text{Very short distance traveled}}{\text{Very short time interval}} \tag{2-1}$$

with the time interval made as short as possible.

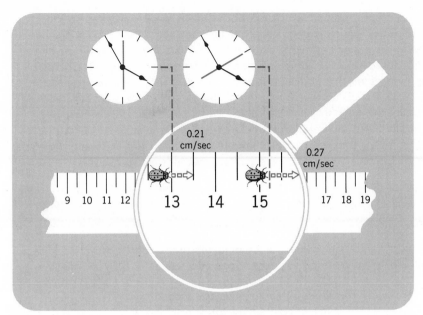

Figure 2-6 *Illustrating the definition of acceleration.*

When the speed of a body is changing, it is of course accelerating. The **acceleration** is the rate at which the speed is changing. Suppose you continue to observe your bug and, using the method just described, you determine that he crosses the 13 cm mark with a speed of 0.21 cm per second (figure 2-6). Ten seconds later you determine his speed again and find that it is 0.27 cm per second. The speed has increased 0.06 cm per second in 10 seconds. It is increasing by 0.006 cm per second every second, and so we say that the acceleration is 0.006 cm per second per second. This is often written 0.006 cm/sec^2 or, better still, 6×10^{-3} cm/sec^2.

The acceleration itself may not be constant. If you slowly move your foot up and down on the accelerator pedal of an automobile, you produce a varying acceleration. We should therefore use the same caution that we did when we defined speed. To determine the acceleration at a particular instant, we should measure the very small change in speed occurring during a very small interval of time.

Instantaneous acceleration,

$$a = \frac{\text{Very small change in speed}}{\text{Very short interval of time}} \cdot \qquad (2\text{-}2)$$

with the time interval made as short as possible.

The simplest case of accelerated motion in a straight line occurs when the acceleration is constant. The most important example of such a motion is a body falling freely under gravity. If the body falls in a vacuum, or if the conditions are such that the resistance of the air can be neglected, and if the total height through which the body falls is small compared to the earth's radius (which is about 6.37×10^8 cm), then, as the body falls its acceleration is always vertically downward and has the same value at all instants. Moreover, at a particular place on the earth's surface this acceleration is the same for all falling bodies, independently of their size, shape, or mass, and it is called the **acceleration due to gravity.** It is usually represented by the symbol g and its value is approximately 980 cm/sec^2 or 9.8 m/sec^2. We shall elaborate these points in Chapter 3.

Figure 2-7 shows a body falling vertically from rest. Its speed increases by g cm/sec every second and so, after t seconds, its speed, v is

$$v = gt \tag{2-3}$$

The student with limited experience of algebraic equations may find it helpful at first to translate all such equations into words.

> The speed in centimeters per second (v)
>   is numerically equal to ($=$)
> the acceleration due to gravity in cm/sec^2 (g)
>   multiplied by
> the time in seconds that has elapsed
> since the body was released from rest (t)

The speed is initially zero and steadily increases to gt in t secs. The average speed v_a during this t seconds is therefore $\frac{1}{2}gt$.

$$\text{Average speed, } v_a = \tfrac{1}{2}gt \tag{2-4}$$

The distance, y, through which the body has then fallen is the average speed multiplied by the time.

$$y = v_a t \tag{2-5}$$
$$\text{or } y = (\tfrac{1}{2}gt)t \tag{2-6}$$
$$\text{and so } y = \tfrac{1}{2}gt^2 \tag{2-7}$$

In words:

> The distance in cm through which the body has fallen (y)
>   is numerically equal to ($=$)
> one half of the acceleration due to gravity in cm/sec^2 ($\frac{1}{2}g$)
>   multiplied by
> the square of the time in secs that has
> elapsed since the body was released from rest (t^2)

Motion

2-5 *A Body Projected Horizontally*

In the previous section we discussed the straight line motion of a body falling vertically downward. We shall now extend the discussion by introducing a horizontal component of the velocity so that the path is curved.

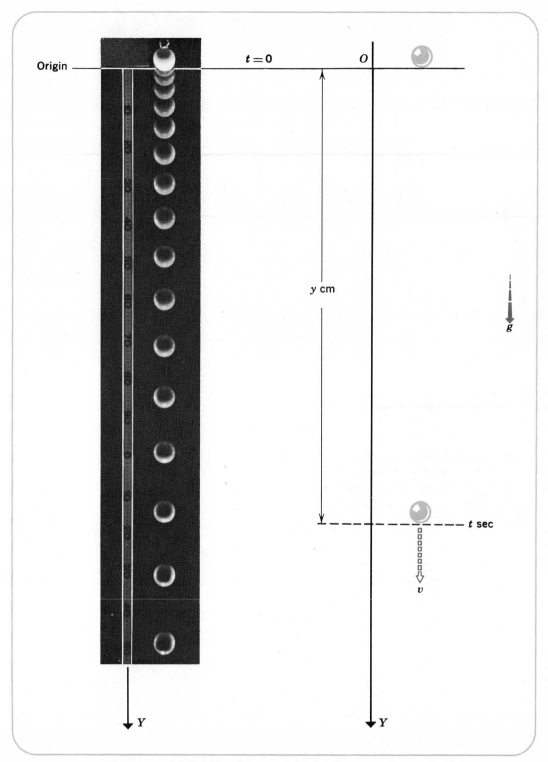

Figure 2-7 A ball falling from rest photographed at successive intervals of 1/30th of a second. (From PSSC Physics, D.C. Heath and Company, Boston, 1965.)

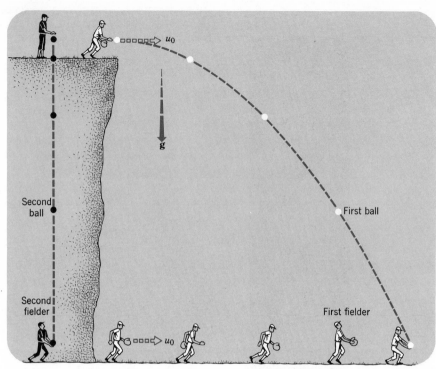

Figure 2-8 A ball projected horizontally over the edge of a cliff.

Consider, for example, a ball thrown over the edge of a cliff with an initial velocity u_0 entirely in the horizontal direction, as in figure 2-8. The acceleration produced by gravity, g, is entirely in the vertical direction and cannot change the horizontal component of the velocity. This implies that a fielder starting from the foot of the cliff at the instant the ball is thrown and running outward with a constant horizontal velocity u_0 always remains directly underneath the ball.

Meanwhile, the ball continues to fall in the vertical direction in exactly the same way as if it had no horizontal velocity. Imagine a second ball dropped from rest at the top of the cliff at exactly the same instant that the first ball is thrown horizontally outward. The two balls fall vertically in exactly the same way and are always at the same height. A second fielder standing stationary at the foot of the cliff catches the second ball at exactly the same instant that the running fielder catches the first ball.

To tackle the problem mathematically, use Cartesian axes with the origin at the top of the cliff. Let the x-axis be horizontal and let the y-axis be vertical (figure 2-9). After a time t, the horizontal distance, x, covered by the first ball is the same as the horizontal distance covered by the running fielder and is the product of the horizontal velocity and the time.

$$x = u_0 t \tag{2-8}$$

Meanwhile, the vertical distance through which either ball has fallen is obtained by using equation 2-7

$$y = \tfrac{1}{2}g t^2 \tag{2-9}$$

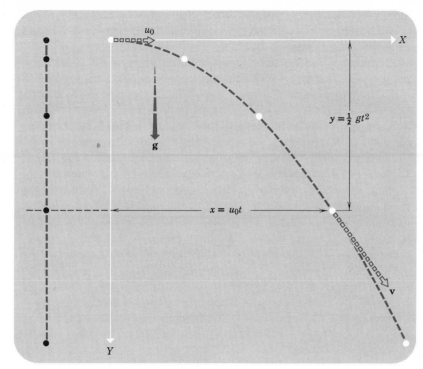

Figure 2-9 The use of Cartesian coordinates to discuss figure 2-8.

The horizontal distance traveled is therefore proportional to the time, t, whereas the vertical distance through which the body has fallen is proportional to the square of the time, t^2. The result is the curved path shown in figure 2-8, which is actually the mathematical curve known as a parabola.

We have been able to discuss this situation by considering the horizontal and vertical motions separately, but suppose we ask what is the velocity at any time, t. In addition to its horizontal velocity u_0, the body has built up a downward vertical velocity gt (see equation 2-3). The two combine in some way to produce a velocity **v** tilted downward from the horizontal (see figure 2-9). The body starts moving horizontally, but as it falls its direction of motion tilts down more and more. The velocity is not adequately specified merely by giving the speed in cm/sec. The direction of motion is equally important. A quantity whose direction is important in addition to its magnitude is called a **vector** and so, before proceeding further, we must consider the nature of vectors.

2-6 *Vectors*

In a three-dimensional space there are many physical quantities that have direction as well as magnitude. Such quantities are called vectors. One of the simplest examples of a vector is the **displacement** of one point from another. The distance between Philadelphia and New York is 90 miles, but you cannot reach New York by traveling 90 miles in a southerly direction from Philadelphia. You must travel 90 miles in exactly the right direction, which is approximately northeast. The displacement of New York from Philadelphia is a vector, and it has both magnitude and direction. Its magnitude is 90 miles and its direction is northeast.

A second example of a vector is **velocity.** From a practical point of view, traveling due west at 60 miles per hour (mph) is very different from traveling due east at 60 mph. To describe completely the velocity of a body, one must give the direction in which it is moving in addition to the distance it travels in a given time. Strictly speaking the word "velocity" should always indicate a vector, whereas the word "speed" should be used to indicate the magnitude of the velocity, its direction being ignored. In this book we shall follow this convention whenever the distinction is important, but in common usage the word "velocity" is often used instead of "speed."

A third example of a vector is a force. Until we have defined force precisely, the reader may rely on his intuitive notion of force as a push or pull. Clearly, if one is trying to push an automobile up a hill, the *direction* in which one pushes is as important as the strength one exerts.

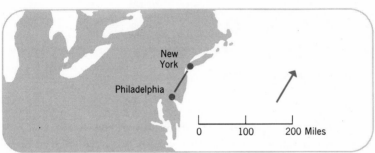

Figure 2-10.

(a) *The locations of Philadelphia and New York on a map.*

(b) *The vectorial representation of the displacement of New York from Philadelphia.*

A quantity that has magnitude but no direction is called a **scalar.** Examples of scalars are time, mass, speed, temperature, electric charge, volume, density, and energy.

In print, the symbol for a vector is set in boldface type, whereas the symbol for a scalar is set in italics. The vector representing a velocity would therefore appear as **v,** whereas the corresponding speed would be v. In handwriting, this distinction is not possible and it is convenient to indicate a vector quantity by placing an arrow above it; for example $\vec{v}$. The vector displacement from point P to point Q might be written $\overrightarrow{PQ}$, with an additional implication that the displacement is in the direction from P to Q, rather than from Q to P. The magnitude of the displacement, a scalar quantity, is written PQ.

A vector is represented pictorially by a line that is drawn in the same direction as the vector and has a length proportional to the magnitude of the vector on some agreed scale. An arrowhead indicates one of the two possible directions along the line. Figure 2-10a is a map of part of the northeastern United States drawn on a scale of 1 inch to 200 miles. Figure

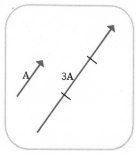

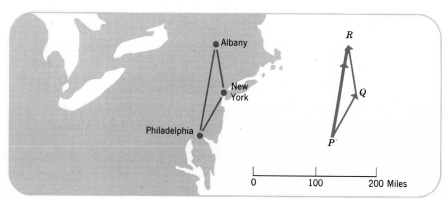

Figure 2-11.

(*a*) *Multiplication of a vector A by a*   (*b*) *The distinction between A and −A.*
scalar quantity of c = 3.

2-10*b* shows how the displacement of New York from Philadelphia can be represented by an arrow drawn in the right direction with a length of 0.45 inches which, according to the scale of the map, represents the 90 miles between the two cities. A velocity of 60 mph due north could be represented by a line with an arrowhead pointing due north and a length of 6 inches, if it had been agreed that 1 inch was to represent 10 mph. A velocity of 40 mph due west would then be represented by a line 4 inches long pointing due west and therefore at right angles to the first line.

Multiplication of a vector by a scalar does not change the direction of the vector but multiplies its magnitude by the scalar in question. The vector **A** has magnitude A and the vector c**A** therefore has magnitude cA and is in the same direction as **A**. Multiplication of a vector by a minus sign reverses its direction so that it points in exactly the opposite direction. Figure 2-11 illustrates these points.

To add vectors together, we proceed as follows: Figure 2-12 is another map of part of the northeastern United States. The displacement from Philadelphia to New York is represented by the vector $\overrightarrow{PQ}$. The displacement from New York to Albany is represented by the vector $\overrightarrow{QR}$. From the point of view of one's eventual location, a trip from Philadelphia to New York followed by a trip from New York to Albany has exactly the same end result as a trip directly from Philadelphia to Albany. The vector $\overrightarrow{PR}$ is therefore ex-

2-6 Vectors

Figure 2-12 Use of a map to illustrate the addition of vectors.

actly equivalent to the sum of the vectors $\overrightarrow{PQ}$ and $\overrightarrow{QR}$. This is written

$$\overrightarrow{PR} = \overrightarrow{PQ} + \overrightarrow{QR} \tag{2-10}$$

Notice that 2-10 is a vector equation and must be clearly distinguished from an ordinary algebraic equation involving scalar quantities, with which the reader is already familiar. It is *not* true, for example, that

$$PR = PQ + QR \tag{2-11}$$

since one covers fewer miles in going directly from Philadelphia to Albany than if one went via New York.

The method used above to add the vectors $\overrightarrow{PQ}$ and $\overrightarrow{QR}$ and obtain the vector $\overrightarrow{PR}$ was to construct a triangle PQR. This method works for the addition of any two vectors, whether they are displacements, velocities, forces, or any other type of vector. The procedure is sometimes referred to as the **triangle of vectors** and is illustrated in a general case in figure 2-13. To add the vectors **A** and **B,** construct a triangle in which **A** and **B** are adjacent sides. The sum of **A** and **B** is then the third side **C** and one can write

$$\mathbf{C} = \mathbf{A} + \mathbf{B} \tag{2-12}$$

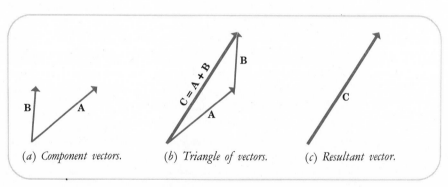

(*a*) *Component vectors.*   (*b*) *Triangle of vectors.*   (*c*) *Resultant vector.*

Figure 2-13 *Addition of two vectors using the "triangle of vectors."*

Notice that the arrowheads on the vectors being added together, **A** and **B,** both proceed in the same direction around the triangle (counterclockwise in figure 2-13*b*), but that the arrowhead on the resultant vector **C** points in the opposite direction (clockwise in figure 2-13*b*). The effect of **C** is exactly the same as the combined effect of the vectors **A** and **B.** **C** is called the **resultant** of **A** and **B,** whereas **A** and **B** are the **components** of **C.** For all practical purposes, the vectors **A** and **B** can be replaced by the single vector **C.** Conversely, if one started with the single vector **C,** it would be permissible to remove it and replace it by the two vectors **A** and **B,** if this should prove convenient for any reason. Let us emphasize again that the addition of vectors is a very special procedure clearly distinct from the addition of the

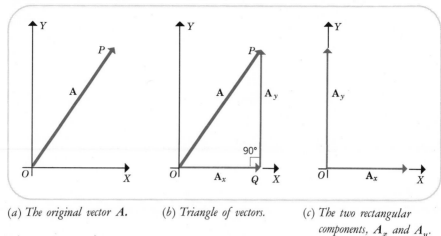

(a) The original vector A. (b) Triangle of vectors. (c) The two rectangular components, A_x and A_y.

Figure 2-14 *Resolving a vector into two rectangular components.*

scalar magnitudes of the quantities. It is not the same thing as $A + B$ and

$$C = A + B \text{ is } not \text{ true} \tag{2-13}$$

Sometimes we need to replace a single vector by two other vectors at right angles, which are then called its **rectangular components.** Consider any vector **A,** as in figure 2-14a. With the starting point of the vector as origin O, draw any two axes OX and OY at right angles to one another. From the end P of the vector **A** drop a perpendicular PQ to the axis OX (the angle PQO is then 90°). Now consider the triangle of vectors OPQ in part b of the figure. Clearly

$$\overrightarrow{OP} = \overrightarrow{OQ} + \overrightarrow{QP} \tag{2-14}$$

$$\text{or } \mathbf{A} = \mathbf{A}_x + \mathbf{A}_y \tag{2-15}$$

The original vector of part a may therefore be removed and replaced by the two vectors $\mathbf{A}_x$ and $\mathbf{A}_y$ of part c. $\mathbf{A}_x$ and $\mathbf{A}_y$ are the two *rectangular components* of **A.** $\mathbf{A}_x$ is the **x-component** and $\mathbf{A}_y$ is the **y-component.**

2-7 Motion in a Circle

Let us apply the concept of vectors to the important case of a body moving in a circle with constant speed. How shall we determine the velocity at the instant when the body is at the point P shown in figure 2-15? At a slightly later time the body has moved on to the point Q and has therefore undergone a displacement $\overrightarrow{PQ}$. The average velocity during this time interval is

$$v_a = \frac{\text{Displacement, } \overrightarrow{PQ}}{\text{Time to go from } P \text{ to } Q} \tag{2-16}$$

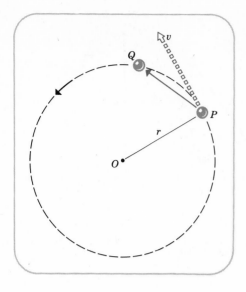

Figure 2-15 The velocity of a body moving in a circle.

This is a vector because it is the vector $\overrightarrow{PQ}$ multiplied by a scalar, the reciprocal of the time. It is in the same direction as $\overrightarrow{PQ}$.

To obtain the instantaneous velocity **v** at P, the time interval must be made very short. As the time interval is made shorter, Q approaches nearer and nearer to P and it is easy to see that the direction of $\overrightarrow{PQ}$ becomes the direction of the tangent to the circle at P. In fact this is one way of defining the tangent. The instantaneous velocity **v** is therefore tangential to the circle and, of course, perpendicular to the radius r. If, for example, a ball is whirled around in a horizontal circle on the end of a string and the string suddenly breaks, then the ball flies off along a straight line tangential to the circle (figure 2-16).

The speed, a scalar quantity v, is the magnitude of the vector **v**. It is the same as the speed with which the body goes round the perimeter of the circle. If T secs is the time for the body to go once round the circle, cover-

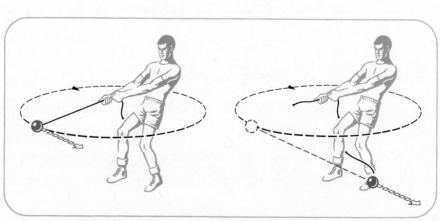

Motion

Figure 2-16 The instantaneous velocity is tangential to the circle.

ing a total perimeter $2\pi r$, then

$$v = \frac{2\pi r}{T} \qquad (2\text{-}17)$$

This discussion of velocity can be generalized to motion along any curved path.

Instantaneous vector velocity,

$$v = \frac{\text{Very small displacement, } \overrightarrow{PQ}}{\text{Very short time to go from } P \text{ to } Q} \qquad (2\text{-}18)$$

with the time interval made as short as possible.

The direction of **v** is always tangential to the path.

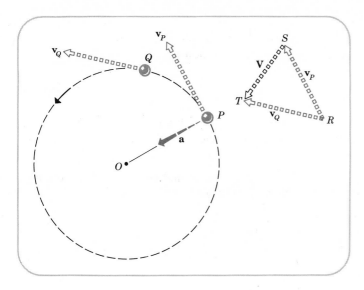

Figure 2-17 *Acceleration of a body moving in a circle with constant speed.*

2-7 Motion in a Circle

Returning to the special case of motion in a circle with constant speed, we must not assume that because the speed v is constant there is no acceleration. As the body moves round the circle, the vector velocity **v** is continually changing its direction, even though its magnitude v remains constant. A change in direction of a velocity is just as real an acceleration as a change in magnitude.

In figure 2-17, the velocity of the body at the point P is represented by the symbol v_P. At a slightly later time the body has moved on to Q and has a new velocity v_Q with the same magnitude v but a different direction. In the triangle of vectors RST, the side $\overrightarrow{RS}$ represents v_P and the side $\overrightarrow{RT}$ represents v_Q. The initial velocity v_P has been changed into the final velocity v_Q

by the addition of a small vector **V** represented by $\overrightarrow{ST}$. **V** is in fact the change in velocity during the time to go from P to Q, and the instantaneous vector acceleration **a** can be defined in the following way.

Instantaneous vector acceleration,

$$\mathsf{a} = \frac{\text{Very small change in vector velocity, } \mathsf{V}}{\text{Very short time to go from } P \text{ to } Q} \tag{2-19}$$

with the time interval made as short as possible.

As Q approaches closer and closer to P, v_Q swings round toward the direction of v_P and the angle at the vertex R of triangle RST becomes very small. Perhaps you can see that **V** then becomes perpendicular to v_P. The acceleration **a** is in the same direction as **V** and is therefore perpendicular to the tangent, pointing in fact along the radius toward the center O of the circle.

With a little mathematical effort, it can be shown that the magnitude of the acceleration is given by the following equation

Acceleration of a body moving in a circle

If the body moves with constant speed v in a circle of radius r, the magnitude of its acceleration is

$$a = \frac{v^2}{r} \tag{2-20}$$

The direction of the acceleration is along the radius, pointing toward the center of the circle.

Figure 2-18 summarizes these facts about motion in a circle with constant speed and should be studied carefully.

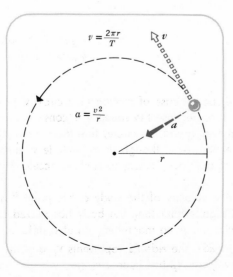

Motion

$$v = \frac{2\pi r}{T}$$

$$a = \frac{v^2}{r}$$

Figure 2-18 *Velocity v and acceleration a of a body moving with constant speed v in a circle of radius r. T is the time for one complete revolution.*

A
1. Which of the following distances is comparable with the length of your index finger?
   (a) 10^6 cm, (b) 10 m, (c) 0.1 m, (d) 10^{-3} cm, (e) 10^{-6} m.
2. Think of an object comparable in size with each of the following distances.
   (a) 10^7 m, (b) 10^5 cm, (c) 10^3 cm, (d) 1 m, (e) 10^{-3} m, (f) 10^{-8} cm, (g) 10^{-12} cm
3. A vector has zero magnitude. Is it necessary to specify its direction?
4. Can a vector have zero magnitude if one of its rectangular components is not zero?

B
5. Several vectors, which are not all in the same plane, add together to give a zero resultant. What is the minimum number of vectors that can satisfy this requirement?
6. If the instantaneous velocity of a body is zero, is its instantaneous acceleration necessarily zero? Give examples.
7. If the acceleration of a body is constant in magnitude and direction, is its path necessarily a straight line? Give examples.

C
8. After a laborious calculation, you conclude that the distance from the center of the earth to the moon is 3.8×10^{10} cm and that the diameter of the moon is 3.1×10^{10} cm. Relying only on your knowledge of the appearance of the moon in the night sky, can you think of a simple reason why these results cannot both be correct?
9. Two racing cars are traveling around a circular track at steady speeds of 90 mph and 100 mph. The driver of the slower car has an elaborate piece of equipment that enables him to measure the velocity of the faster car *relative to himself*. Does this relative velocity ever become zero? Does it matter whether the cars are going round in the same or opposite directions? If the two cars were going around at the same speed in opposite directions, would the relative velocity ever be zero?
10. Is it possible to have a straight line motion with **v** and **a** always in opposite directions? (Hint: Consider a rocket launched vertically with a very high velocity.)
11. If the acceleration of a body is always constant in magnitude, but not in direction, is the path necessarily either a straight line or a circle? Explain.
12. Is it possible for a body to move in a circle with a speed that varies in such a way that the acceleration points *outward* along a radius?
13. Suppose that very accurate measurements suggest that the diameter of the earth is gradually shrinking. How could we be sure that this was not really due to a gradual increase in the length of the standard meter?

Problems

1. The top speed of a certain aircraft is 420 mph. An international commission decides to define a new mile that is three times as long as the present one. What is the top speed of the aircraft in "new-miles" per hour?

2. Convert the following quantities to the CGS system of units. You may use the table of conversion factors on the inside of the cover.
(a) 5 miles, (b) 50 millimeters, (c) 30 m, (d) 5 kg, (e) 5×10^{-3} kg, (f) 18 months, (g) 6 minutes, (h) 1000 mph, (i) 5 kilometers per hour, (j) 0.2 mm/sec, (k) 15 m/sec.

3. Convert the quantities in the previous question to the MKS system of units.

4. Add together the three vectors shown in the diagram.

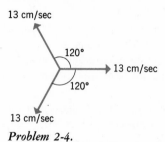

Problem 2-4.

5. Draw diagrams to find the magnitude and direction of the sum (**A** + **B**) of each pair of vectors in the figure.

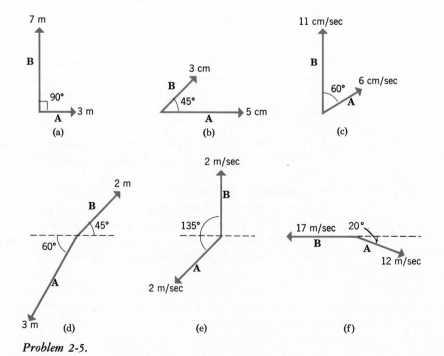

Motion

Problem 2-5.

6. A body falls vertically from rest. (a) What is its velocity after it has been falling for 3 sec? (b) How far has it then fallen?

7. A car traveling at 60 mph applies its brakes and comes to rest in 0.8 sec. Assuming the acceleration (or, if you prefer, deceleration) was constant during this time, calculate its value in the CGS system of units.

8. A body moves around a circle of radius 300 cm with a constant speed of 30 cm/sec. What is its radial acceleration?

B

9. The speed limit on a certain expressway is 60 mph. An international commission decides that the hour shall be redefined so that it takes the earth only 20 hours to rotate once on its axis. What should the new speed limit be?

10. About how many seconds are there in the lifetime of an average man? Express your answer in the form $n \times 10^m$, where n and m are integers.

11. A famous traveler took 80 days to go around the world. At the same average speed, how long would it take him to go to (a) the moon, (b) the sun, (c) Pluto (6×10^{14} cm away), (d) the nearest star, Alpha Centauri (4×10^{18} cm away)?

12. A rocket is able to achieve speeds in the vicinity of 30,000 mph. At this speed, how long would it take to encircle the earth at the equator? How long would it take to go to the celestial objects mentioned in the previous question?

13. A ball is thrown vertically downward with an initial speed of 10^3 cm/sec. What is its speed after it has been falling for 2 sec?

14. A skater is gliding over the ice with a velocity of 2.6 m/sec. Another skater pushes him from behind and produces a steady acceleration of 1.3 m/sec^2 until his velocity has been raised to 18.2 m/sec. How long did the push last?

15. A mailman's round is made up as follows: (a) $\frac{1}{2}$ mile due east, (b) $\frac{1}{4}$ mile due north, (c) $\frac{3}{4}$ mile northwest, (d) $\frac{1}{2}$ mile due south, (e) 1 mile southwest. When he has finished, what is his displacement relative to his starting point? Use graph paper and a protractor.

16. The second hand of a wrist watch is 1.2 cm long from the center of the dial to the tip of the hand. Calculate the velocity and acceleration of the tip.

17. A simple theory of the hydrogen atom concludes that the electron moves around the proton with a velocity of 2.2×10^8 cm/sec in a circular orbit of radius 5.3×10^{-9} cm. Calculate the acceleration of the electron and compare it with the acceleration due to gravity, g.

18. A phonograph record with a diameter of 12 inches (30.5 cm) rotates at $33\frac{1}{3}$ rpm. What is the velocity of a point on its rim? What is the acceleration of this point?

C

Problems

19. The Empire State Building is 102 stories high. Working in centimeters, place a lower limit on the height of the building. Make this lower limit as large as you can, but be prepared to advance a completely convincing argument to prove that the height could not possibly be less than this.

20. A student calculates that the velocity that must be given to a rocket to enable it to escape completely from the earth is 721 cm/sec. Is this reasonable? Why or why not?

21. A car traveling along a straight highway maintains a speed of 60 mph for 1 hour and then 30 mph for 2 hours. What was its average velocity over the 3-hour interval? What would its average velocity have been if it had turned round at the end of the first hour?

22. A car traveling along a straight highway maintains a speed of 60 mph for the first 20 miles and then 30 mph for the next 40 miles. What was its average velocity over the whole 60-mile run?

23. Two bodies are dropped simultaneously from rest at heights of 125 m and 35 m above the ground. What is their difference in height after they have been falling for (a) 2.0 sec, (b) 3.0 sec, (c) 10.0 sec? When one of these bodies hits the ground, it does not bounce but sticks to the surface.

24. On a certain straight highway the speed limit is 60 mph. A policeman times a car with a stop watch that may introduce an error of 0.5 sec and concludes that the speed of the car is 62 mph. What is the minimum distance in yards over which the car must be timed before the policeman can be quite certain that it was exceeding the speed limit?

25. A stone is dropped from rest at the top of a building 150 m high. How long will it take to reach the ground? With what velocity will it strike the ground?

26. An automobile crashes into a tree at 60 mph. From what height would it have to be dropped to hit the ground at the same speed? About how many stories would there be in a building this high?

27. A cannon is mounted on a hill 150 m high, and its barrel points horizontally. It is trying to hit the gate of a city on the plain below, a horizontal distance of 350 m away. What should be the velocity of the cannonball as it emerges from the muzzle?

28. A body is at rest on the ground at the equator. Calculate its velocity and its acceleration due to the rotation of the earth.

Motion

3 *The Laws of Motion*

3-1 *Newton's First Law of Motion*

Within the framework of classical physics, let us pursue our plan of visualizing the universe as a collection of particles in motion. In the two preceding chapters we devoted our attention to the description of this motion and we considered carefully such concepts as velocity and acceleration. We shall now turn to the basic question of why each particle moves along the particular path it does and how we might predict the details of its path. The underlying idea is that the details of the motion are determined by the interactions between the particle and the other particles in the universe. To use a language that is already familiar to the reader, but the precise significance of which will have to be described in detail, the other particles exert forces on the particle which push and pull it along the twists and turns of its path.

Before pursuing this point, a still more basic question must be answered. How would the particles behave if they did not interact with one another, if they were entirely oblivious of one another's existence? Would they all remain at rest, move in straight lines or circles, or oscillate backward and forward about fixed centers? The answer is not obvious and can only be found by carefully observing the universe and discovering which assumption best fits the facts. The answer is contained in Newton's first law of motion.

Newton's first law of motion:

In the absence of any interaction with the rest of the universe, a body would either remain at rest or move continually in the same straight line with a constant velocity. That is, the velocity, regarded as a vector, would remain constant; and rest would be the special case of zero velocity. Alternatively, one can say that the acceleration would be zero.

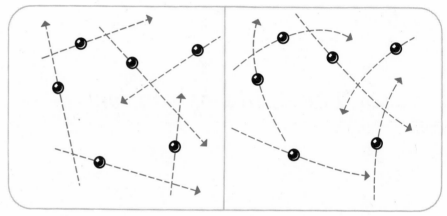

Figure 3-1.

(a) *If the particles did not interact they would move in straight lines with constant velocities.*

(b) *When they interact they accelerate one another and distort one another's paths.*

In the absence of interactions, the particles in the universe would move in straight lines with constant velocities as in figure 3-1a. The role of the interactions is to change the velocities, to produce accelerations and hence complicated paths, as in figure 3-1b. A large part of classical physics consists of formulating the laws needed to calculate the acceleration produced in one particle by a second particle. The procedure is somewhat as follows. Suppose that, at a fixed instant of time, we know the positions and velocities of all the particles, which are labeled 1, 2, 3, 4, 5, etc., as in figure 3-2. To calculate the acceleration of particle 1, let us consider first its interaction with particle 2. If we know the distance between 1 and 2, their velocities, and certain properties of the two bodies such as their masses and electric charges, the fundamental laws enable us to calculate the acceleration a_{12} induced in 1 by its interaction with 2. Similarly we can calculate the acceleration a_{13} induced in 1 by its interaction with 3 and so on to obtain the contribution to the acceleration of 1 from all the other bodies in

The Laws of Motion

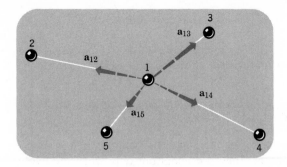

Figure 3-2 *The acceleration of body 1 is the vector sum of contributions resulting from interactions with each of the other bodies in the universe.*

the universe. The vectorial sum $a_{12} + a_{13} + a_{14} + a_{15} + \cdots$ then gives us the total acceleration of body 1 at the fixed instant of time we are considering. The accelerations of all the other bodies at this same instant of time can be calculated in a similar way. It can be shown that this procedure enables us to describe the path of each particle in complete detail and hence calculate the positions, velocities, and accelerations of all the particles at all future instants of time and all past instants of time.

> ## Determinism in classical physics
>
> In classical physics, if the positions and velocities of all the particles are known at one instant of time, then the fundamental laws describing the interactions of the particles with one another can be used to calculate in precise detail the behavior of the particles at all past, present, and future times.

This implication of classical physics clearly has profound philosophical significance. It would seem to imply that the future of the universe is completely determined by its past and that there is no such thing as free will or human intervention. Suppose that I take the extreme point of view that my thoughts are merely a consequence of the positions and velocities of the electrons, protons, and neutrons in my brain. Then, if I know their positions and velocities at 12:30 P.M. and the positions and velocities of certain other particles in my immediate vicinity, with sufficient effort and the use of sufficiently complicated computers, I can presumably predict that at 12:55 P.M. the positions and velocities of these particles in my brain will be such as to cause me to make a decision to go out to lunch. Any feeling I have that this decision was made voluntarily is pure delusion. If, in the process of crossing the road to the lunch counter, I am struck by an automobile and killed, this could presumably have been predicted from the previous positions and velocities of the particles in the universe and was quite inescapable. These conclusions are very difficult to avoid if one accepts without question the basic approach of classical physics. Fortunately, modern quantum mechanics provides a loophole, as we shall discover in a later chapter, although it is not clear that it provides us with an alternative that is any more comforting.

The richness and variety of physical theory is partly a result of the fact that the interaction between two particles has several aspects, which are enumerated below.

A. Gravitational forces. These are always attractive and exist between any pair of particles in the universe.

B. Electromagnetic forces. These are present if the two particles have electric or magnetic properties.

C. Nuclear forces. These are the forces that hold the neutrons and protons together in the nucleus of an atom. They are important only when the two particles are very close together, within about 10^{-12} cm of one another.

D. Various other interactions are currently being studied at the frontier of physics and are not yet properly understood. These interactions are not conspicuous in everyday experience.

During our presentation of Newtonian mechanics we shall concentrate on gravitational and electromagnetic forces. The other forces are more complicated and will have to be carefully explained at the appropriate places later in the book.

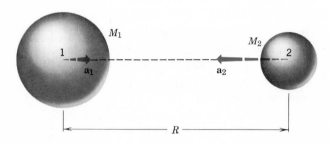

Figure 3-3 *The gravitational interaction of two bodies.* M_1 *and* M_2 *are their gravitational masses.*

3-2 *Gravitation, Electromagnetism, and Mass*

Imagine any two bodies 1 and 2 at a distance R apart, as in figure 3-3. They might be two fundamental particles, or the earth and the moon, or the earth and a falling body, or the sun and a planet, or two stars. If they are not electrically charged and not magnetized, there are no electromagnetic forces between them. If they are much farther apart than 10^{-12} cm, the nuclear forces can also be neglected. Then only their gravitational interaction need be considered. (The gravitational interaction is still present even when there are other types of force, but sometimes these other forces are much stronger and completely obscure the weak gravitational interaction.) In order to avoid mathematical complications of a nonessential nature, we shall assume that the distance between the bodies is very large compared with their size. As we have already explained in section 2-1, this requirement is frequently satisfied, especially in the solar system.

Gravitational phenomena, such as the motion of the planets and their moons and the behavior of bodies falling to the earth, can be explained in terms of the following assumption. The gravitational effect of body 2 on body 1 is such as to give body 1 an acceleration pointing directly toward body 2 of magnitude

$$a_1 = \frac{GM_2}{R^2} \tag{3-1}$$

Similarly, interchanging the roles of the two bodies, the gravitational effect of body 1 on body 2 is such as to give body 2 an acceleration pointing directly toward body 1 of magnitude

$$a_2 = \frac{GM_1}{R^2} \tag{3-2}$$

The first thing to notice about these formulas is that a_1 and a_2 vary inversely with the square of R. As the distance between the two bodies is increased, the gravitational accelerations they produce in one another decrease rapidly. If the distance is doubled, the accelerations are divided by four. G is a "universal constant of nature" called the **gravitational constant.** It should not be confused with the acceleration due to gravity g, which is not a universal constant but has a value that depends on the accidental circumstance of being located at a particular spot near the earth's surface and even varies from place to place over the surface of the earth. The numerical value of G, which is assumed to be the same for any pair of bodies anywhere in the universe, is

$$G = 6.67 \times 10^{-8} \text{ in the CGS system} \tag{3-3}$$

The quantities M_1 and M_2 are the **gravitational masses** of the bodies 1 and 2, and equations 3-1 and 3-2 serve to define these masses. This means that we can associate with any body an invariable number M which, when inserted into an equation such as 3-1, enables us to calculate the gravitational acceleration that the body produces in any other body. The procedure for determining the value of M will be described in a simplified form in the next paragraph. Gravitational mass is always positive. As we explained in section 2-3, the unit of mass is arbitrarily defined in terms of the mass of a particular platinum-iridium cylinder. This arbitrary choice determines the numerical value of the constant G. This can be illustrated by the following sequence of imaginary experiments. Let M_2 be a standard 1 gm mass. Place a small mass M_1 at a measured distance R cm from it, and then measure the acceleration a_1 of M_1 due to M_2, say in some remote region of space where the effect of all other masses can be neglected. Then

3-2 Gravitation, Electromagnetism, and Mass

$$a_1 = \frac{G \times 1}{R^2} \text{ if } M_2 = 1 \text{ gram} \tag{3-4}$$

$$G = a_1 R^2 \tag{3-5}$$

Since R and a_1 have both been measured, this experiment measures G, which will be found to have the value given in equation 3-3. However, it has this particular value only because the standard gram is as large as it is.

Now take an unknown mass M, place a small mass m at a measured distance R away from it, and measure the acceleration a of the mass m, all other masses in the universe being too far away to influence the result. Then

$$a = \frac{GM}{R^2} \tag{3-6}$$

$$M = \frac{aR^2}{G} \tag{3-7}$$

Since a and R have been measured and G has been determined by the first experiment, this second experiment measures the unknown mass M. Notice that the measurement of a and R involves only basic measurements of length and time.

A careful examination of equations 3-1 and 3-2 will convince you that the gravitational mass of a body is a measure of its ability to produce an acceleration in *another* body. As long as we confine ourselves to gravitational interactions, there is no need to introduce any other definition of mass. Equations 3-1 and 3-2 are completely adequate to describe all purely gravitational phenomena, such as the behavior of the solar system. However, once we introduce electromagnetic interactions we are led to the concept of **inertial mass,** which rather seems to measure the ability of a body to resist the production of an acceleration in itself. Gravitational mass and inertial mass appear to be entirely different concepts and yet, to the best of our present knowledge, they are always exactly equal to one another for any body. This is a very important fact, but we do not properly understand why it should be so.

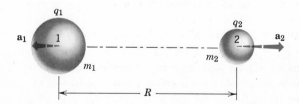

Figure 3-4 *The electrostatic interaction of two bodies. q_1 and q_2 are their electrical charges, and m_1 and m_2 are their inertial masses.*

Electromagnetism is a complicated subject that we shall discuss more fully later. For our present purpose, consider the particularly simple case of two electrically charged bodies 1 and 2 that are instantaneously at rest at a distance R apart, as in figure 3-4. We can associate with the bodies two numbers q_1 and q_2, which are called their **electric charges.** Unlike gravitational mass, electric charge can be either positive or negative. If the two charges are both positive or both negative they repel one another and accelerate directly away from one another. But if one is positive and the other negative they attract one another and accelerate directly toward one another. In addition to the gravitational acceleration, which is always present, the electrical interaction gives body 1 an acceleration

$$a_1 = \frac{1}{m_1}\frac{q_1 q_2}{R^2} \tag{3-8}$$

and body 2 an acceleration

$$a_2 = \frac{1}{m_2}\frac{q_1 q_2}{R^2} \tag{3-9}$$

m_1 and m_2 are the inertial masses of bodies 1 and 2 respectively. The equations take this particularly simple form only in the CGS system, but, since the complications introduced into the MKS formulation are not directly related to the physical behavior of the universe, we shall ignore them here. In the CGS system the units of charge are **statcoulombs**. Notice that electromagnetic accelerations also vary inversely with the square of the distance between the two bodies.

Notice that in equation 3-8 for the electromagnetic acceleration of body 1 the inertial mass of body 1 appears in the denominator, whereas in equation 3-1 for the gravitational acceleration of body 1 the gravitational mass of body 2 appears in the numerator. This is the reason why we say that the gravitational mass produces gravitational accelerations in other bodies, whereas the inertial mass resists the production of an acceleration in the body itself.

Equations 3-8 and 3-9 imply that

$$\frac{a_1}{a_2} = \frac{m_2}{m_1} \tag{3-10}$$

The accelerations are in inverse ratio to the masses. If we measure the two accelerations, we can deduce the ratio of the two inertial masses. If one of the bodies is the standard one kilogram cylinder of platinum-iridium, its inertial mass is defined to be 1000 in the CGS system, and so the mass of the other body can be deduced. Comparing masses in pairs in this way, the inertial masses of all bodies can be determined.

3-3 Force, and a Spate of New Laws

The simple procedures that we have described for measuring gravitational and inertial masses are not, of course, used in practice. A complete description and analysis of the procedures actually used would be very lengthy and might, by virtue of its complexity, prove more confusing than helpful. However, when properly understood, these procedures are equivalent in principle to the direct procedures we have described. The standard gram is taken to be the CGS unit of both gravitational and inertial mass, but in the case of

$\mathbf{F} = m\mathbf{a}$

$\mathbf{a}$

m

Figure 3-5 *Illustrating the definition of force.*

other bodies there seems to be no initial reason why the measured gravitational mass should turn out to be equal to the measured inertial mass. In fact, however, the gravitational mass has never been found to differ from the inertial mass by more than 1 part in 10^4, which is the limit of the experimental accuracy. The physical significance of this is still not properly understood, but it is essential to the development of Newtonian mechanics, since it makes possible the introduction of the concept of **force** and leads to Newton's second and third laws of motion.

Newton's second law and the definition of force are the same thing. If a body of mass m has an acceleration $\mathbf{a}$ (figure 3-5), then the force acting on it is defined as the product of its mass and its acceleration.

Newton's second law of motion

Force = Mass × Acceleration

$\mathbf{F} = m\mathbf{a}$ (3-11)

Dynes = Grams × cm/sec^2

We are no longer making any distinction between gravitational and inertial mass and so the mass m is a uniquely defined quantity. Notice that equation 3-11 is a vector equation. Force is a vector and its direction is the same as the direction of the acceleration $\mathbf{a}$. Its magnitude is equal to the product of the scalar mass and the scalar magnitude of $\mathbf{a}$. The unit of force is so important that it has a special name. In the CGS system it is called a **dyne**. If m is measured in grams and a in cm/sec^2, then F is measured in dynes.

Let us now take another look at gravitation and electromagnetism, using this concept of force. Figure 3-6 is similar to figure 3-3, except that the emphasis is now on the forces that the two bodies exert on one another. The force F_{21} exerted by body 2 on body 1 is obtained with the help of equation 3-1.

$$F_{21} = M_1 a_1 \tag{3-12}$$

$$= \frac{GM_1 M_2}{R^2} \tag{3-13}$$

(Remember that it does not now matter whether we use m or M.) Similarly,

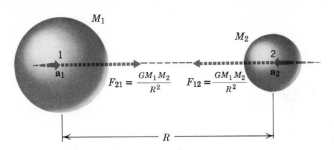

$$F_{21} = \frac{GM_1M_2}{R^2} \qquad F_{12} = \frac{GM_1M_2}{R^2}$$

Figure 3-6 *Gravitational forces between two bodies.*

the other force F_{12}, exerted by body 1 on body 2 is obtained with the help of equation 3-2.

$$F_{12} = M_2a_2 \qquad\qquad (3\text{-}14)$$

$$= \frac{GM_1M_2}{R^2} \qquad\qquad (3\text{-}15)$$

Notice that the right-hand side of equation 3-15 is exactly the same as the right-hand side of equation 3-13. The forces exerted by the two bodies on one another have the same magnitude. However, they have exactly opposite directions, since gravitation is always attractive and the force on either body points directly toward the other body.

The gravitational interaction between two bodies can now be expressed in a particularly elegant form using the concept of force, as was first realized by Newton.

Newton's universal law of gravitation

Any two bodies in the universe have a gravitational attraction for one another. If their masses are M_1 and M_2 and their distance apart, R, is large compared with the size of either, then the force on either body points directly toward the other body and has a magnitude

$$F = \frac{GM_1M_2}{R^2} \qquad\qquad (3\text{-}16)$$

3-3 Force,
and a Spate of
New Laws

Now consider the electrostatic forces between two stationary electric charges, as in figure 3-7. The electrostatic force exerted by 2 on 1 is obtained with the help of equation 3-8.

$$F_{21} = m_1a_1 \qquad\qquad (3\text{-}17)$$

$$= \frac{q_1q_2}{R^2} \qquad\qquad (3\text{-}18)$$

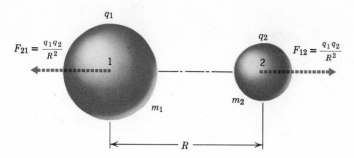

$$F_{21} = \frac{q_1 q_2}{R^2}$$

$$F_{12} = \frac{q_1 q_2}{R^2}$$

Figure 3-7 *Electrostatic forces between two bodies.*

Similarly, the electrostatic force exerted by 1 on 2 is obtained with the help of equation 3-9.

$$F_{12} = m_2 a_2 \tag{3-19}$$

$$= \frac{q_1 q_2}{R^2} \tag{3-20}$$

Again the two forces are equal in magnitude, but opposite in direction. The fundamental law of electrostatics, discovered by the French physicist Coulomb, can be stated in the following form.

Coulomb's law for electrostatic forces

If a stationary electric charge q_1 is at a distance R from a stationary charge q_2, the force on either charge due to the other has magnitude

$$F = \frac{q_1 q_2}{R^2} \tag{3-21}$$

If the charges are both positive or both negative, the forces are repulsive. If one charge is positive and the other negative, the forces are attractive.

The word "electrostatic," incidentally, implies that the charges are stationary. If they are in motion, the forces are much more complicated, as we shall discover later.

A very important fact that has emerged from the above discussion of both gravitational and electrostatic interactions is that the forces two bodies exert on one another are equal in magnitude but opposite in direction. This is one reason why the concept of force is so useful. It is called Newton's third law of motion. It played a very important role in the highly successful mathematical development of Newtonian mechanics in the eighteenth and nineteenth centuries. Notice particularly that the two forces referred to in Newton's third law act on different bodies. They never act on the same body.

> *Newton's third law of motion*
>
> If body 1 exerts on body 2 a force $\mathbf{F}_{12}$, then body 2 exerts on body 1 a force $\mathbf{F}_{21}$ which is equal in magnitude but exactly opposite in direction.
>
> $$\mathbf{F}_{12} = -\mathbf{F}_{21} \qquad\qquad (3\text{-}22)$$

Newton believed that his law was universally valid for all forces. In a later chapter, when we are discussing magnetic forces, we shall encounter a clear-cut case where the third law is not true.

3-4 Some Comments on All This

In everyday life many forces are encountered that seem to have no simple relation to either gravitational or electrostatic interactions. A few examples are: the pull of a muscle when we lift a weight; the tug of a tractor dragging a log along the ground; the upward force exerted by the top of a desk to support a book; the buoyancy of the air raising a balloon. These forces are now known to be due to the forces that atoms exert on one another and they are basically electromagnetic in character. The atom as a whole is electrically neutral because each negatively charged electron is balanced by a positively charged proton. However, as the electron moves around inside the atom, the electromagnetic force it exerts on an electron moving around inside a neighboring atom does not quite average out to zero. This cannot be completely understood without recourse to quantum mechanics. It is found that when the atoms are far apart they attract one another, but when they get too close they begin to repel one another. In a solid, the atoms are arranged in an orderly array at such a distance apart that, if they are pulled further apart and then released, they attract each other back into position, but if they are pushed closer together they repel each other back into position.

Interatomic forces between electrically neutral atoms always obey Newton's third law. The force exerted on atom 1 by atom 2 is equal in magnitude but opposite in direction to the force exerted on atom 2 by atom 1. When the small interatomic forces are added together for the whole body, the third law is found to be obeyed on a large scale and this is one important reason why the third law is so obviously valid in everyday experience. Those instances where the third law is violated are more subtle and do not have obvious consequences in everyday life.

To illustrate these points, consider a man trying to push a crate that is too heavy to move (figure 3-8). From a macroscopic point of view we would say that the man's hand is exerting a force $\mathbf{F}$ on the crate, while the crate is exerting an equal and opposite force $-\mathbf{F}$ on the man's hand. The origin of all this is that the atoms in the skin of the man's hand are in close proximity to certain atoms on the surface of the crate and these atoms exert repulsive electromagnetic forces on one another. The force exerted by an atom of the man's skin on a nearby atom of the crate is equal and opposite to the force exerted by the atom of the crate on the atom of the man's

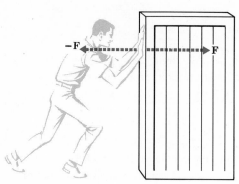
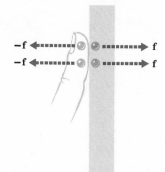

Figure 3-8.

(a) *A man attempting to push a crate*
    illustrates Newton's third law.

(b) *The forces are caused by the*
    electromagnetic interaction between
    atoms in the crate and atoms in
    the man's hand.

skin. When these forces are added together for all the atoms, the resultant force **F** on all the atoms of the crate is equal and opposite to the resultant force −**F** on all the atoms in the skin of the man's hand.

Now consider our intuitive concept of the nature of inertial mass. It is much easier to push one's way through a light bamboo curtain than through a heavy revolving door. We can easily distinguish between being hit on the head by a table tennis ball and being hit on the head by a baseball, even though they have the same velocity. These experiences all involve interatomic forces. As the hand pushes on the revolving door, the atoms of the door exert forces on the atoms of the skin. These atoms in turn move closer to the atoms underneath them and exert forces which are thereby transmitted deeper and deeper into the hand. The forces are passed on from atom to atom until they reach the atoms of our muscles, which readjust themselves in some complicated way in order to produce the necessary forces. When the baseball strikes the head, atoms on the outside of the ball first push against atoms in the skin. The force is transmitted inward from atom to atom, displacing the atoms from their rightful positions and thereby producing a certain amount of damage. The physiological details are complicated, but the basic processes are interatomic and the forces are electromagnetic.

Mass is often defined as "the quantity of matter in a body." To be of value in physics this concept has to be made more precise and more quantitative. An obvious approach is to say that mass is a measure of the number of atoms in a body. This cannot cope with the fact that there are many different kinds of atoms with different masses. It is more unpleasant to be hit on the head by a lead brick than by a lump of charcoal containing the same number of atoms. An atom of lead is about seventeen times more massive than an atom of carbon.

The difference between various atoms is related to the number of protons, neutrons, and electrons they contain. Can we say that the mass of a body is a measure of the number of fundamental particles it contains? Unfortunately not, because fundamental particles do not all have the same mass. The neutron has a mass 1.0014 times as large as the mass of the pro-

The Laws of
Motion

46

ton, and the proton has a mass 1836.1 times the mass of the electron. A deuterium atom containing three particles (one proton, one neutron, and one electron) is not $\frac{3}{2}$ times as massive as a hydrogen atom containing two particles (one proton and one electron). It is approximately twice as massive.

The basic problem is this: imagine a proton and electron exerting electrostatic forces on one another. In figure 3-4, suppose that body 1 is an electron and body 2 is a proton. In equation 3-8, for the acceleration of the electron, we insert the mass of an electron. In equation 3-9, for the acceleration of the proton, we insert the mass of the proton, which is 1836.1 times larger. But what is the significance of this peculiar number 1836.1, which is not even an integer? This can justifiably be said to be the major unsolved problem of contemporary physics. It is complicated by the following facts, which we shall discuss in more detail later in the book. According to the theory of relativity, the mass of a body is found to vary with its velocity. Mass and energy are similar quantities and can be converted into one another. The mass of an atom is not even the sum of the masses of the fundamental particles of which it is composed!

The major obstacle to a proper understanding of mass is probably our intuitive feeling that we know what it is before we start. Rather than attempting to justify this intuitive feeling, we would do better to realize how little we really understand the concept. At this stage, it is probably best to regard mass as a number that has to be inserted in equations such as equation 3-8 in order to get the right answer.

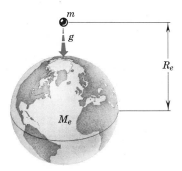

Figure 3-9 A body falling near the surface of the earth.

3-5 The Earth's Gravitation; Weight and Mass

Figure 3-9 shows a small mass m falling near the surface of the earth. This situation is similar to figure 3-3 in which two bodies are shown exerting gravitational attractions on one another, except that in figure 3-3 we assumed that the diameters of the bodies were small compared with their distance apart. Clearly the radius of the earth is not small compared with the distance between the falling body and the center of the earth. However, it may be shown that a perfectly spherical body with uniform density behaves as though its total mass were concentrated at the center of the sphere. If we assume that the earth is a perfect sphere of uniform density, we may therefore apply equation 3-1 to the freely falling body m.

The acceleration of a freely falling body near the surface of the earth is called the **acceleration due to gravity** and is denoted by the symbol g. If

M_e is the mass of the earth and R_e is the radius of the earth, which is practically the same thing as the distance of the falling body from the center of the earth, then equation 3-1 takes the form

$$g = \frac{GM_e}{R_e{}^2} \qquad\qquad (3\text{-}23)$$

Equation 3-23 embodies a feature of gravitational effects that is well worth emphasizing. The acceleration g depends on the universal constant G, the mass of the earth M_e, and the radius of the earth R_e, but it does not depend on any property of the falling body. At a particular place on the earth's surface all bodies fall with the same acceleration (if the effect of the air on the motion is eliminated or allowed for). This important fact was first realized by Galileo and is immortalized in the famous story of how he dropped two very different bodies from the top of the Leaning Tower of Pisa and observed that they reached the ground at the same time. Since then the point has been tested by experiments of increasing complexity and sophistication, and we now know that g for various bodies of different size and chemical composition does not differ by more than 1 part in 10^{11}. More generally, as equations 3-1 and 3-2 imply, when a body is made to accelerate by the gravitational attraction of other bodies, the acceleration does not depend on any property of the body that is being accelerated. This peculiar aspect of gravitational forces is not true of any other type of force. It is one of the fundamental principles underlying Einstein's general theory of relativity.

The earth is not a perfect sphere and its density is not uniform throughout, so equation 3-23 is only approximately true. In fact g varies from place to place on the earth's surface. At the top of a high mountain g is less than at sea level because one is farther from the center of the earth and the inverse square law comes into play. There is a tendency for g to increase as one goes from the equator to the poles. There are also local variations in g caused by variations in the nature of the earth's crust. As we might expect, g changes when we approach a mountain, because of the gravitational attraction of the mountain, but the effect can be quite complicated because the mountain may have a ''root'' below sea level. Characteristic local variations in g have been used to help locate oil deposits. The approximate value of g is

$$g \fallingdotseq 980 \text{ cm/sec}^2 \qquad\qquad (3\text{-}24)$$

and it varies by about 0.5% over the earth's surface. It can now be measured with such great accuracy that it is possible to observe the effect of the tides. The gravitational attraction of the ocean is slightly different at high and low tide, so that the value of g oscillates very slightly up and down twice a day.

So far we have concentrated on the *acceleration* of the falling body. The gravitational *force* of attraction exerted on the body by the earth is called the **weight** of the body, **W**. Applying Newton's second law to a freely falling body, the gravitational force on it is equal to the product of its mass and its acceleration.

> *Weight*
>
> In terms of the vector forces:
>
> $$\mathbf{W} = m\mathbf{g} \qquad\qquad (3\text{-}25)$$
>
> In terms of the scalar magnitudes:
>
> $$W = mg \qquad\qquad (3\text{-}26)$$

Since the weight of a body is the gravitational force that the earth exerts on that body, it can be obtained from Newton's universal law of gravitation by inserting the appropriate symbols into equation 3-16:

$$W = \frac{GM_e m}{R_e{}^2} \qquad\qquad (3\text{-}27)$$

This equation could also be obtained by inserting into equation 3-26 the value of g given by equation 3-23.

It is important to distinguish carefully between weight and mass. Mass is measured in grams, whereas weight is a force measured in dynes. The mass of a body is an invariant property of that body and remains the same wherever the body is and whatever it is doing. (In the theory of relativity, mass varies with velocity, but we are ignoring this at present.) However, the weight of a body depends on the local value of g and varies from place to place on the earth's surface. If the body is taken to the moon or one of the planets, its mass remains unaltered, but its weight suffers a drastic change, since the acceleration due to gravity is very different. Strictly speaking, it is incorrect to talk about a weight of 10 grams. One should refer to a mass of 10 grams, which has a weight of 10g dynes, the exact value of this weight depending on the local value of g.

3-6 A Falling Book and a Book Resting on a Desk

If I lift up a book of mass m and then let it fall to the floor, while it is falling it has a downward acceleration **g**. The only force acting on it is its weight $\mathbf{W} = m\mathbf{g}$, which is the force of gravitational attraction exerted on the book by the earth. In accordance with Newton's third law, the book must exert on the earth an upward force $\mathbf{W}'$, which is equal in magnitude but opposite in direction to $\mathbf{W}$ (figure 3-10a).

$$\mathbf{W}' = -\mathbf{W} \qquad\qquad (3\text{-}28)$$

Also, while the book is falling downward toward the earth, the earth is falling upward toward the book with an acceleration

$$g_e = \frac{Gm}{R_e{}^2} \qquad\qquad (3\text{-}29)$$

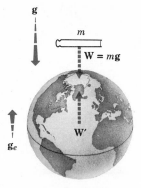
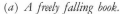

(a) A freely falling book.

(b) A book resting on a desk.

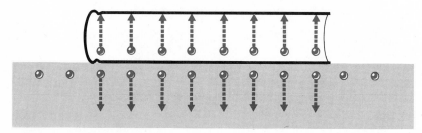

(c) The interatomic forces.

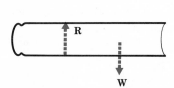

(d) A system consisting of the book only.

(e) A system consisting of the desk and the earth only.

Figure 3-10 Newton's laws applied to a book falling freely or resting on a desk. For the sake of clarity some of the forces have been displaced sideways.

The ratio of this upward acceleration of the earth to the downward acceleration of the book is

$$\frac{g_e}{g} = \frac{m}{M_e} \tag{3-30}$$

This equation can be derived from equations 3-23 and 3-29. It may also be obtained by applying Newton's third law to equate the magnitudes of the forces on the earth and book and Newton's second law to express these forces as mass times acceleration:

$$M_e g_e = mg \qquad (3\text{-}31)$$

Since the earth is much more massive than the book by a factor of about 10^{25}, the acceleration of the earth g_e is a factor 10^{25} smaller than g and is therefore negligible in practice, although it is present in principle.

If the book is placed on top of a desk it is still subject to the gravitational attraction of the earth, but it no longer accelerates downward because there is a compensating upward force exerted on it by the top of the desk (figure 3-10b). The atoms at the bottom of the book are in close proximity to the atoms on the top of the desk and they exert interatomic forces on one another (figure 3-10c). When the book is first placed on the desk its weight causes it to sink down, the atoms come closer and closer together and repel one another with stronger and stronger forces until the total upward repulsive force on the atoms at the bottom of the book is just sufficient to counteract the weight of the book. The interatomic forces obey Newton's third law and so, corresponding to the upward force **R** exerted on the book by the desk, there is an equal but opposite force **R'** exerted on the desk by the book:

$$\mathbf{R'} = -\mathbf{R} \qquad (3\text{-}32)$$

Now consider the application of Newton's second law to this situation. To apply the second law correctly, we must first concentrate our attention on some well defined part of the universe. We then add together all the forces exerted on this part by the whole of the rest of the universe and equate the resultant force to the total mass of the separated part multiplied by its acceleration. Let us do this for the book (figure 3-10d). Since the book is at rest and has zero acceleration, the resultant force on it must be zero. This force is composed of two parts, a downward force **W** due to the gravitational attraction of the earth and an upward force **R** exerted by the atoms on top of the desk, so

$$\mathbf{W} + \mathbf{R} = 0 \qquad (3\text{-}33)$$
$$\mathbf{W} = -\mathbf{R} \qquad (3\text{-}34)$$

Similarly, if we isolate a system consisting of the earth and the desk, but not the book (figure 3-10e), this also has zero acceleration and the resultant force on it is zero. This resultant force has two parts, an upward force **W'** on the earth due to the gravitational attraction of the book and a downward force **R'** on the top of the desk due to the atoms at the bottom of the book, so

$$\mathbf{W'} + \mathbf{R'} = 0 \qquad (3\text{-}35)$$
$$\mathbf{W'} = -\mathbf{R'} \qquad (3\text{-}36)$$

3-6 A Falling Book and a Book Resting on a Desk

This analysis may seem trivial and obvious, but it is important to distinguish between the roles played by the third law and the second law. The third law tells us that **W** is equal but opposite to **W'** and that **R** is equal but opposite to **R'**. It is the second law that tells us that **W** is equal but opposite to **R** and that **W'** is equal but opposite to **R'**, and then only because the accelerations are zero. It would be incorrect to adduce the third law to prove that **W** is equal and opposite to **R** because in the third law the forces that are equal and opposite act on different bodies, not on the same body. Notice also that **W** is a gravitational force and **R** is an electromagnetic force.

3-7 A Satellite in Orbit

A man-made satellite will continue to orbit about the earth for many revolutions even though it is not powered by any fuel, whereas an airplane must use up fuel if it is to maintain its altitude. There are some important differences between the satellite and the airplane. The presence of the air is essential to the flight of an airplane, but this implies that its motion is inevitably opposed by the frictional resistance of the air and it must use up fuel to overcome this resistance. A satellite, however, must be placed in orbit at a very high altitude where the atmosphere is so rarefied that it exerts very little frictional drag on the satellite. (Even so, it is this small frictional drag that eventually brings down the satellite after it has completed many revolutions around the earth.) Once it has reached this rarefied part of the atmosphere, the satellite moves in a stable orbit only if its velocity has exactly the right magnitude and direction for the orbit in question. The required magnitude of the velocity is greater than can be achieved by an airplane.

To avoid complicated mathematics and yet still illustrate the physical principles involved, consider the particularly simple case of a circular orbit of radius R (figure 3-11). Let M_e be the mass of the earth, M_s the mass of the satellite, and v the velocity that the satellite must have if it is to remain in this orbit. The situation will be described from the point of view of an observer in space watching the satellite revolve around the earth (and also watching the earth rotate on its axis, although this is not relevant to the present problem). The gravitational attraction of the earth gives the satellite

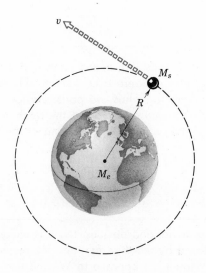

Figure 3-11 A satellite in orbit.

an acceleration in accordance with equation 3-1:

$$a_s = \frac{GM_e}{R^2} \tag{3-37}$$

By virtue of the fact that it is moving in a circle of radius R with velocity v, the satellite has an acceleration toward the center of the circle (section 2-7 and equation 2-20).

$$a_s = \frac{v^2}{R} \tag{3-38}$$

The circle is a possible stable orbit only if the velocity has exactly the right value to make these two expressions for the acceleration equal:

$$\frac{GM_e}{R^2} = \frac{v^2}{R} \tag{3-39}$$

Whence

$$v = \sqrt{\frac{GM_e}{R}} \tag{3-40}$$

The radius of the orbit of Sputnik I was about 4300 miles or 7×10^8 cm. Its velocity must therefore have been

$$v = \sqrt{\frac{6.67 \times 10^{-8} \times 6.0 \times 10^{27}}{7 \times 10^8}} \tag{3-41}$$

$$= 7.5 \times 10^5 \text{ cm/sec} \tag{3-42}$$
$$= 17,000 \text{ miles per hour} \tag{3-43}$$

Since the perimeter of its orbit was $2\pi R$, or about 4.4×10^9 cm, the time taken for a complete revolution was

$$T = \frac{2\pi R}{v} \tag{3-44}$$

$$= \frac{4.4 \times 10^9}{7.5 \times 10^5} \text{sec} \tag{3-45}$$

$$= 5.9 \times 10^3 \text{ sec} \tag{3-46}$$

3-7 A Satellite in Orbit

$$= 1.64 \text{ hour} \tag{3-47}$$

Reverting to the layman's question "Why does the satellite not fall down to the earth?" the only requirement that the earth's gravitational field imposes on the satellite is that it shall always have an acceleration directed toward the center of the earth with the magnitude given by equation 3-37. A body released from rest at a height achieves this acceleration by falling

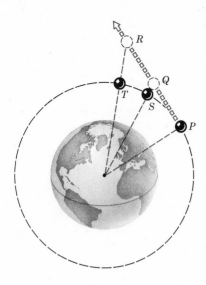

Figure 3-12 *In a certain sense, the satellite is falling toward the earth.*

directly downward. A satellite can achieve it equally well by moving continually in the same circle with just the right speed, since motion in a circle always involves an acceleration directed toward the center of the circle and given by equation 3-38. There is a sense in which the satellite *is* falling toward the earth. In the absence of the earth's gravitational attraction it would continue to move in a straight line tangential to the circle, in accordance with Newton's first law of motion. It would then occupy the successive positions, *P, Q, R* shown in figure 3-12. Actually, when it moves in a circular orbit it occupies the successive positions *P, S, T*, and this can be imagined to be the result of having fallen through the distances *QS* and *RT*.

We are now in a position to understand **weightlessness** inside a spaceship in orbit. Suppose that an astronaut holds up an object and then releases it. Just before and immediately after being released, the object has the same velocity **v** as the spaceship. However, this is exactly the right velocity to keep it moving in the same orbit as before. Notice that equation 3-40 does not contain the mass M_s of the orbiting object and therefore works just as well for the released object as for the spaceship. The object and the ship therefore continue to move as before with the same velocity in the same orbit and stay in the same positions relative to one another. The object remains suspended in mid-air. It does not need to fall up or down because it can acquire the necessary acceleration in the gravitational field merely by continuing in orbit as before.

Finally, let us realize that the arguments applied above to the motion of a man-made satellite around the earth are equally applicable to the motion of the moon around the earth, to the motion of any moon around any planet, or to the motion of any planet around the sun, as long as the appropriate radius and the appropriate mass are used. The ideas outlined above are the general principles underlying the theory of the motions of the planets and their moons in the entire solar system.

The Laws of Motion

Questions

A

1. Argue as strongly as you can in favor of the theory that a body will come to rest unless a force is exerted on it to keep it in motion. Having established a good case, refute it.
2. Taking into account the whole of your experience, do you believe that the future of the universe is completely predetermined?
3. If a body is instantaneously at rest, is the force on it necessarily zero?
4. In figure 3-8, why do the forces shown not produce a forward acceleration of the crate and a backward acceleration of the man? Remember that the crate is too heavy for the man to move.

B

5. How are interatomic forces involved in (a) the pull of a muscle when we lift a weight, (b) the tug of a tractor dragging a log along the ground, (c) the upward force exerted by the top of a desk to support a book, (d) the buoyancy of the air raising a balloon.
6. A satellite is moving around the earth once every three hours in a circular orbit that passes over both the north and south poles. A hydrogen bomb hangs from a string below the satellite. When the satellite is over the north pole, the pilot cuts the string. The bomb is timed to explode 90 minutes later. Where will the explosion take place and who will suffer most from it?
7. When a baseball strikes your head, why are you aware of the total mass of the baseball, rather than just the mass of a few atoms in contact with your skin?
8. A body of mass m is being whirled in a circle on the end of a string. The string suddenly snaps. Immediately afterward, what are the velocity and acceleration of the mass and what is the force on it, if this takes place (a) in outer space in a gravity-free region, (b) on the surface of the earth? In the latter case, does it matter whether the circle is horizontal or vertical?
9. A trapeze artist is hanging from a swinging trapeze at the end of a rope held between her clenched teeth. According to Newton's third law, which force is equal but opposite to her weight?

C

10. Is it possible to deduce the mass of a planet by a careful study of its motion around the sun? Is it possible to deduce the mass of a planet by a careful study of the motion of the other planets?
11. Does the time taken for a satellite to go once round the earth in a stable circular orbit increase or decrease as the radius of the orbit is made larger?
12. The force of friction due to the very rarefied outer atmosphere gradually slows down a satellite. According to equation 3-40 a smaller velocity corresponds to a larger orbit. In fact, however, the satellite does not move outward but spirals inward toward the earth and eventually burns out in the more dense lower atmosphere. Resolve this dilemma.

A

1. What is the magnitude of the force that produces an acceleration of 3.2 cm/sec^2 in a mass of 5 gm?
2. When a force of 25 dynes acts on a body it produces an acceleration of 150 cm/sec^2. What is the mass of the body?
3. A mass of 2 gm moves in a circle of radius 5 cm with a constant speed of 7 cm/sec. What is the magnitude of the force acting on the mass?
4. A mass of 5 gm is 20 cm away from a mass of 9 gm. What is the magnitude of the gravitational force exerted by one mass on the other?
5. A speck of dirt is 200 cm away from a stone with a mass of 100 gm. What is the gravitational acceleration of the speck due to the stone?
6. Using the data on the inside of the cover of this book, find the weight of one standard kilogram. Would it have exactly this weight on top of Mount Everest?
7. The mass of Jupiter is 1.9 $\times$ 10^{30} gm and its diameter is 1.4 $\times$ 10^{10} cm. What is the acceleration due to gravity on the surface of Jupiter?
8. What is the acceleration due to gravity at a distance from the center of the earth equal to the diameter of the earth?
9. At what distance from the center of the earth is the acceleration due to gravity equal to 490 cm/sec^2?
10. A mass of 2 $\times$ 10^3 gm has a weight of 1.6 $\times$ 10^6 dynes on a certain planet. What is the acceleration due to gravity on this planet?
11. A planet moves around a star in a circular orbit of radius 5 $\times$ 10^{13} cm with a speed of 3 $\times$ 10^7 cm/sec. What is the mass of the star?

B

12. What is the gravitational acceleration of the earth due to the attraction of (a) the moon and (b) the sun? Compare them with one another and with the acceleration due to gravity on the surface of the earth.
13. What is the gravitational force exerted on the sun by the earth?
14. If the electron in a hydrogen atom moves around the proton in a circle of radius 5.3 $\times$ 10^{-9} cm, what is the gravitational force exerted on the electron by the proton?
15. A book with a mass of 800 gm is projected across the top of a bench with an initial velocity of 60 cm/sec. It comes to rest after traveling 15 cm. What is the force of friction exerted on the book by the top of the bench?
16. A planet of mass 3 $\times$ 10^{29} gm moves around a star with a constant speed of 2 $\times$ 10^6 cm/sec in a circle of radius 1.5 $\times$ 10^{14} cm. What is the gravitational force exerted on the planet by the star?

C

17. A mass of 12 gm on the end of a thread is being whirled around in a circle of radius 2 m. If the thread breaks when the force on it is 2000 dynes, what is the maximum speed the mass can have? Ignore gravity.
18. At what point in space does the gravitational force of the moon on a spaceship exactly counteract the gravitational force of the earth? Do not assume that this point is on the line joining the earth to the moon, without first proving that it could not possibly lie to one side of this line. Could the point be on the far side of the moon from the earth, or on the far side of the earth from the moon? Why or why not?

19. Two astronauts have deserted their spaceship in a region of space far from the gravitational attraction of any other body. Each has a mass of 10^5 gm and they are 10^4 cm apart. They are initially at rest relative to one another. How long will it be before the gravitational attraction between them brings them 1 cm closer together? Express your answer in units which make it obvious how long a time this is (for example, is it about 1 minute? 1 day? 1 year?).

20. If the mass of the earth were increased by a factor of 9, by what factor would its radius have to be changed to keep g the same?

21. A satellite of mass 10^6 gm moves around a planet with a speed of 5×10^5 cm/sec in a circular orbit of radius 1.6×10^9 cm. If the radius of the planet is 7×10^8 cm, what is the acceleration due to gravity on its surface?

22. Calculate the radius of the orbit of the satellite that always stays vertically above the same place on the earth's surface.

4 Momentum, Angular Momentum, and Energy

4-1 Momentum

As we have already pointed out, it is more painful to be hit on the head with a baseball than with a table tennis ball. The mass is not the only thing that matters, though. The velocity is also important. To be hit on the head by a baseball dropped from a height of one inch is much less painful than if the baseball were dropped from the roof of the house at a height of 20 feet and thereby given time to build up a greater velocity. The direction of the velocity is equally important. The real danger arises from a fast ball coming straight at you, whereas a glancing blow from the same ball can be comparatively harmless.

What matters is the product of the mass m and the velocity $\mathbf{v}$. This is called the **momentum** and is represented by the symbol $\mathbf{p}$.

Momentum,

$$\mathbf{p} = m\mathbf{v} \tag{4-1}$$

Since the velocity **v** is a vector, the momentum **p** is also a vector and is in the same direction as the velocity. The plural of momentum is momenta.

Newton's second law of motion can be expressed in a different form, using momentum. Suppose that a body of mass m has a velocity $\mathbf{v}_1$ at a time t_1 sec, and that it has a slightly different velocity $\mathbf{v}_2$ at a slightly later time t_2. The change in velocity is the final velocity minus the initial velocity, $(\mathbf{v}_2 - \mathbf{v}_1)$, and this change has taken place during a time interval $(t_2 - t_1)$ sec. The acceleration **a** is therefore, according to equation 2-19,

$$\text{Acceleration} = \frac{\text{Change in velocity}}{\text{Time taken}} \tag{4-2}$$

$$\mathbf{a} = \frac{(\mathbf{v}_2 - \mathbf{v}_1)}{(t_2 - t_1)} \qquad \text{when } (t_2 - t_1) \text{ is very small} \tag{4-3}$$

Newton's second law of motion is

$$\text{Force} = \text{mass} \times \text{acceleration} \tag{4-4}$$

$$\mathbf{F} = m\mathbf{a} \tag{4-5}$$

$$= m\frac{(\mathbf{v}_2 - \mathbf{v}_1)}{(t_2 - t_1)} \tag{4-6}$$

$$= \frac{(m\mathbf{v}_2 - m\mathbf{v}_1)}{(t_2 - t_2)} \tag{4-7}$$

At the time t_1 the momentum $\mathbf{p}_1$ is $m\mathbf{v}_1$ and at time t_2 the momentum $\mathbf{p}_2$ is $m\mathbf{v}_2$. Therefore,

$$\mathbf{F} = \frac{(\mathbf{p}_2 - \mathbf{p}_1)}{(t_2 - t_1)} \qquad \text{when } (t_2 - t_1) \text{ is very small} \tag{4-8}$$

The new form of Newton's second law is

$$\boxed{\text{Force} = \text{Rate of change of momentum with time} \tag{4-9}}$$

In classical mechanics equations 4-4 and 4-9 are logically equivalent. However, in the theory of relativity the mass is no longer constant, but varies with the velocity. It is then found that the definition of force as mass multiplied by acceleration is unsatisfactory and that the definition as rate of change of momentum with time is much to be preferred. Moreover, as our exposition of physical theory gradually unfolds, it will be seen that the concept of momentum becomes more and more useful and develops more and more physical significance.

One reason why the concept of momentum is so useful is that, under the right circumstances, the total momentum of a system of interacting bodies remains constant as the motion proceeds. Consider a system of bodies that is well defined in the sense that it is clearly understood which bodies belong to

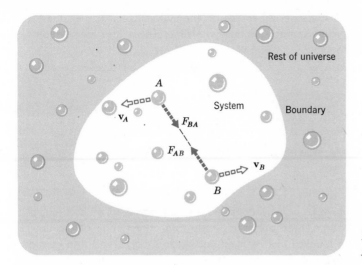

Rest of universe

System

Boundary

A

$\mathbf{v}_A$

$\mathbf{F}_{BA}$

$\mathbf{F}_{AB}$

$\mathbf{v}_B$

B

Figure 4-1 *A system of bodies
separated from the rest of the Universe.*

the system and which do not (figure 4-1). Let A and B be any two of these
bodies. Then, according to Newton's third law of motion, the force exerted
on A by B is equal and opposite to the force exerted on B by A:

$$\mathbf{F}_{AB} = -\mathbf{F}_{BA} \tag{4-10}$$

Since force is equal to the rate of change of momentum with time, the
change in the momentum of A produced by B is always exactly counterbalanced by the change in the momentum of B produced by A. Therefore,
the vectorial sum $(m_A\mathbf{v}_A + m_B\mathbf{v}_B)$ of the momenta of A and B is not
changed by the mutual forces between A and B, any change in $m_A\mathbf{v}_A$ produced by B being exactly compensated by an opposite change in $m_B\mathbf{v}_B$ produced by A. This argument is clearly applicable to any pair of bodies in the
system.

The **total momentum P** of the system is defined as the *vectorial* sum of
the momenta of the individual bodies in the system:

Total momentum, $\mathbf{P} = m_A\mathbf{v}_A + m_B\mathbf{v}_B + \cdots$

. . . summed for all the bodies in the system   (4-11)

Since all the *internal* forces between bodies inside the system occur in equal
and opposite pairs, these internal forces cannot change the value of the
total momentum, even though changes may occur in the individual terms of
the sum. Any change in the total momentum must be caused by *external*
forces exerted by bodies outside the system on bodies inside the system. In

fact:

The rate of change of the total momentum **P** with time is equal to
the resultant of all the *external* forces exerted on the system by
the rest of the universe.

If the resultant of all the external forces on the system is zero, then the total momentum of the system cannot change with time. This is the law of conservation of momentum.

Law of conservation of momentum

If the resultant force exerted on a system of bodies by the rest of the universe is zero, then the vectorial sum of the momenta of all the bodies in the system is a constant vector which does not change with time.

If the whole universe can be considered to be a system of bodies subject to no external force, then presumably the total momentum of the universe is constant. A quantity like momentum which is conserved and cannot be created or destroyed is of particular interest in physics. Later in the present chapter we shall consider two other such quantities, angular momentum and energy.

The proof of the law of conservation of momentum is critically dependent on the validity of Newton's third law of motion. In electromagnetism the third law is not always true and it is not obvious that momentum is conserved in this case. However, conservation of momentum is reestablished by introducing the concept of the momentum of electromagnetic radiation in free space (see section 8-4) and it appears that conservation of momentum is more basic than the third law.

4-2 Some Examples of Conservation of Momentum

Figure 4-2 illustrates a lecture demonstration apparatus which has recently become a popular toy. It can even be found on the desks of tired executives, who play with it so that the rationality of the laws of physics may soothe their nerves which have been frayed by the irrationality of human affairs. Five steel balls, all of the same size and mass, hang side by side in a row. Imagine that the end ball on the right is pulled outward, raised through a height h and then released from rest (part a of figure 4-2). As it swings downward its velocity increases and just before it strikes the other balls (part b) it has acquired a velocity **v** which depends on h. All five balls then collide, but immediately after the collision it is observed that the four balls on the right are stationary and only the end ball on the left is moving (part c).

Moreover it is moving with the same velocity **v** that the right-hand ball had before the collision because the left-hand ball rises through a height of exactly h before coming to rest (part d). It is easy to see why this follows by considering a single ball swinging like a pendulum as in figure 4-3. It falls through a height h and acquires a velocity **v** (a to b), but it is well known that this velocity is just right to enable it to rise through an equal height h on the other side (b to c). Returning to figure 4-2, the fact that the left-hand ball rises to a height h in part d therefore implies that it must have had a velocity **v** in part c.

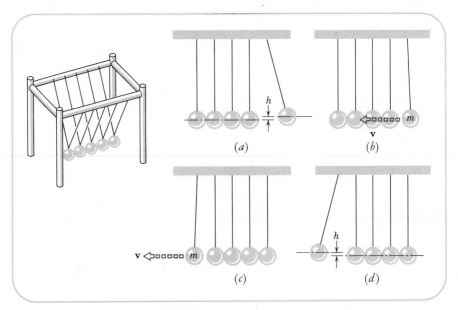

Figure 4-2 *A popular toy demonstrates the law of conservation of momentum.*

Figure 4-2 therefore illustrates conservation of momentum in a simple, direct fashion. Just before the collision (part *b*) the momentum is entirely due to the right-hand ball which has a mass *m* and a velocity **v**, and consequently a momentum *m***v**. Immediately after the complicated five ball collision, all the momentum is in the left-hand ball which also has a mass *m*, a velocity **v**, and a momentum *m***v** (part c). If the use of algebraic symbols confuses you, interpret these remarks as meaning that *m* has a definite numerical value (say 20 gm), which is the same for all the balls, and **v** is a definite velocity (say 70 cm/sec toward the left). In both *b* and *c* the momentum is 20 × 70 = 1400 gm cm/sec toward the left. Momentum is therefore conserved during the collision.

The more complicated collisions of figure 4-4 can now be readily understood. In figure 4-4*a* two right-hand balls are released from rest at a height *h* and strike the other balls with a velocity **v**. The total momentum just before the collision is therefore 2*m***v**. After the collision the three right-

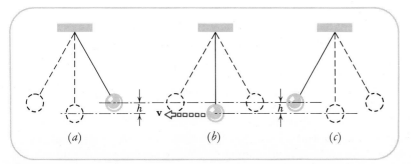

Figure 4-3 *During the swing of a simple pendulum a fall through a height h from (a) to (b) produces a velocity **v**. This velocity causes the ball to rise through an equal height h on the other side during the motion from (b) to (c).*

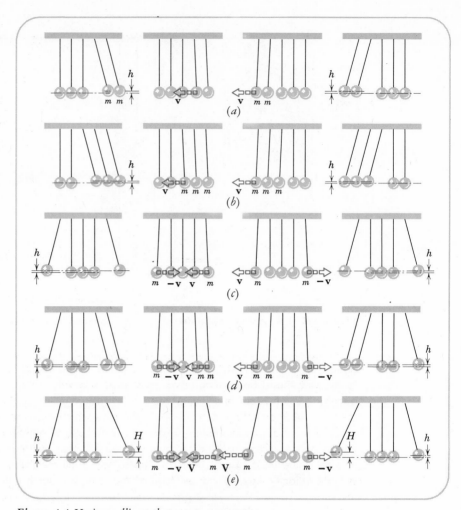

Figure 4-4 *Various collisons that conserve momentum.*

hand balls are stationary and *two* left-hand balls fly away with velocity **v** and total momentum 2m**v**, exactly equal to the total momentum just before the collision.

In figure 4-4b three balls are dropped and three balls fly off on the other side. The interesting aspect of this is that the middle ball keeps going.

In figure 4-4c both end balls are simultaneously released from a height *h*. The right hand ball strikes with a velocity **v** and a momentum m**v**. However, we must remember that velocity and momentum are vectors, and although the left-hand ball strikes with the same speed *v* it is moving in the opposite direction and its velocity is actually −**v** and its momentum is −m**v**. The total momentum before the collision is therefore m**v** − m**v** = 0, because the two momentum vectors are in opposite directions and cancel one another. After the collision the three central balls remain stationary but the end balls are observed to rebound with their velocities exactly reversed. The right-hand ball has velocity −**v** and momentum −m**v**, whereas the left-hand ball has velocity +**v** and momentum +m**v**. The total momentum is then −m**v** + m**v** = 0, the same as before the collision.

In figure 4-4*d* two right-hand balls are released simultaneously with one left-hand ball at the same height. The two right-hand balls strike with a momentum $+2m\mathbf{v}$ and the left-hand ball strikes with a momentum $-m\mathbf{v}$. The total initial momentum is therefore $+2m\mathbf{v} - m\mathbf{v} = +m\mathbf{v}$. After the collision *one* right-hand ball flies off with a momentum $-m\mathbf{v}$ and *two* left-hand balls fly off with momentum $+2m\mathbf{v}$ and so the total momentum remains $-m\mathbf{v} + 2m\mathbf{v} = +m\mathbf{v}$.

Finally, figure 4-4*e* introduces a new feature. Whereas the left-hand ball is still released from a height *h* and acquires a velocity $-\mathbf{v}$, the right-hand ball is released from a greater height *H* and acquires a velocity $\mathbf{V}$. The initial momentum is therefore $m\mathbf{V} - m\mathbf{v} = m(\mathbf{V} - \mathbf{v})$. The reader should by now be able to see that momentum is conserved if the left-hand ball flies off with a velocity $\mathbf{V}$ and rises to a height *H*, while the right-hand ball flies off with a velocity $-\mathbf{v}$ and rises to a height *h*.

A final comment should be made for the benefit of thoughtful students. In figure 4-2 the momentum in part *b* is equal to the momentum in part *c*. However, the total momentum in parts *a* and *d* is zero, because all the balls are then at rest. Thus, momentum is not conserved during the motion from *a* to *b* or from *c* to *d*. The reason for this is that during the motions from *a* to *b* and from *c* to *d* there are *external* forces, the gravitational forces responsible for the weights of the balls, and these forces have plenty of time to change the momenta of the swinging balls. However, during the transition from *b* to *c*, the multiball collision takes place very rapidly and the external forces are not given sufficient time to change the momentum by any significant amount.

4-3 Angular Momentum

The momentum $m\mathbf{v}$ that we have just discussed is sometimes called *linear* momentum because it is associated with the motion of a body along a line. There is a similar quantity, called *angular* momentum, which is associated with rotation about an axis. In the simple case of a particle of mass *m* rotating with a speed *v* in a circle of radius *r*, the angular momentum A is defined as *mvr* (see figure 4-5).

Angular momentum of a particle moving in a circle

$$A = mvr \tag{4-12}$$

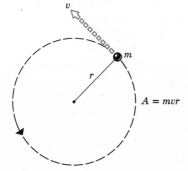

$A = mvr$

Figure 4-5 *The angular momentum of a particle moving in a circle is mvr.*

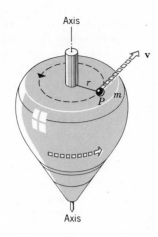

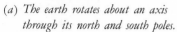

(a) *The earth rotates about an axis through its north and south poles.*

(b) *A spinning top.*

Figure 4-6.

Now consider a rotating solid body, such as the earth rotating about an axis through its north and south poles, or a spinning top (figure 4-6). Any very small part P of the spinning top moves around the axis of rotation in a circle and its angular momentum is mvr. However, v and r are different for different parts of the top. Nevertheless, the total angular momentum of the top may be defined simply by adding together the contributions mvr from all the small parts such as P which make up the whole top. Notice that m is not the mass of the whole top, but of a small part P of it. Moreover, v and r refer to the small part P and have no meaning for the top as a whole.

Angular momentum of a rotating solid body

A is the sum of contributions of the form mvr from a large number of small parts that together make up the whole body.

To change the angular momentum by changing the rotational velocities v, it is necessary to apply forces that have a "twisting" effect. To see how this can be done, grasp the bottom right-hand corner of this book in your right hand and grasp the bottom left-hand corner in your left hand. With the right hand push the book directly away from you and at the same time pull it directly toward you with the left hand, using a force equal in magnitude but exactly opposite in direction to the force of the right hand (figure 4-7). Since the two forces exactly compensate one another, the resultant force on the book is zero, its acceleration is zero, and so this procedure cannot start the book moving bodily away from you. What happens, of course, is that the book rotates in a counterclockwise direction. The reason is that, although the forces are equal but opposite, they are displaced sideways from one another and therefore produce a turning effect.

Momentum,
Angular
Momentum,
and Energy

66

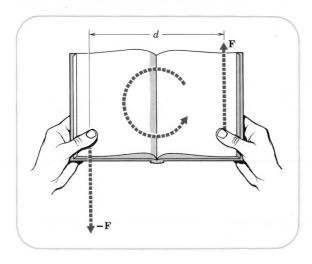

Figure 4-7 *A couple rotating a book.*

A pair of equal but opposite forces displaced sideways from one another is called a **couple.** If the magnitude of each force is F and their perpendicular distance apart is d, the **moment** of the couple is defined in the following way.

Moment, or torque, of a couple,

$$\tau = Fd \tag{4-13}$$

It will be represented by τ, which is the Greek letter tau. An alternative name for the moment of a couple is the **torque** exerted by the couple.

The moment, or torque, plays the same role for rotational motion that force plays for motion along a straight line. The resultant external force **F** acting on a system of bodies is equal to the rate of change of the total momentum **P** of the system. Similarly, the external *torque* τ is equal to the rate of change of the *angular* momentum A.

Torque = Rate of change of angular momentum with time (4-14)

4-4 *Conservation of Angular Momentum*

If there is no external torque, the total angular momentum remains constant. This is the **law of conservation of angular momentum,** which can be demonstrated in the following way. A man stands on a slowly rotating turntable mounted on ball bearings that introduce very little friction (figure 4-8*a*). Initially his arms are outstretched and each hand holds a mass m at a distance R from the axis of rotation. The two masses make a contribution $2mvR$ to the angular momentum of the rotating system. He then bends his arms and pulls in the two masses until their distance from the axis has the much smaller value r (figure 4-8*b*). Their contribution to the angular momentum is thereby reduced. Since the moment of the couple exerted on the sys-

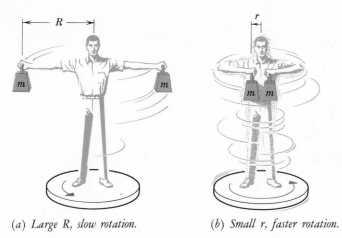

(a) Large R, slow rotation.   (b) Small r, faster rotation.

Figure 4-8 A demonstration to illustrate the law of conservation of angular momentum.

tem by the frictional forces due to the ball bearings is negligibly small, the total angular momentum of the system cannot change. The only way in which this can be achieved is by a sudden increase in the velocities of all parts of the rotating system. The result of the maneuver is that the man suddenly rotates much faster. If he extends his arms again, the rotation immediately slows down. This principle is used by a figure skater executing a spin (figure 4-9). If she starts her spin slowly with outstretched arms, she can increase her rate of rotation suddenly by folding her arms across her chest. She can slow down by extending her arms again.

Linear momentum is a vector with direction as well as magnitude. *Angular* momentum can also be associated with a direction, the direction of the axis about which the body is rotating. The vector representing angular momentum therefore has a length A and a direction parallel to the axis of rotation. There are two such directions, opposite to one another, and it is necessary to establish a convention for choosing between them. The convention is that the body shall be seen to be rotating clockwise if we look along the axis of rotation in the direction of the arrowhead on the angular momentum vector. This is illustrated in figure 4-10.

The vectorial nature of angular momentum can be demonstrated in the

Figure 4-9 A figure skater increasing her rate of rotation.

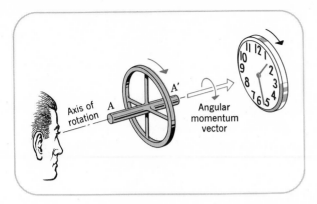

Figure 4-10 *Representation of angular momentum by a vector.*

following way. The man on the turntable holds a bicycle wheel which is rotating about a horizontal axis (figure 4-11a). Initially the turntable is not rotating and the total angular momentum of the system is represented by a *horizontal* vector. The vertical component of the angular momentum is zero. Since the ball bearings cannot exert an appreciable couple about a vertical axis, this vertical component must subsequently remain zero. If the man turns the axle of the wheel so that it has an angular momentum about a vertical axis (figure 4-11b), then the man and the turntable start to rotate in the opposite direction to the wheel. The vertical vector representing the angular momentum of the man and the turntable must be exactly equal and opposite to the vector representing the angular momentum of the wheel. If the man then turns the axle of the wheel through 180°, he reverses his own direction of rotation so that his angular momentum still cancels that of the wheel (figure 4-11c).

Angular momentum has a very basic significance in physics. Some of the fundamental particles, in particular electrons, protons, and neutrons, are found to be spinning like tops. Moreover, the angular momentum is the same for each of these particles and is equal to $h/4\pi$, where h is a fundamental constant known as Planck's constant, which plays a basic role in quantum mechanics. Angular momentum is therefore a distinctive property of the fundamental particles and we shall have occasion to consider it later in this book.

4-4 Conservation of Angular Momentum

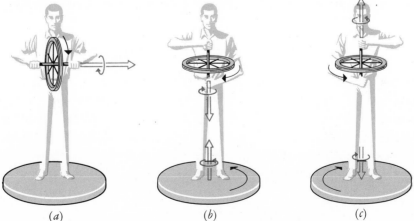

(a) (b) (c)

Figure 4-11 *A demonstration of the vectorial nature of angular momentum.*

4-5 Energy

Because the concept of energy is so familiar to us, its physical nature is correspondingly more difficult to understand. We know, for example, that gasoline contains energy and if we pour gasoline into a car this energy can be converted into energy of motion of the car. It is very tempting to visualize energy as though it were a sparkling fluid that is added to a body in order to make it move. Conservation of energy is then a simple consequence of the fact that the amount of fluid does not change. This attitude is too simpleminded and frequently very misleading. The best way to combat it is to emphasize the precise mathematical definition of the various forms of energy, although unfortunately it will not be possible to give rigorous mathematical proofs in a book at this level.

The point that must be made at the very beginning is that energy is another quantity like momentum which is conserved for a system of bodies subject to no external interference. The energy of a system of bodies is the sum of several parts, such as kinetic energy, gravitational potential energy, electromagnetic potential energy, and so on, which are quite different from one another in character. The problem is to decide how precisely to define the various forms of energy so that the law of conservation of energy will apply. The energy of the gasoline is then understood when it is realized that gasoline is composed of atoms and atoms are composed of negatively charged electrons and positively charged nuclei. The energy of the gasoline is ultimately discovered to be the kinetic energy and electromagnetic potential energy of these electrons and nuclei. So let us first consider the definitions of kinetic energy and potential energy.

Kinetic energy is energy of motion. If a particle has mass m and velocity v, its kinetic energy is $\frac{1}{2}mv^2$.

$$\text{Kinetic Energy} = \tfrac{1}{2}mv^2 \qquad\qquad (4\text{-}15)$$

Energy is a scalar quantity. There is no direction associated with it. Kinetic energy is always positive. In the CGS system, when m is in gm and v in cm/sec, the unit of energy is the **erg.** Thus, a mass of 1 gm moving with a velocity of 1 cm/sec has a kinetic energy of $\frac{1}{2}$ erg.

Kinetic energy is the energy that a single body possesses by virtue of its motion. **Potential energy** is a mutual property of a *pair* of bodies resulting from their proximity to one another and the fact that they interact with one another. Suppose, for example, that a body of mass m_1 is at a distance R from a body of mass m_2. Then their mutual *gravitational* potential energy is represented by Φ_G (the Greek letter capital phi with a subscript G to remind us that we are talking about gravitation). It is

Gravitational potential energy,

$$\Phi_G = -\frac{Gm_1m_2}{R} \qquad\qquad (4\text{-}16)$$

Notice that, in contradistinction to the gravitational *force*, the potential energy varies inversely as the distance apart, R, not as the square of R. Notice also that gravitational potential energy is negative, whereas kinetic energy is positive. When the two bodies are infinitely far apart, their mutual potential energy is zero (when the R in the denominator of equation 4-16 is very large, the fraction is very small). As they approach closer together, their mutual potential energy becomes more and more negative.

Gravitational potential energy is associated with gravitational interactions. There are corresponding types of potential energy associated with other interactions. For example, if two stationary electric charges, q_1 and q_2, are at a distance R apart, their mutual *electrostatic* potential energy is

Electrostatic potential energy,

$$\Phi_e = \frac{q_1 q_2}{R} \tag{4-17}$$

in the CGS system

If the charges are in motion, the situation is more complicated. However, we shall ignore such complications since the electrostatic potential energy just defined is often by far the most important part of the electromagnetic energy. Remember that the electric charges q_1 and q_2 can be positive or negative. If they are both positive or both negative, Φ_e is positive. But if one is positive and the other is negative, Φ_e is negative. Like gravitational potential energy, electrostatic potential energy is zero when the charges are infinitely far apart.

Now imagine a system of bodies in motion exerting gravitational and electrical forces on one another, but not subjected to any appreciable interference by the rest of the universe (figure 4-12). The bodies have masses m_1, m_2, m_3, and so on, and may also have electric charges q_1, q_2, q_3, and so on. At a particular instant of time they have velocities v_1, v_2, v_3 and so on, but these velocities change as the bodies are accelerated by the forces they exert on one another. The exact way in which this happens is determined by the laws of motion discussed in Chapter 3. These laws enable us to write down for each body an equation in which the product of its mass and its acceleration is put equal to the sum of all the forces that the other bodies in the system exert on it. The mathematical solution of these equations enables us to predict the future motion of all the bodies.

However, in addition to this prediction it is also possible to manipulate the equations mathematically in such a way as to prove that a certain algebraic expression has a constant numerical value that always remains the same as the motion proceeds. This number is the total energy of the system and the fact that it is constant is the law of conservation of energy. The algebraic expression is complicated and contains several different kinds of terms. It contains the term $\frac{1}{2}m_1 v_1^2$ and a similar term of the form $\frac{1}{2}mv^2$ for each body in the system. This part of the expression represents the positive kinetic energy. In addition, there is a term of the form $(-Gm_1 m_2/R_{12})$ for each *pair* of bodies in the system. R_{12} is the distance between the masses

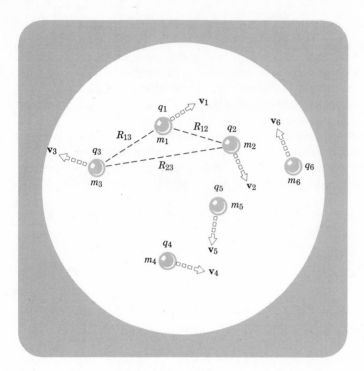

Figure 4-12 *A system of bodies exerting forces on one another, but isolated from the rest of the universe.*

m_1 and m_2, and this part of the expression represents the negative gravitational potential energy. Also, for each pair of charged bodies in the system there is a term of the form (q_1q_2/R_{12}), and this part of the expression represents the electrostatic potential energy, which may be positive or negative. The whole expression is the sum of the kinetic energies, the gravitational potential energies, and the electrostatic potential energies. It is this sum of three different kinds of energy that remains constant as the motion proceeds.

Energy always has this significance. In a particular physical situation, which might be much more complicated than the one we have considered, the equations of motion can always be manipulated to yield a mathematical expression that retains a constant numerical value throughout the motion. The various different parts of this expression can then be identified with various types of energy.

4-6 *A Body Near the Surface of the Earth*

Momentum, Angular Momentum, and Energy

Figure 4-13*a* shows a body of mass m resting on the surface of the earth. Its distance from the center of the earth is R_e, the radius of the earth. As before, the mass of the earth is denoted by M_e. The mutual gravitational potential energy of the earth and the body m is

$$\Phi_G = -\frac{GM_em}{R_e} \tag{4-18}$$

Now suppose that the body is raised through a height h (figure 4-13*b*) so that its distance from the center of the earth becomes $(R_e + h)$. Then the

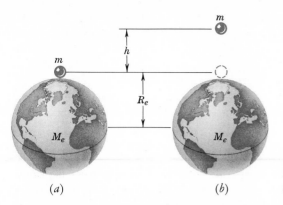

(a)   (b)

Figure 4-13 *Increasing the potential energy of a body by raising it above the surface of the earth.*

new potential energy is

$$\Phi'_G = -\frac{GM_e m}{(R_e + h)} \qquad (4\text{-}19)$$

The increase in potential energy is the new value minus the old value, or

$$\Phi'_G - \Phi_G = -\frac{GM_e m}{(R_e + h)} - \left(-\frac{GM_e m}{R_e}\right)$$

$$= GM_e m \left(\frac{1}{R_e} - \frac{1}{R_e + h}\right)$$

$$= GM_e m \frac{[(R_e + h) - R_e]}{R_e(R_e + h)}$$

$$= \frac{GM_e mh}{R_e(R_e + h)} \qquad (4\text{-}20)$$

In many problems involving bodies moving in the vicinity of the surface of the earth the height h is very small compared with the radius of the earth, R_e. Then very little error is introduced if the expression $(R_e + h)$ in the denominator is replaced by R_e. The increase in potential energy then becomes

$$\Phi'_G - \Phi_G = \frac{GM_e}{R_e^2} mh \qquad (4\text{-}21)$$

4-6 A Body Near the Surface of the Earth

In section 3-5 of Chapter 3 it was proved that the acceleration due to gravity near the surface of the earth is

$$g = \frac{GM_e}{R_e^2} \qquad (4\text{-}22)$$

Therefore, equation 4-21 for the increase in potential energy may be changed to

$$\Phi'_G - \Phi_G = gmh \qquad (4\text{-}23)$$

Changes in gravitational potential energy near the
surface of the earth

When a body of mass *m* near the surface of the earth is *raised*
through a height *h* small compared with the radius of the earth,
the *increase* in gravitational potential energy is *mgh*.

The mutual gravitational potential energy of the earth and the body is
negative, but positive energy must be added to the system to raise the body.
Let us look into this in more detail. When the body is resting on the surface
of the earth, let us suppose that the potential energy, which is given by the
expression $(-GM_em/R_e)$, has the value -1000 ergs. When the body is
raised, the potential energy is given by $(-GM_em)/(R_e + h)$. Increasing the
denominator makes the fraction smaller, and its new value may be supposed
to be -999 ergs. To raise the body and change the potential energy from
-1000 ergs to -999 ergs, it is necessary to *add* a positive energy of $+1$
erg, which is given by the expression $+mgh$.

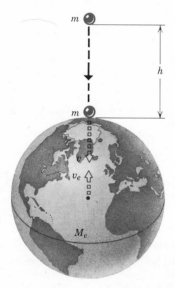

Figure 4-14 *Conservation of energy and momentum for a falling body.*

Momentum,
Angular
Momentum,
and Energy

Now suppose that the body is released from rest at a height *h* and
falls to the ground (figure 4-14). We may apply the law of conservation of
energy to a system consisting of the earth and the body. As a result of the
fall through a height *h*, the potential energy has decreased by an amount
mgh. However, the body is continually accelerating as it falls and hits the
ground with a velocity *v* and a kinetic energy $\frac{1}{2}mv^2$. When it was at rest at
a height *h*, *v* was zero and the kinetic energy was zero. The increase in
kinetic energy is therefore $\frac{1}{2}mv^2$ and this must exactly compensate for the
decrease in potential energy, *mgh*, so that the total energy may remain constant.

$$\tfrac{1}{2}mv^2 = mgh \qquad\qquad (4\text{-}24)$$

Cancelling m and multiplying both sides by 2,

$$v^2 = 2gh \qquad\qquad (4\text{-}25)$$

So,

$$v = \sqrt{2gh} \qquad\qquad (4\text{-}26)$$

The law of conservation of energy therefore enables us to calculate very easily the velocity that a body acquires after falling through a given height.

We have cheated slightly because as the body falls downward the earth must fall upward toward the body (see section 3-5). In fact the law of conservation of momentum must be obeyed, and the total momentum must always be zero, since it is initially zero when both the earth and the body have zero velocity just before the body is released at the height h. The downward momentum mv acquired by the body just before it hits the ground must be compensated by an upward momentum $M_e v_e$ of the earth.

$$mv = M_e v_e \qquad\qquad (4\text{-}27)$$

Therefore, as the body strikes the earth, the earth is moving upward to meet it with a velocity

$$v_e = \frac{m}{M_e} v \qquad\qquad (4\text{-}28)$$

However, the mass m of the body is usually very much smaller than the mass M_e of the earth. The fraction m/M_e is therefore very small and the velocity v_e of the earth is very much smaller than the velocity v of the body. Moreover, the kinetic energy of the earth is

$$\frac{1}{2} M_e v_e{}^2 = \frac{1}{2} M_e \left(\frac{mv}{M_e}\right)^2 \qquad\qquad (4\text{-}29)$$

$$= \frac{m}{M_e} \frac{1}{2} mv^2 \qquad\qquad (4\text{-}30)$$

The kinetic energy of the earth is equal to the kinetic energy of the body multiplied by the very, very small fraction m/M_e. It follows that a negligible part of the decrease in potential energy is used to give kinetic energy to the earth.

A closer look at some aspects of the situation convinces us that conservation of energy is not always realized in so simple a way. The original increase mgh in the potential energy might be achieved by a man who carries the body up the stairs to the top of a tall building. How is the man able to supply this energy? That is a complicated physiological question, but the original source of the energy is the food that the man has eaten. During his metabolic processes there are chemical reactions that release **chemical energy.** The atoms taking part in these chemical reactions are composed of electrons, with negative electric charge, and nuclei with positive electric charge. During a chemical reaction, the electrons and nuclei are rearranged and change their motions. The total energy of the atoms is made up of the kinetic energies of the electrons and nuclei and the electrostatic potential energies of pairs of these charged particles. As a result of the rearrangement produced by the chemical reaction, there is a decrease in this total energy and energy is made available to be used in some other way. The complicated physiological processes going on in the man's muscles enable him to use part of this available energy to raise the body and increase its gravitational potential energy.

Another interesting situation arises when the body strikes the ground and imbeds itself into the soft earth. The kinetic energy suddenly disappears, but without any corresponding increase in gravitational potential energy. Where has the lost energy gone? As the body imbeds itself in the ground, the body and the earth warm up and the lost kinetic energy reappears as **heat energy.** Actually, as we shall explain in the next chapter, when the temperature of a body is increased, its atoms are made to move faster. The lost energy therefore reappears as kinetic energy of the random motion of the atoms in the body and the earth.

4-7 Rocketry

The launching of a rocket provides a good example of the application of the laws of conservation of momentum and energy. Figure 4-15 illustrates the basic principles underlying the·operation of a rocket. The two fuels are mixed in the combustion chamber, where they undergo a violent chemical

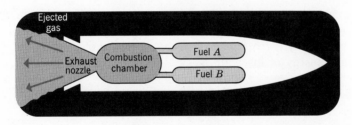

Figure 4-15 Rocket propulsion.

reaction. The products of this reaction are hot, high pressure gases that escape through the exhaust nozzle with a high velocity and thereby acquire a large backward momentum. The total momentum of the system must remain constant and so the backward momentum acquired by the ejected gases must be compensated by an equal and opposite forward momentum given to the rocket. The forward velocity of the rocket therefore continually increases as long as the fuel is being burnt, but not, of course, after the fuel has been used up.

Consider the firing of a rocket from the point of view of a system which includes the rocket, the fuel, and the earth. The advantage of this system is that the gravitational forces between the rocket and the earth are *internal* to the system, whereas the external gravitational forces exerted by the other heavenly bodies are too weak to matter. It is therefore permissible to apply the law of conservation of momentum to this system. Assume, as is usually the case, that the fuel is all spent before the rocket leaves the earth's atmosphere. During the firing the rocket acquires an upward momentum and the ejected gases a downward momentum. However, the ejected

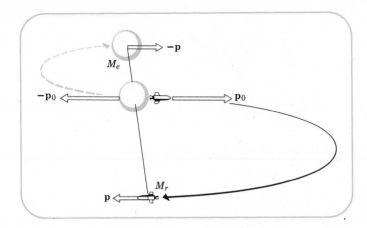

Figure 4-16 *Conservation of momentum applied to the system that includes the rocket and the earth. The relative size of the earth's orbit has been exaggerated for the sake of clarity. It is actually very small indeed compared with the orbit of the rocket.*

gases mingle with the earth's atmosphere and eventually transfer their momentum to the earth. Therefore, when the firing is complete and the rocket has escaped from the earth's atmosphere, the earth has recoiled in the opposite direction to the motion of the rocket and has a momentum $-\mathbf{p}_0$ exactly equal and opposite to $\mathbf{p}_0$, the momentum of the rocket. The total momentum of the system earth plus rocket remains unchanged (figure 4-16).

Subsequently the rocket moves in a curved trajectory as the earth's gravitational force acts on it. However, according to Newton's third law, the force exerted by the earth on the rocket is exactly equal but opposite to the force exerted by the rocket on the earth. The total momentum of the combined system of earth plus rocket therefore remains zero. At any subsequent instant when the rocket has a momentum $\mathbf{p}$, the earth has an equal but opposite momentum $-\mathbf{p}$. The earth therefore describes an orbit that is a sort of inverted copy of the orbit of the rocket, but greatly reduced in size (figure 4-16). It is reduced in size because, although the momentum of the earth is always equal in magnitude to the momentum of the rocket, the mass of the earth is so much greater than the mass of the rocket that the velocity of the earth must be very much smaller than the velocity of the rocket. This is similar to the situation discussed in the previous section in connection with a falling body, and an argument similar to the one used there can convince us that the kinetic energy of the earth is always negligible compared with the kinetic energy of the rocket.

The law of conservation of energy will now be used to derive the "escape velocity," which is the smallest speed with which the rocket must be launched in order to escape completely from the earth's gravitational attraction. Situation 1 of figure 4-17a is the instant just before the firing starts. The total *momentum* is then zero because the earth and the rocket are at rest. Situation 2 (figure 4-17b) is the instant just after the firing is complete and all the fuel is spent. The rocket then has a velocity V_r and the earth is recoiling with a velocity V_e. Momentum is conserved during the firing and, if M_r is the mass of the rocket,

$$M_r V_r = M_e V_e \tag{4-31}$$

The kinetic energy of the earth may be neglected and so the kinetic

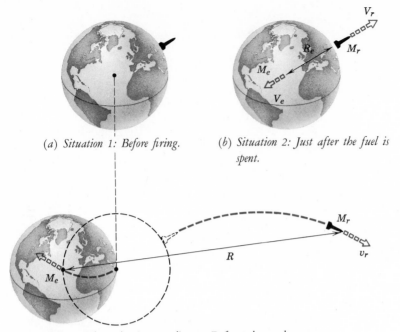

(a) *Situation 1: Before firing.*

(b) *Situation 2: Just after the fuel is spent.*

(c) *Situation 3: The rocket is at a distance R from the earth.*

Figure 4-17 *A rocket escaping from the earth's gravitational attraction.*

energy in situation 2 is

$$\text{Kinetic energy in situation 2} = \tfrac{1}{2}M_r V_r^2 \tag{4-32}$$

The burning of the fuel takes place quickly and the rocket has not traveled very far, so the distance of the rocket from the earth is still very nearly R_e and the mutual gravitational potential energy of the rocket and the earth in situation 2 is

$$\text{Potential energy in situation 2} = -\frac{GM_e M_r}{R_e} \tag{4-33}$$

The total energy is the sum of the kinetic and potential energies.

Total energy in situation 2

$$= \frac{1}{2}M_r V_r^2 - \frac{GM_e M_r}{R_e} \tag{4-34}$$

As the rocket climbs away from the earth, the earth's gravitational attraction tries to pull it back and reduces its velocity. If it has been launched with too small an initial velocity V_r, the pull of the earth eventually turns it around and it starts to fall back again. In situation 3 (figure 4-17c) the rocket has climbed to a distance R from the center of the earth and its velocity has been reduced to v_r. Under these circumstances,

$$\text{Kinetic energy in situation 3} = \tfrac{1}{2}M_r v_r^2 \tag{4-35}$$

$$\text{Potential energy in situation 3} = -\frac{GM_eM_r}{R} \tag{4-36}$$

Total energy in situation 3

$$= \frac{1}{2}M_rv_r^2 - \frac{GM_eM_r}{R} \tag{4-37}$$

During the motion that converts situation 2 into situation 3 the earth and the rocket constitute a system subject to no external interference and the total energy of this system must be conserved. The total energy in situation 2 must equal the total energy in situation 3.

$$\frac{1}{2}M_rV_r^2 - \frac{GM_eM_r}{R_e} = \frac{1}{2}M_rv_r^2 - \frac{GM_eM_r}{R} \tag{4-38}$$

To understand what is going on, let us put in some numbers. Suppose that the initial kinetic energy $\frac{1}{2}M_rV_r^2$ is $+700$ ergs and the initial potential energy $(-GM_eM_r/R_e)$ is -1000 ergs. (These numbers are chosen for convenience of exposition. Realistic numbers for actual rockets would be very much larger.) Then the initial total energy is $+700 - 1000 = -300$ ergs. When the rocket has doubled its distance from the earth, $R = 2R_e$, and the negative potential energy is halved and becomes -500 ergs. Since the total energy must remain -300 ergs, the kinetic energy $\frac{1}{2}M_rv_r^2$ must have been reduced to $+200$ ergs, because $+200 - 500 = -300$. In general, as the rocket moves away v_r becomes smaller, the positive kinetic energy $\frac{1}{2}M_rv_r^2$ becomes smaller and the positive energy thus made available is needed to make the potential energy less negative.

If the rocket is to escape completely, it must be possible for R to become infinitely large. The potential energy $(-GM_eM_r/R)$ is then 0, whereas in situation 2 just after launching it was $(-GM_eM_r/R_e)$. The potential energy therefore increases by $(+GM_eM_r/R_e)$ and this increase is achieved at the expense of the kinetic energy. For this to be possible the initial kinetic energy must be at least $(+GM_eM_r/R_e)$. The minimum speed of launching before the rocket can escape is therefore given by the equation

$$\frac{1}{2}M_rV_r^2 = \frac{GM_eM_r}{R_e} \tag{4-39}$$

Whence

$$V_r^2 = \frac{2GM_e}{R_e} \tag{4-40}$$

$$V_r = \sqrt{\frac{2GM_e}{R_e}} \tag{4-41}$$

The values of the gravitational constant G, the mass of the earth M_e, and the radius of the earth R_e may be found inside the cover of this book.

Inserting them into equation 4-41 we find that

$$\text{Escape velocity} = \sqrt{\frac{2 \times 6.67 \times 10^{-8} \times 6.0 \times 10^{27}}{6.4 \times 10^8}} \quad (4\text{-}42)$$

$$= 1.11 \times 10^6 \text{ cm/sec} \quad (4\text{-}43)$$

$$= 24{,}800 \text{ mph} \quad (4\text{-}44)$$

Finally, notice that we cannot relate situation 1 to situation 2 by applying the law of conservation of energy in the form of an equation similar to equation 4-38. This is because, during the burning of the fuel, the chemical energy of the fuel is being converted into kinetic energy of the rocket. We then have a situation involving forms of energy other than kinetic energy and gravitational potential energy.

Questions

1. Two identical lumps of clay traveling with the same speed in exactly opposite directions collide and stick together. What can you conclude from the law of conservation of momentum? What about conservation of energy?
2. Consider each of the collisions in figures 4-2 and 4-4 from the point of view of conservation of *energy*.
3. Can you imagine circumstances under which the total momentum of a system of bodies is conserved, but its total energy is not?
4. A boy sleds down a slope and ends up in a snow bank. Where did his kinetic energy come from and where does it go to?

5. How does the momentum of a satellite in a circular orbit vary with time? How does its magnitude vary with the radius of the orbit?
6. How does the law of conservation of momentum apply to a car accelerating along a straight horizontal highway?
7. If the angular momentum of the earth due to its daily rotation is represented by a vector, does this vector point from the south toward the north or from the north toward the south?
8. An automobile is traveling at a steady speed along a straight horizontal road. What happens to the energy released from the gasoline that is used up?

9. A firework is projected vertically upward. At its highest position it explodes into three pieces with equal masses and equal speeds. What can be said about the directions of their velocities?
10. How does the law of conservation of momentum apply to the motion of the earth around the sun? (Hint: Is the sun really stationary at the center of the orbit? Suppose the sun and the earth had the same mass?)
11. Why is it difficult to design a helicopter with only one propeller?
12. If you lift a book from the floor and place it on a desk, its gravitational potential energy increases. Where did the energy come from? Trace back the source of the energy as far as you can, in the hope of finding the ultimate source of all energy on the surface of the earth.

Problems

A

1. If a proton has a velocity of 3×10^6 cm/sec, what is its momentum?
2. If an electron has a momentum of 2×10^{-22} gm cm/sec, what is its velocity?
3. An electron moves in a circle of radius 4 cm. What must its speed be if its angular momentum about the center of the circle is $h/2\pi = 1.05 \times 10^{-27}$ CGS units?
4. If a mass of 3 gm has a velocity of 4 cm/sec, what is its kinetic energy?
5. If a mass of 0.7 gm has a kinetic energy of 140 ergs, what is its velocity?
6. If an electron has a kinetic energy of 10^{-10} erg, what is its velocity?
7. What is the mutual gravitational potential energy of two stars of masses 5×10^{32} gm and 3.5×10^{33} gm when the distance between them is 2×10^{14} cm?
8. A 200-gm book is carried from the street to a second floor room 4.5×10^2 cm higher. What is the change in its gravitational potential energy?

B

9. A spaceman is stationary in a gravity-free region of space. His total mass, including all his equipment, is 10^5 gm. He takes off his oxygen tank, which has a mass of 10^4 gm, and hurls it away with a velocity of 200 cm/sec. With what velocity does he recoil?
10. Two spacemen are both floating with zero velocity in a gravity-free region of space. Spaceman A has a mass of 1.2×10^5 gm and spaceman B has a mass of 9×10^4 gm. A pushes B away from him toward the door of their spaceship. If B's velocity is 50 cm/sec, with what velocity does A recoil?
11. Two skaters are stationary in the center of a circular rink. They then push on one another so that they fly apart. One of the skaters has a mass of 9×10^4 gm and acquires an initial velocity of 80 cm/sec. If the other skater has a mass of 7.5×10^4 gm, what is his initial velocity?
12. Referring to the diagram for this problem, the two small 3 gm balls are connected by a stout wire of negligibly small mass. A large 2×10^3 gm ball exerts a gravitational force of attraction on a nearby small ball. The gravitational force that it exerts on the small ball furthest away from it can be neglected. Calculate the couple on the system consisting of the two small balls and the wire joining them.

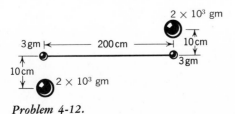

Problem 4-12.

13. A phonograph record is rotating at $33\frac{1}{3}$ rpm. A fly of mass 0.2 gm is

stuck to the record at a distance of 8 cm from the center. What is the angular momentum of the fly?

14. If a proton has a momentum of 10^{-18} gm cm/sec, what is its kinetic energy?

15. A mass of 400 gm is dropped from rest at a height of 1.5×10^3 cm. What is its kinetic energy just before striking the ground? (It is not necessary to calculate the velocity of the body.)

16. If the momentum of a body is doubled, what is the ratio of its final to its initial kinetic energy? If the kinetic energy of a body is doubled, what is the ratio of its final to its initial momentum?

17. Referring to figure 4-2, find a relationship between the speed v with which a falling ball strikes the other balls and the height h through which it has fallen.

18. A lump of clay with a mass of 80 gm and a velocity of 200 cm/sec overtakes another lump of clay with a mass of 120 gm and a velocity of 150 cm/sec in exactly the same direction. What is their common velocity after they coalesce?

19. If a body of mass 300 gm is dropped from rest at a height of 200 cm, what is its momentum just before it strikes the ground?

20. What is the angular momentum about an axis through the north and south poles of a 1 kg mass resting on the ground at the equator?

21. Niagara Falls is approximately 5×10^3 cm high and about one billion grams of water flows over it per second. How much potential energy is lost per second? An electric light bulb requires 10^9 ergs of energy per second. How many such bulbs could be lit if all the energy of Niagara Falls could be utilized? At an average of one bulb per person, what sort of community would be supplied with electric power for lighting?

22. A rubber ball of mass 75 gm is dropped from a height of 150 cm and bounces to a height of 120 cm. What is the ratio of the magnitudes of its momenta just before and just after it strikes the ground?

23. What is the escape velocity from the moon?

24. When a satellite is in a stable circular orbit, what is the ratio of its kinetic energy to its potential energy.

25. A rocket is fired vertically upward with a velocity of 8×10^5 cm/sec. To what height does it rise?

Momentum,
Angular
Momentum,
and Energy

5 *Atoms and Heat*

5-1 *Atoms*

The laws of mechanics were discovered in the seventeenth century. The eighteenth century was devoted mainly to developing the mathematical techniques that enabled these laws to be applied to various problems. This was the age of the great mathematical physicists d'Alembert, Bernoulli, Euler, Lagrange, Laplace, and many others. They were interested in such problems as the detailed behavior of the solar system, the motion of a spinning top, and the flow of liquids. The mathematics was very difficult and the success of their solutions produced overwhelming confidence in the validity of Newtonian mechanics. However, all these successes were concerned with the behavior of macroscopic bodies, and many important questions were left unanswered. The laws of mechanics predicted that a lead ball would fall to the earth at the same rate as an iron ball, but the difference between lead and iron was not understood and it was not known why one cubic centimeter of lead has a mass of 11.37 grams, whereas one cubic centimeter of iron has a mass of 7.87 grams. The first major step toward providing answers to questions of this kind was the emergence of the atomic theory of matter at the beginning of the nineteenth century.

The idea that matter is composed of atoms was proposed by the Greek philosophers Empedocles and Democritus between 500 and 400 B.C., but the establishment of a scientific theory involves more than a bright idea. It is necessary to show that this idea is useful in explaining known phenomena and in predicting new phenomena. In the case of atoms, the initial evidence came mainly from chemistry. The evidence was first presented by Dalton at the beginning of the nineteenth century and the situation was finally clarified by Cannizzaro in the middle of the century. This interval of 50 years between the presentation of an important idea and its proper understanding is a striking illustration of the slow progress of science in the past compared with its present explosive development. Once the chemical evidence had convinced scientists of the existence of atoms, the atomic theory was found to provide a very successful explanation of the properties of gases and later of solids and liquids also. In this chapter we shall give a very brief review of the chemical evidence and then proceed to discuss the properties of gases.

Hydrogen and oxygen combine to form water. Hydrogen is a **chemical element** and is therefore composed of atoms that are all of the same kind and are called hydrogen atoms. Actually the hydrogen atoms join together in pairs to form hydrogen **molecules.** Since a hydrogen molecule contains two atoms it is said to be *diatomic.* Oxygen is also a chemical element and is also diatomic, but its molecule is composed of two oxygen atoms that are different in many respects from hydrogen atoms. In particular, an oxygen atom is approximately 16 times more massive than a hydrogen atom. When hydrogen and oxygen react chemically, two hydrogen molecules come to-

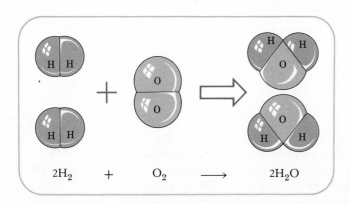

$2H_2$ $+$ O_2 $\longrightarrow$ $2H_2O$

Figure 5-1 Two molecules of hydrogen combine with one molecule of oxygen to form two molecules of water.

gether with one oxygen molecule and the atoms are reshuffled to form two water molecules, each of which contains one oxygen atom and two hydrogen atoms (figure 5-1). Since the molecule of water contains atoms of more than one kind, water is a **chemical compound.** Four atoms of hydrogen combine with two atoms of oxygen and the mass of the oxygen atom is 16 times that of the hydrogen atom. This provides a ready explanation of the experimental observation that a given mass of hydrogen combines with 8 times its mass of oxygen to give 9 times its mass of water.

At the time of this writing, 105 different chemical elements are known,

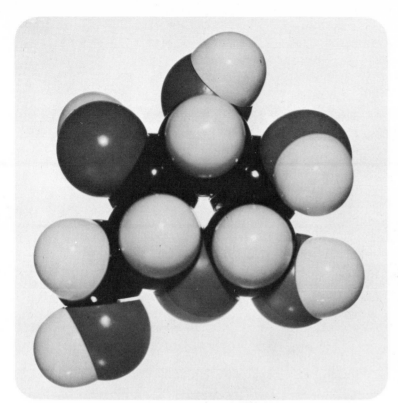

Figure 5-2 *A model of a sugar molecule. This particular sugar molecule is composed of six carbon atoms, twelve hydrogen atoms, and six oxygen atoms. Its formula is therefore $C_6H_{12}O_6$. The carbon atoms are represented by the black balls, the hydrogen atoms by the white balls, and the oxygen atoms by the gray balls. (Courtesy of the Physical Science Study Committee and D. C. Heath and Company.)*

corresponding to 105 different kinds of atoms. A molecule is obtained by selecting some of these atoms and joining them together in some spatial arrangement. The number of possible selections is large and the number of possible spatial arrangements is large, thus explaining the large variety of chemical compounds in nature. Figure 5-2 shows a model of a molecule of sugar.

A chemical reaction occurs when several molecules come together and reshuffle their atoms to produce different molecules. As in the case of the combination of hydrogen and oxygen to form water, if it is known which atoms are involved in the composition of each molecule and if the relative masses of the atoms are known, it is easy to calculate the proportions by mass in which the reactants must be mixed in order to react completely and to calculate the mass of each product formed. It was this success of the atomic theory in predicting the masses involved in chemical reactions that first convinced chemists and physicists of the existence of atoms.

Atoms are very small. The diameter of an atom is only about 10^{-8} cm, which means that if atoms were placed side by side along a line touching one another, it would take about one hundred million of them to form a line only one centimeter long. An atom is in fact too small to be seen di-

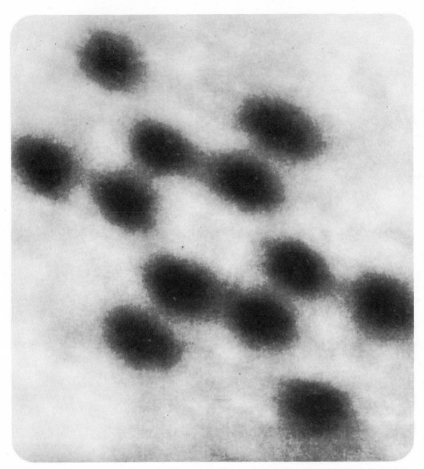

Figure 5-3 *A "photograph" of a hexamethylbenzene molecule. The chemical formula is $C_6(CH_3)_6$ or $C_{12}H_{18}$. The dark black blobs are the twelve carbon atoms: the hydrogen atoms do not show up clearly. This is not a photograph of an individual molecule. It was obtained by gathering x rays from a large number of molecules and electronically converting the x-ray signal into a visible signal. (From Jay Orear, Fundamental Physics, John Wiley & Sons, Inc. The photograph was obtained by Dr. M. L. Huggins of the Kodak Research Laboratory.)*

Atoms and
Heat

rectly, even with a microscope. A fundamental rule of microscopy is that two objects cannot be distinguished from one another if their distance apart is much less than the wavelength, λ, of the light illuminating them. (The wavelength was defined in section 1-1. See also figure 1-1.) An object smaller than λ is seen as a blur of light with a diameter of about λ. This is a consequence of the diffraction phenomena, which will be discussed in section 8-6. Since the smallest wavelength of visible light is 4×10^{-5} cm and the diameter of an atom is about 10^{-8} cm, we obviously cannot hope to look at an atom and examine its structure. (The student who is still not familiar with powers of ten should pause and consider that 4×10^{-5} is 4000 times as big as 10^{-8}.)

Nevertheless, there are special techniques that come very close to providing pictures of individual atoms. X-rays are similar to light, but they have much shorter wavelengths that can be much smaller than the diameter of an atom. It is therefore possible to design an x-ray "microscope" that will reveal the structure of the atom. It is still not possible to see an individual atom, but if the x-rays are scattered by an orderly array of atoms in a solid, it is possible to deduce from the detailed nature of the scattering the spatial arrangement of the atoms in the solid. Figure 5-3 is a photograph of a complicated organic molecule reconstructed from such an x-ray experiment.

One of the most impressive ways of "seeing" atoms is the *field ion microscope* invented by E. W. Müller. The explanation of how it works is complicated, but figure 5-4 is easily appreciated. It is a "photograph" of the tip of a tungsten needle and each bright dot is caused by an atom, or a group of two or three atoms, on the tip.

To illustrate the vast number of atoms in even a small particle of matter, let us suppose that a microscope had been invented that could distinguish the individual atoms. Suppose that you attempted to use this microscope to count the atoms in a rain drop only 0.1 cm in diameter. The psychological limit on the rate of counting could hardly be in excess of 5 per second. At this rate, it would take 3×10^{11} years to count all the atoms! Many astronomers believe that the universe was created about 10^{10} years ago. Therefore, if you had started to count these atoms the very instant the universe came into being, you would still be far from having completed the job.

5-2 The Internal Energy of an Ideal Monatomic Gas

The ideal monatomic gas is the simplest state of matter and can be discussed with relatively simple mathematics. **Monatomic** means that the molecules are merely single atoms. This includes helium (He), neon (Ne), argon (Ar), mer-

Figure 5-4 *Atoms on the tip of a tungsten needle in a field ion microscope. (Courtesy of Dr. Erwin Müller.)*

cury vapor (Hg), sodium vapor (Na), potassium vapor (K), but it excludes diatomic gases such as oxygen (O_2), nitrogen (N_2), hydrogen chloride (HCl), and also excludes triatomic gases such as ozone (O_3), steam (H_2O), and carbon dioxide (CO_2). A gas is said to be **ideal** if there are no forces between its molecules. Actually all molecules exert forces on one another, but these forces decrease very rapidly as the distance between the molecules is increased. If the gas is expanded into a very large volume so that its density becomes very small and the molecules are very far apart, the forces between the molecules become negligibly small. An ideal gas is therefore the limiting case of an actual gas when its volume becomes very large and its density becomes very small.

An ideal monatomic gas can be visualized as a collection of small hard spheres (the atoms) that are separated by distances much larger than their diameters and that exert no forces on one another. These atoms are in rapid motion and move in straight lines until they rebound from the wall of the container, or very occasionally collide with one another (figure 5-5). The dis-

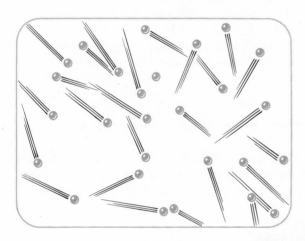

Figure 5-5 The kinetic theory picture of an ideal monatomic gas. The tails are intended to create an impression of the motion of the atoms and do not imply that the atom leaves any sort of wake behind it.

tribution of the atoms in space is perfectly random and the velocities are uniformly distributed over all possible directions. The atoms do not all have the same speed. They have all possible speeds from zero upward. However, there is an average speed, v_a, which we shall soon define more precisely. Most of the atoms have speeds that are not very different from v_a and there are only a few atoms with speeds very much smaller than v_a or very much larger than v_a. This average speed v_a is related to the temperature of the gas. The higher the temperature, the greater is the average speed.

Since the atoms of an ideal gas exert no forces on one another there is no mutual potential energy of these atoms. The only source of energy is the kinetic energy of motion of the atoms. An atom of mass m moving with a velocity v has a kinetic energy $\frac{1}{2}mv^2$ and so the total kinetic energy, U, of all the atoms is

Atoms and Heat

Total kinetic energy,
$$U = \text{Sum of } \tfrac{1}{2}mv^2 \text{ for all the atoms}$$
$$\text{of the ideal monatomic gas} \quad (5\text{-}1)$$

U is called the **internal energy** of the gas. It is a form of energy present within the gas that is not immediately obvious until we realize that the gas consists of atoms in motion. If we take an object such as a baseball with mass M and throw it with a velocity V, then in mechanics its kinetic energy is said to be $\frac{1}{2}MV^2$. However, if the baseball is visualized as a collection of atoms in rapid motion in all directions, the total kinetic energy of all these atoms is greatly in excess of $\frac{1}{2}MV^2$. In fact, the velocity V is unlikely to be much greater than 5×10^3 cm/sec, whereas the internal velocities could easily be as large as 5×10^5 cm/sec. The internal kinetic energy is therefore at least 10,000 times as great as the "obvious" term $\frac{1}{2}MV^2$.

Fortunately, the total kinetic energy can be divided into two parts, a part $\frac{1}{2}MV^2$ that is due to the motion of the baseball as a whole, and an "internal kinetic energy" that is the same as the total kinetic energy of the random motion of the atoms when the baseball as a whole is not moving. As long as the temperature remains constant the "internal kinetic energy" remains constant and all our attention can be directed toward changes in the term $\frac{1}{2}MV^2$, as in the previous chapter. But if energy is added in the form of heat the internal kinetic energy increases and the temperature rises. If, for example, the baseball strikes the ground and embeds itself in a patch of loose earth, the kinetic energy $\frac{1}{2}MV^2$ is destroyed and at least part of it is converted into extra internal kinetic energy, so that the atoms move faster and the baseball warms up.

There are some complications involved in the internal energy of a solid such as a baseball, so let us return to the ideal monatomic gas. The average speed v_a can be defined in such a way that the internal energy U would be the same if all the atoms had this speed. If there are N atoms in the gas, v_a is *defined* by

$$U = N \tfrac{1}{2}mv_a{}^2 \tag{5-2}$$

$$N \tfrac{1}{2}mv_a{}^2 = \text{Sum of } \tfrac{1}{2}mv^2 \text{ for all the atoms} \tag{5-3}$$

Cancelling $\frac{1}{2}m$ on both sides,

$$Nv_a{}^2 = \text{Sum of } v^2 \text{ for all the atoms} \tag{5-4}$$

Notice that v_a is a rather special sort of average. It is really the kinetic energy that is being averaged.

For an ideal monatomic gas there is a direct relationship between the internal energy and the concept of absolute temperature, T.

> *Internal energy and absolute temperature*
>
> $$U = N \tfrac{1}{2}mv_a{}^2 \tag{5-5}$$
>
> $$U = N \tfrac{3}{2}kT \tag{5-6}$$

The constant k is a fundamental constant called

> *Boltzmann's constant*
>
> $k = 1.380 \times 10^{-16}$ CGS units $\hspace{2cm}$ (5-7)

Equation 5-6 is one way of defining the temperature T, which is called the **absolute temperature** and is written $T°K$. This is referred to verbally as "degrees absolute" or "degrees Kelvin" (in honor of the physicist Lord Kelvin who was responsible for clarifying the concept of temperature during the middle years of the nineteenth century). The absolute scale of temperature is much more commonly used in physics than the Centigrade or Fahrenheit scales because the zero of the absolute scale is really the lowest possible temperature that can ever be attained. Equation 5-6 implies that, when $T = 0$, then $U = 0$ and $v_a = 0$. The **absolute zero of temperature**, $0°K$, is therefore the temperature at which the ideal monatomic gas has lost all its internal energy and its atoms have come to rest. Clearly, we cannot do any better than this.

There is a simple relationship between the temperature $T°K$ measured on the absolute scale and the temperature $t°C$ measured on the more familiar Centigrade scale.

$$T°K = t°C + 273.15 \hspace{2cm} (5\text{-}8)$$

It follows that the absolute zero of temperature is

$$0°K = -273.15°C \hspace{2cm} (5\text{-}9)$$

The temperature at which ice melts under a pressure of one standard atmosphere (the ice point) is

$$0°C = 273.15°K \hspace{2cm} (5\text{-}10)$$

The temperature at which water boils under a pressure of one standard atmosphere (the steam point) is

$$100°C = 373.15°K \hspace{2cm} (5\text{-}11)$$

The most familiar scale of temperature is the Fahrenheit scale, because of its use in meteorology and medicine. On the Fahrenheit scale, the ice point is $32°F$, the steam point is $212°F$ and the absolute zero is $-459.67°F$. However, the Fahrenheit scale is very rarely used in physics and the absolute scale is much to be preferred to it or the Centigrade scale.

Finally, it is important to emphasize that the simple expression of equation 5-6 for the internal energy is true only for an ideal monatomic gas. For all other substances the relationship between internal energy U and temperature T is much more complicated.

Atoms and Heat

90

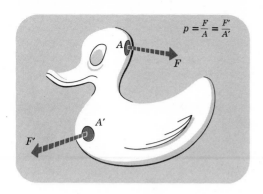

$$p = \frac{F}{A} = \frac{F'}{A'}$$

Figure 5-6 *The pressure inside a balloon with a complicated shape.*

5-3 The Pressure of an Ideal Monatomic Gas

The atoms of a gas are in rapid motion and are continually striking the walls of the container. This steady bombardment of atoms is equivalent to a force pushing the wall outward, and is the origin of the pressure of the gas. More precisely, the pressure is the force exerted on unit area of a plane wall. However complicated the shape of the container (figure 5-6), select a small area of the wall that is small enough to be essentially plane and that has an area A. The bombardment of atoms exerts on it an outward force F that is perpendicular to the small portion of the wall. The pressure p of the gas is then

---

Pressure = force per unit area

$$p = F/A \tag{5-12}$$

Assuming that the temperature is constant throughout the whole of the gas, and ignoring some very small effects due to the earth's gravitational field, the pressure has the same value on all parts of the wall. Notice that pressure is measured in dynes per square centimeter. For the benefit of readers who have encountered pressures measured in cm of mercury or standard atmospheres, let us emphasize that such units are secondary and nonfundamental. The equations given in this book are true only when the pressure is expressed in dyne/cm². The conversion factors are

$$1 \text{ cm of mercury} = 1.33 \times 10^4 \text{ dyne/cm}^2 \tag{5-13}$$

$$1 \text{ standard atmosphere} = 76 \text{ cm of mercury}$$
$$= 1.01 \times 10^6 \text{ dyne/cm}^2 \tag{5-14}$$

When an atom of the gas strikes the wall it rebounds with its velocity and momentum pointing in a different direction. Since the total momentum must be conserved, some momentum is transferred to the wall. As illustrated in figure 5-7, the momentum **P** of the atom just before striking the wall must be equal to the vector sum of the momentum **P'** of the rebounding atom

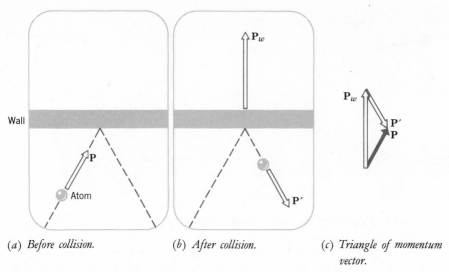

(a) Before collision.   (b) After collision.   (c) Triangle of momentum
                                                     vector.

Figure 5-7 *Conservation of momentum when a gas atom strikes the wall.*

plus the momentum $\mathbf{P}_w$ given to the wall. The continual bombardment of the wall by the gas atoms therefore produces a steady transfer of momentum to the wall. However, as explained in section 4-1, the momentum given to the wall in one second is the same thing as the force exerted on the wall by the gas atoms. To calculate the pressure p we must therefore calculate the momentum transferred in *one second* to *unit area* of the the wall. This is a laborious calculation when done properly, but the result is comparatively simple.

Pressure of an ideal monatomic gas

$$p = \frac{1}{3}\frac{N}{V}mv_a{}^2 \tag{5-15}$$

---

p is the pressure of the gas in dyne/cm². N is the number of atoms of the gas. V is the volume occupied by the gas in cm³. m is the mass of an atom in gm. v_a is the average speed of the atoms in cm/sec as defined by equation 5-4.

This result may be understood in the following way. In a single collision the momentum transferred to the wall is proportional to mv. The colliding atoms have all velocities from zero upward but the average value of mv is approximately mv_a. The number of collisions per second is proportional to the number of atoms per unit volume in the vicinity of the wall, that is to N/V. Also, the faster the atoms are moving, the more often they collide with the wall and the number of collisions per second is proportional to v_a. Putting these three factors together, the momentum transferred to the wall per second is proportional to

$$(mv_a) \times \frac{N}{V} \times (v_a) = \frac{N}{V}mv_a{}^2$$

This differs from equation 5-15 only by a factor of $\frac{1}{3}$, which comes from making the correct averaging procedure.

The absolute temperature of an ideal monatomic gas was introduced by equations 5-5 and 5-6. From these equations it follows very simply that

$$\tfrac{1}{3}mv_a{}^2 = kT \tag{5-16}$$

Substituting this into equation 5-15 we obtain

$$p = \frac{N}{V}kT \tag{5-17}$$

Multiplying both sides by V we finally obtain

> *The ideal gas equation*
>
> $$pV = NkT \tag{5-18}$$

Now let us pause and consider what we have achieved. We started with a model of an ideal monatomic gas as a collection of atoms in random motion. The pressure of the gas was then calculated by considering the bombardment of gas atoms on the walls of the container. The pressure was found to depend on the number of atoms per unit volume and the average kinetic energy of the atoms (equation 5-15). The concept of temperature was introduced as a measure of the average kinetic energy (equation 5-6). Finally the ideal gas equation was obtained in a form involving macroscopically measurable quantities such as the pressure p, the volume V, and the absolute temperature T.

Actually, we have reversed the historical order of events. Originally, crude temperature scales were based on the fact that most substances expand as the temperature is raised, as in the familiar mercury-in-glass thermometer. On the basis of such crude temperature scales, the macroscopic form of the ideal gas equation was discovered experimentally. It was then realized that a more fundamental temperature scale could be based on the properties of gases at very low densities. Finally, the atomic theory provided a microscopic explanation of the ideal gas equation and also revealed the relationship between absolute temperature and internal energy for an ideal monatomic gas.

5-3 The Pressure of an Ideal Monatomic Gas

If a fixed mass of gas is kept at a fixed temperature, everything on the right-hand side of equation 5-18 is constant and so

$$pV = K = \text{a constant} \tag{5-19}$$

$$p = \frac{K}{V} \tag{5-20}$$

$$V = \frac{K}{p} \tag{5-21}$$

This is **Boyle's Law** and was discovered experimentally by the English physicist Robert Boyle as early as the seventeenth century, a few years before Newton discovered the laws of motion. It tells us that if we keep the temperature of a sample of gas constant and decrease its volume, the pressure increases in inverse proportion to the volume. To obtain a vivid picture of what is happening, the reader may imagine the atoms to be analogous to a swarm of angry mosquitoes, trapped inside a rubber balloon and continually flying from side to side and battering themselves against the rubber in an attempt to escape (figure 5-8). When the balloon has a large volume (figure 5-8a) the mosquitoes spend most of their time flying across the balloon and only occasionally hit the rubber; but when the volume is small (figure 5-8b) they hit the rubber much more frequently and exert a greater outward pressure on it.

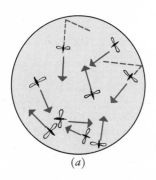

(a) (b)

Figure 5-8 *An analogy between Boyle's law and the behavior of a swarm of mosquitoes inside a rubber balloon.*

If the volume of the gas is kept constant and its temperature is varied, then

$$p = \frac{Nk}{V}T \tag{5-22}$$

$$p = K'T \tag{5-23}$$

where K' is another constant. The pressure of the gas is directly proportional to the absolute temperature. This is essentially **Charles' Law,** although the French physicist Jacques Charles did not express it in quite this way, since he was not familiar with the concept of absolute temperature. He proposed this law at the end of the eighteenth century, more than a century after the discovery of Boyle's law and a few years before Dalton's atomic theory. Its pictorial interpretation is that raising the temperature is equivalent to making the mosquitoes angrier, so that they fly faster, hit the rubber more often and with a greater impact and therefore exert a greater pressure on it.

Equation 5-18 may also be rewritten

$$N = \frac{pV}{kT} \tag{5-24}$$

Consider two different ideal monatomic gases corresponding to two different chemical elements. If they have the same pressure p, the same temperature T, and occupy equal volumes V, then everything on the right hand side of equation 5-24 is the same for the two gases. They must therefore contain the same number of atoms N. This is **Avogadro's Law.** Equal volumes of two ideal monatomic gases at the same temperature and pressure contain the same number of atoms.

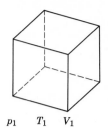 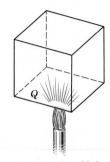 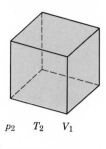

$p_1 \quad T_1 \quad V_1$ $\qquad\qquad\qquad\qquad\qquad\qquad p_2 \quad T_2 \quad V_1$

(a) *Initial conditions.*   (b) *Heat is added.*   (c) *Final conditions.*

Figure 5-9 *Addition of heat to an ideal monatomic gas at constant volume.*

5-4 Heat and the First Law of Thermodynamics

Heat is a form of energy. The heat added to a system is the same thing as the energy added to the system, but a little care must be taken not to neglect some part of this energy.

Consider first the simplest case of heat added to an ideal monatomic gas in a container of fixed volume V_1 (figure 5-9). Suppose that the initial temperature and pressure are T_1 and p_1 and that the addition of an amount of heat Q raises the temperature and pressure to T_2 and p_2. Initially the total kinetic energy of the gas atoms is

$$U_1 = \tfrac{3}{2}NkT_1 \tag{5-25}$$

The addition of heat raises the total kinetic energy to

$$U_2 = \tfrac{3}{2}NkT_2 \tag{5-26}$$

The heat added is equal to the increase in the energy of the system:

$$Q = U_2 - U_1 \tag{5-27}$$

or

$$Q = \tfrac{3}{2}Nk(T_2 - T_1) \tag{5-28}$$

During the above procedure, raising the temperature of the gas while keeping its volume constant results in an increase in the pressure of the gas. Let us now consider a different arrangement in which the pressure is kept constant, but the gas is allowed to expand as its temperature rises. In the device of figure 5-10 the lower part of the cylinder contains the gas and the upper part is a vacuum. Separating the gas from the vacuum is a piston, which fits tightly enough to prevent any gas from escaping, but is well lubricated so that it can move freely up and down without friction. If the cross-sectional area of the piston is A and the pressure of the gas is p_1, the gas exerts an upward force p_1A on the lower face of the piston. The space above the piston is a vacuum and there is no similar force on the upper face of the piston. However, weights are placed on top of the piston until the combined weight mg of the piston and added weights is just sufficient to balance the upward thrust of the gas and the piston remains stationary.

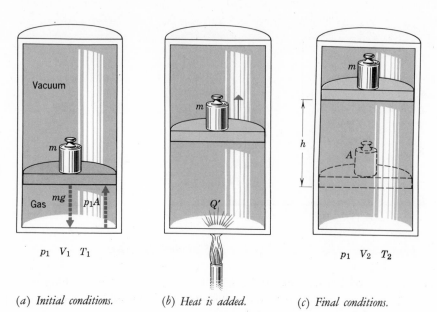

(a) Initial conditions.   (b) Heat is added.   (c) Final conditions.

Figure 5-10 *Addition of heat to an ideal monatomic gas at constant pressure.*

Then

$$mg = p_1 A \qquad (5\text{-}29)$$

Initially, the pressure, volume and temperature are p_1, V_1, and T_1 respectively. An amount of heat Q' is then slowly added to the gas. The temperature starts to increase and the pressure also starts to increase, so that the upward force $p_1 A$ exceeds the downward force mg and the piston starts to rise. However, the resulting increase in volume of the gas tends to lower the pressure again. If the heat is added very, very slowly, the piston rises very slowly and never acquires an appreciable acceleration, so the two forces on it must always approximately cancel one another. The volume and temperature therefore slowly increase but the pressure remains constant with the value

$$p_1 = \frac{mg}{A} \qquad (5\text{-}30)$$

When all the heat has been added, the volume and temperature have increased to V_2 and T_2, the piston has risen through a height h, but the pressure remains p_1. The internal energy of an ideal monatomic gas depends only on the temperature and not on the pressure or volume, so the increase in internal energy is still given by equation 5-28

$$U_2 - U_1 = \tfrac{3}{2} Nk(T_2 - T_1) \qquad (5\text{-}31)$$

The interesting feature of the situation, though, is that there is also an increase in the gravitational potential energy of the piston, which has a mass m and is raised through a height h in the earth's gravitational field.

Therefore,

$$\text{Increase in gravitational potential energy} = mgh \tag{5-32}$$

$$= p_1 Ah \tag{5-33}$$

because of equation (5-29). Notice that the added volume, $(V_2 - V_1)$, is a cylinder of base area A and height h, and so

$$(V_2 - V_1) = Ah \tag{5-34}$$

An alternative form of equation 5-33 is therefore

$$\text{Increase in gravitational potential energy} = p_1(V_2 - V_1) \tag{5-35}$$

The heat Q' added to the gas is seen to be used up in two ways, partly to increase the kinetic energy of the atoms and partly to increase the gravitational potential energy of the piston.

$$Q' = (U_2 - U_1) + mgh \tag{5-36}$$

Alternatively,

$$Q' = \tfrac{3}{2}Nk(T_2 - T_1) + p_1(V_2 - V_1) \tag{5-37}$$

Behind all this is the law of conservation of energy. What we are saying is that the energy of the system has increased by an amount given by the right-hand side of equation 5-37, and we call this "the heat added to the system." Since energy must be conserved, some other system must have *lost* an equal amount of energy. Suppose, for example, that the gas is warmed up by a flame. The system that loses energy is the fuel plus the oxygen in the air with which it reacts chemically inside the flame. As we have already explained in section 4-6, a chemical reaction involves a rearrangement of electrons and nuclei and the chemical energy released comes from a reduction in the kinetic energy and mutual electrostatic potential energy of these particles. This reduction is the heat supplied by the flame, its numerical value is Q' and it is equal to the energy gained by the gas and the piston.

Thermodynamics is the sophisticated name for the present subject, which is concerned with heat, temperature, and such properties of matter as pressure and volume. The first law of thermodynamics is merely another form of the law of conservation of energy.

5-4 Heat and the First Law of Thermo-dynamics

The first law of thermodynamics

The heat added to a system is equal to the increase in energy of the system. Care must be taken not to overlook some part of the energy of the system.

It is traditional to measure heat in calories. One **calorie** is the heat required to raise the temperature of one gram of pure water from 14.5°C to 15.5°C. Since heat is really energy and we have already defined the unit of energy, the erg, it is more satisfying from a fundamental point of view to measure heat in ergs. The conversion factor is

$$1 \text{ calorie} = 4.186 \times 10^7 \text{ ergs} \tag{5-38}$$

5-5 Order and Disorder

Heat is *disorganized* energy. Consider again the case of a baseball of mass M thrown through space with a velocity V. Its kinetic energy $\frac{1}{2}MV^2$ would not normally be called heat. However, when it strikes the ground and warms up, the kinetic energy of its bodily motion through space has been converted into kinetic energy of the random disorganized motion of its individual atoms, which is rightly called heat. This concept of randomness, or disorder, is very important in thermodynamics.

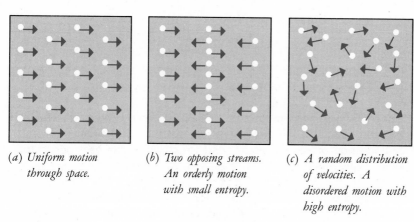

(a) *Uniform motion through space.*

(b) *Two opposing streams. An orderly motion with small entropy.*

(c) *A random distribution of velocities. A disordered motion with high entropy.*

Figure 5-11 *Three ways of achieving a total kinetic energy $\frac{3}{2}NkT = \frac{1}{2}Nmv_a{}^2$ in an ideal monatomic gas.*

For an ideal monatomic gas at temperature T, we have so far only specified that the total kinetic energy shall be $\frac{3}{2}NkT$. This could be realized in many different ways. For example, all the atoms could be moving with the same velocity v_a in exactly the same direction (figure 5-11a). This case, however, would be more correctly described as a gas at 0°K moving bodily through space with a velocity v_a. (The container would also have to move through space with a velocity v_a, otherwise collisions with the walls would soon upset the situation.) Alternatively, we might divide the atoms into two equal streams, one moving to the right with speed v_a and the other moving to the left with speed v_a (figure 5-11b). Each atom would then bounce backward and forward between the right-hand and left-hand walls of the container and there would be no motion of the gas as a whole. It is intuitively obvious that a gas does not really look like this. Why should there not be streams up and down, or back to front, or for that matter in any direction? Even if all the atoms started out with exactly the same speeds, would not collisions between the atoms soon increase the speeds of some and

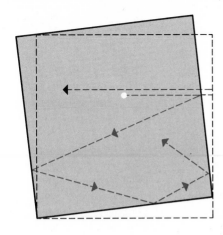

Figure 5-12 *A very slight tilt of the container makes an ordered motion degenerate rapidly into a disordered one.*

lower the speeds of others? It seems much more reasonable to assume that the atoms are moving in all possible directions and have various possible speeds, subject to the restriction that the total kinetic energy must be $\frac{3}{2}NkT$ (figure 5-11c).

Suppose that we started with the situation of figure 5-11b and the container then received a small jolt which rotated it through a small angle (figure 5-12). Then the atoms would no longer bounce straight back from the wall, but would bounce off at an angle. There is good reason to believe that the simplicity of the motion would soon be destroyed and it would rapidly revert to complete randomness as in figure 5-11c. Can the same thing happen in reverse? Can a disordered motion develop into an ordered motion? It obviously can because of a fundamental principle of Newtonian mechanics known as **time reversal,** which says that for any possible motion, there is another possible motion that would be obtained by running the first motion backward in time. Suppose I drop a body from rest at a height *h* and it hits the ground with a speed *v*. The time-reversed motion corresponds to throwing the body *upward* from the ground with speed *v* so that it rises to a height *h* before it comes to rest. If I were to film the falling motion, then by running the film *backward* through the projector I could reproduce in all details the time-reversed motion in which the ball was thrown upward.

Now return to the above case of the ordered motion of figure 5-11b degenerating into the disordered motion of figure 5-11c. Halt the process at some instant of time when disorder has been reached, and then reverse the velocity of each atom without changing its magnitude. The subsequent motion will be an exact time reversal of the preceding motion and will continue until the ordered motion of figure 5-11b has been established.

In practice this never happens. The reason is that a disordered motion can be realized by assigning positions and velocities to the atoms in an enormously large number of ways. Of this very large number of disordered motions only a few are of the very special kind that would develop into an ordered motion such as that in figure 5-11b within a reasonably short time. It is conceivable that any random motion, if given sufficient time, would eventually develop into an ordered motion. However, in the case of an overwhelming majority of random motions the time taken would be enormously large, greatly in excess of the "age of the universe." Moreover, even when the ordered motion was realized it would persist for only a fraction of a second. The situation is similar to taking a new pack of cards that is carefully ordered into suits, and shuffling it. The initial order is very soon destroyed, but it would be a very long time indeed before the shuffling reproduced the initial order again. There are very many ways of arranging a pack in a random fashion, but there is only one way of arranging it in the order in which it comes from the manufacturer.

What we are saying then is that the ordered arrangement of figure 5-11*b* is quite consistent with the laws of mechanics, but it is very unlikely to occur in practice. On the other hand, the disordered arrangement of figure 5-11*c*, or *some similar random arrangement,* is very likely to be found in practice. Moreover, a random arrangement is *overwhelmingly* more probable. A student confronted with these ideas for the first time is likely to argue that, although the ordered arrangement is improbable, it is possible and he may ask why it does not sometimes occur. Such a student should sit down with a pack of cards and try shuffling them until they are arranged into suits. His patience with his own argument will soon become exhausted.

5-6 Entropy; the Second and Third Laws of Thermodynamics

It is possible to put these ideas on a firm mathematical basis and to define a quantity S, called the **entropy,** which is a measure of the probability that a particular type of motion will occur. Since a disordered situation is also a more probable situation, S is also a measure of disorder. The situation in figure 5-11*b* is an ordered situation and an improbable situation, so it corresponds to a small value of the entropy S. The situation in figure 5-11*c* is a disordered situation, but a more probable situation, so it corresponds to a larger value of the entropy S. Subject to the slightest perturbation, the ordered situation would rapidly develop into the disordered situation with larger entropy. The chance of the disordered situation ever developing into the ordered situation is so very small that it can be neglected. This is an example of a very general law.

> ### The second law of thermodynamics
>
> When a system containing a large number of particles is left to itself, it assumes a state with maximum entropy, that is, it becomes as disordered as possible.

Since entropy is a measure of probability, all that this says is that "what will occur is what is most likely to occur." The probability of what is likely to occur is so overwhelmingly large compared with any other possibility that this other possibility can be completely discounted.

Let us now concentrate on the value of the entropy when equilibrium has been reached and ignore nonequilibrium states. Start with a system at temperature T with an entropy S and add a small quantity of heat q so that the temperature increases to $T + t$ and the entropy of the new equilibrium state is $S + s$. The amount of heat added must be very small, so that t is very small compared with T, and it must be added slowly and carefully so that the system passes through a sequence of equilibrium states. It can be shown that:

The practical definition of entropy

$$s = \frac{q}{T} \tag{5-39}$$

$$\text{Increase in entropy} = \frac{\text{Heat added}}{\text{Temperature}}$$

Thus, although entropy is a subtle mathematical concept related to the laws of probability, as long as we confine ourselves to equilibrium states this equation provides us with a direct macroscopic method of measuring changes in entropy. It also shows that, as the temperature is increased, the entropy increases. The system is more disordered at higher temperatures. Conversely, as the temperature is lowered the entropy decreases and the system becomes more ordered. When the system is perfectly ordered, it is not possible to go any further and the absolute zero of temperature has been reached.

The third law of thermodynamics

A system in equilibrium at the absolute zero of temperature is in a state of perfect order and has zero entropy.

Let us add, though, that the third law, when stated in this particular form, must be applied with caution. It is very easy to obtain a system at a temperature near 0°K which is not really in equilibrium, but which appears to be so because the rate at which it is changing in the direction of equilibrium is extremely slow. Such a system might be very disordered and have a high entropy. Given enough time, it would eventually come into equilibrium with its low temperature surroundings. It would become perfectly ordered and would have zero entropy. However, this might take a very long time indeed.

Perfection is difficult to achieve and the absolute zero of temperature is impossible to achieve. It can be shown that an inescapable consequence of the laws of thermodynamics is that it is impossible to make an apparatus that will reach the absolute zero of temperature in a finite number of operations.

5-7 The
Arrow of
Time and the
Fate of the
Universe

5-7 The Arrow of Time and the Fate of the Universe

The second law of thermodynamics can be illustrated by a familiar example. Suppose that a body at a temperature T_1 is connected to a body at a lower temperature T_2 by a rod which can conduct heat (figure 5-13). This situation is a departure from equilibrium and it is well known that heat flows from

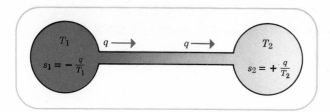

the hot body to the cold body until their two temperatures are equal. Suppose that a small amount of heat q leaves the hot body and enters the cold body. Then the entropy of the hot body decreases by q/T_1 and the entropy of the cold body increases by q/T_2. The change in the total entropy of the system is therefore:

$$\text{Change in entropy} = -\frac{q}{T_1} + \frac{q}{T_2}$$

$$= q\left(\frac{1}{T_2} - \frac{1}{T_1}\right) \tag{5-40}$$

Since q is positive and T_2 is less than T_1, it is easy to convince yourself that this change is positive. Heat flows from a body at a high temperature to a body at a lower temperature because this increases the total entropy of the system, as required by the second law of thermodynamics. If the heat q were to flow from the cold body to the hot body, the change in entropy would be $q\left(\frac{1}{T_1} - \frac{1}{T_2}\right)$, which is negative. The combined entropy of the two bodies would therefore decrease and the second law would be violated.

In principle, of course, we can imagine a film running backward through a projector and showing a flow of heat from the cold body to the hot body. Such an "uphill" flow of heat is dynamically possible, but the exact set of initial circumstances that would make it happen is so very unlikely to occur in practice that it can be completely discounted.

Everyday experience contains many examples of processes that are very unlikely to occur in reverse. Milk poured into coffee soon distributes itself uniformly by diffusion, but it is inconceivable that it should later separate out again into a stream of pure milk leaping up into the jug. A log readily burns to form ash and various gases, which mix with the air, but it is inconceivable that these gases should return from the atmosphere to the flame and recombine with the ash to form a new log. Can you imagine any process that would reconvert an omelet into an egg capable of hatching into a chicken? Wind, rain, and frost can reduce an ancient city to rubble, but are unlikely ever to rebuild it again.

All these examples involve the interaction of a large number of atoms, and they are all instances of systems that are initially not in equilibrium but that move toward equilibrium. The process that actually occurs starts from a state of order with low entropy and develops into a state of greater disorder with higher entropy, in accordance with the second law of thermodynamics. The time-reversed process, which never occurs in nature, produces an increase in order from a state of high entropy to a state of lower entropy.

A film running backward will therefore not deceive us for long. We shall soon notice the custard pie reforming on the face of the comedian and returning to the hand of the thrower. The correct forward direction of time is unambiguously determined as the direction in which order diminishes and entropy increases. For this reason, entropy has been described as "time's arrow," pointing from the past toward the future.

More important than the direction of time is the implication that the universe is heading toward a death of disordered chaos. Life as we know it requires the existence of states of high order and departures from equilibrium such as temperature differences. A living organism is itself a highly ordered entity with low entropy which cannot come into equilibrium with its surroundings except by dying. The functioning of any machine relies on the availability of sources of energy not in equilibrium with their surroundings. A fuel is a chemical substance of low entropy, but the burning of the fuel produces chemical substances with higher entropy which are of no further use as fuels.

When we have used up all the fuels available to us on earth, we might make use of the energy of the tides. The tides are caused by the gravitational forces exerted by the moon on the oceans. The frictional drag of the tides is gradually slowing down the rotation of the earth. Eventually the length of the day will be equal to the length of the month, the moon will always stay in the same position above a fixed point on the earth's surface, and there will be no more lunar tides.

We might then use the energy of sunlight or harness the winds, which derive their energy from sunlight. Ultimately, however, the sun will burn itself out and will probably end up as some cold inactive object at the same temperature as the space around it. This, in fact, may well be the ultimate fate of the whole universe, which may well become a uniform mixture of lifeless objects all in equilibrium with one another and with the radiation filling the space around them. Nothing that we would consider to have any interest or any value could then occur except as a highly improbable, short-lived, small fluctuation away from equilibrium.

There is a possible reprieve. We do not yet understand the behavior of the universe as a whole. There may be some aspect of this behavior that overrides the second law of thermodynamics. We still have too poor an understanding of these matters to predict the distant future with any confidence.

Questions

A

1. An ideal monatomic gas is compressed in such a way that its temperature does not change. How does its internal energy vary with its volume?
2. Criticize the following argument. Atmospheric pressure exerts a force of 15 pounds on every square inch of surface. The surface area of a man is about 4000 square inches. Therefore, the force of the atmosphere on a man is 15 × 4000 pounds or 30 tons.
3. Comment on the following definitions of heat. (a) Another name for energy. (b) The change in the total kinetic energy of random motion of the atoms of a body. (c) The change in the total internal energy of a

body. (d) Energy in motion from a high temperature to a low temperature.

4. What are the units of entropy?

B

5. Why, do you think, was the factor 3/2 introduced into equation 5-6? (A reasonable alternative might be to define a different Boltzmann constant $k' = \frac{3}{2}k$.)

6. Imagine a container of gas isolated in gravity-free outer space. If the molecules of the gas transfer momentum to the walls of the container, does the container thereby acquire a steady acceleration?

7. Describe qualitatively how the first law of thermodynamics can be applied to (a) warming a poker in a fire, (b) boiling away a saucepan of water on a stove, (c) the boy scout method of lighting a fire by rubbing together two sticks.

8. Would a universe in complete equilibrium be an interesting place to live in? List all the ways in which it would differ from the actual universe.

C

9. Give several examples familiar in everyday life of entropy increasing toward its maximum possible value.

10. In a device such as the one shown in figure 5-10, the piston is slowly pushed downward with a steady velocity. As the gas is compressed it warms up. Explain this by considering the collisions of the gas atoms with the moving piston.

11. Can mankind invalidate the second law of thermodynamics by intervening and attempting to create order out of chaos?

12. A balloon containing helium gas is punctured and the helium diffuses throughout the whole room. How does this illustrate the second law of thermodynamics?

13. A container is divided into two parts by a thin partition. One side contains hydrogen gas and the other side contains helium gas. If the partition is broken, does the entropy of the system increase? If both parts contain helium, does the entropy increase?

Problems

A

1. If the same temperature is represented by $t_F°$F on the Fahrenheit scale, by $t°$C on the Centigrade scale, and by $T°$K on the absolute scale, prove that

$$t_F = \tfrac{9}{5}t + 32$$
$$t_F = \tfrac{9}{5}T - 459.67$$

Atoms and

Heat

2. A gas at a pressure of 5×10^4 dyne/cm^2 is contained inside a cubical box. If the length of a side of the box is 25 cm, what is the force exerted by the gas on one face of the box?

3. If the cylinder of figure 5-10 has an inside radius of 5 cm and the total mass of the piston is 150 gm, what is the pressure of the gas?

4. Neon gas is contained inside a box of volume 500 cm^3 at a pressure of 5×10^4 dyne/cm^2 and a temperature of 150°K. What is the average kinetic energy per atom? (Neon is monatomic).

5. An ideal monatomic gas at 400°K exerts a pressure of 10^5 dyne/cm^2. If it is compressed to half its original volume and, as a result, its temperature changes to 480°K, what is its new pressure?

6. If an ideal monatomic gas occupies a volume of 5×10^4 cm^3 and exerts a pressure of 3×10^3 dyne/cm^2 at a temperature of 250°K, how many atoms are there?

B

7. An ideal monatomic gas is at a temperature T_1, exerts a pressure p_1 and occupies a volume V_1. If its volume is changed to V_2 and its pressure to p_2, find an algebraic expression for its new temperature T_2.

8. Most gases liquefy long before they can be cooled to a temperature very near 0°K. The most favorable case is helium, which can probably be studied in its gaseous form down to about 0.2°K. What is the average speed of the helium atoms at this temperature?

9. Assuming that interstellar space contains one atom per cubic centimeter and is at a temperature of 10°K, what is the pressure there?

10. Show that equation 5-28 may also be written

$$Q = \tfrac{3}{2}V_1(p_2 - p_1)$$

11. Show that equation 5-37 may also be written

$$Q' = \tfrac{5}{2}p_1(V_2 - V_1)$$

C

12. What are the extreme temperatures produced by your local climate? Suppose that you inflate your automobile tires on the coldest day in winter to the correct pressure of 26 pounds per square inch as recommended by the manufacturer. Assuming that no air leaks out and that the volume remains constant, what will the pressure be on the hottest day in summer? Ignore heating of the tire by frictional contact with the road. (When the gauge reads 0 pounds per square inch, the pressure inside the tire is one atmosphere, which is 14.7 pounds per square inch.)

13. Some helium gas is contained in a box of negligible weight at the ambient temperature of the room, which may be taken to be 300°K. If all the internal energy could be abstracted from the gas and used to lift the box, through what height could it be raised? Express your answer in miles.

14. At what temperature would the average speed of the molecules of the air be equal to the escape velocity from (a) the earth, (b) the moon? Discuss carefully the question of why the earth has retained its atmosphere, but the moon has not.

15. Imagine that an international commission has defined a new temperature scale on which the absolute zero is 0°N and the ice point is 100°N. What is the new value of Boltzmann's constant, k, in ergs per degree N? What is the steam point on this new scale?

16. An astronomical object is believed to be a rarefied cloud of hydrogen. It contains both monatomic and diatomic molecules. What is the ratio of their average speeds?

17. At 0.01°K helium vapor in contact with liquid helium has a pressure of about 4×10^{-312} dyne/cm^2. Calculate the average distance between helium atoms in light years.

6 Electricity

6-1 Stationary Electric Charges

In the previous chapters most of the emphasis has been placed on gravitational forces and very little has been said about electromagnetic forces. Whenever interatomic forces had to be introduced, we were reminded of the existence of electromagnetic forces, but the details of the electromagnetic character of interatomic forces cannot be explained properly without quantum mechanics. As we shall soon see, electromagnetic forces are more complicated than gravitational forces and our discussion of them will lead us to important new concepts such as field theory, the special theory of relativity, and eventually quantum mechanics. Although interatomic forces play an important part in everyday experience, there are no common everyday experiences that make us directly aware of the basic character of electromagnetic forces in the same way that the behavior of falling bodies makes us directly aware of the character of gravitational forces. To discover the basic facts about electromagnetic forces we must do some deliberate experimentation such as the following.

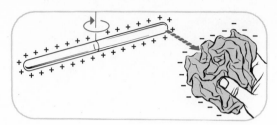

(*a*) *The silk attracts the glass rod.*

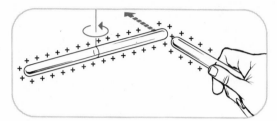

(*b*) *The two glass rods repel one another.*

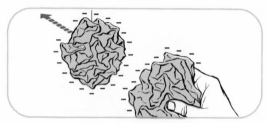

(*c*) *The two silk cloths repel one another.*　　　　**Figure 6-1** *The basic experimental facts of electricity.*

Rub a glass rod with a silk cloth. Suspend the rod at the end of a length of string, as in figure 6-1a, and place the silk cloth near one end of it. The rod will be attracted to the silk. Rubbing the glass and silk together has conditioned them in such a way that they exert additional forces on one another. Moreover, an estimate of the magnitude of these forces shows that they are comparable with the weight of the glass rod. This is an important point, because the weight of the glass rod results from the gravitational attraction of an extremely large body, the earth, whereas the new "electrical force" is produced by a comparatively small body, the silk cloth. Electrical forces, once they have been produced, seem to be very strong.

Now rub a second glass rod with a second silk cloth and bring the second rod up to the first suspended rod. This time there will be repulsion (figure 6-1b). Similarly, if one of the silk cloths is wadded into a ball and suspended, and the second cloth is brought up to it, again there is repulsion (figure 6-1c). Gravitational forces are always attractive, but this new force seems to be sometimes attractive and sometimes repulsive.

Electricity

These facts were known to the ancient Greeks, but the first attempt at a correct explanation of them was made by Benjamin Franklin and his contemporaries in the eighteenth century. They postulated that electricity is some sort of fluid which exists in two forms, positive and negative. Most bodies are *electrically neutral* because they contain equal amounts of positive and negative electricity that exactly cancel one another. When the silk cloth is rubbed against the glass rod, electricity flows from one to the other in such

108

a way that the glass rod is left with an excess of positive electricity and the silk cloth therefore has an equal excess of negative electricity. It is assumed that each type of fluid can be neither created nor destroyed. Since the glass and the silk each contained equal amounts of positive and negative electricity before they were rubbed together, it is easy to see that, if a certain amount of excess positive electricity appears on the glass, an equal amount of negative electricity must appear on the silk, assuming that suitable experimental precautions have been taken to prevent any of the electricity from leaking away.

In figure 6-1 these ideas have been represented by placing plus signs near the glass rod and minus signs near the silk cloth. The experimental facts illustrated in this figure prove that positive electricity repels positive electricity, negative electricity repels negative electricity, but positive and negative electricity attract one another.

Modern physics explains all this in the following way. Atoms are composed of neutrons, protons, and electrons. Neutrons are electrically neutral, so we may ignore them at the moment. A proton is a unit of positive electricity and an electron is a unit of negative electricity. These two units have equal but opposite electrical effects, so that a negative electron placed exactly on top of a positive proton would cancel it electrically. An atom contains a certain number of protons in a central nucleus and an equal number of electrons orbiting around this nucleus. Except in the immediate vicinity of the atom, this turns out to be equivalent to putting the electrons on top of the nucleus and the atom as a whole is therefore electrically neutral. When the silk is rubbed against the glass, some of the electrons near the outside of the glass atoms (silicon and oxygen) are rubbed off and migrate into the silk. Electrons prefer to be in silk rather than glass for reasons that are now understood but are too complicated to describe here. The silk cloth therefore acquires an excess of electrons and a net negative charge. The glass rod has a deficiency of electrons, or, if you prefer, an excess of protons, and this gives it a net positive charge.

This model gives a facile explanation of the nature of electricity, but let us extract from it the following basic facts, which are still not properly understood. As far as electrical effects are concerned, fundamental particles come in three different kinds, neutral, positive, and negative. Neutral particles show no electrical effects. Positive particles repel positive particles. Negative particles repel negative particles. A positive particle and a negative particle attract one another. The effect of a negative particle is exactly equal but opposite to the effect of a positive particle.

The idea that electricity cannot be created or destroyed may seem to imply that electrons and protons cannot be created or destroyed. We now know that there are processes in nature during which particles are created or destroyed. However, particles are always created in pairs, one negative and one positive. Conversely, in a destruction process a positive particle and a negative particle come together and destroy one another. The sum of positive and negative electric charge therefore remains constant. This is known as **conservation of charge.** We might even go a little further than what we are quite certain of, and guess that the number of positive particles in the universe is always exactly equal to the number of negative particles.

Suppose that a body contains n_+ fundamental particles with positive electric charge and n_- fundamental particles with negative electric charge.

Then the total electric charge, q, on the body is

$$q = (n_+ - n_-)e \qquad (6\text{-}1)$$

In the CGS system the unit of charge is called the **statcoulomb,** or some-times the **electrostatic unit of charge,** abbreviated to esu. The quantity e is a fundamental constant of nature. It is a measure of the electric charge on any single charged fundamental particle, but for historical reasons it is called **the charge on the electron.** Its value is

$$e = 4.802 \times 10^{-10} \text{ statcoulomb} \qquad (6\text{-}2)$$

For definiteness, we shall always take e to be a positive number. Then, in the case of a positively charged body n_+ is greater than n_- and the total charge q is a positive quantity. In the case of a negatively charged body n_+ is less than n_- and q is a negative quantity.

We shall now repeat some of the basic facts about electric forces that have already been mentioned in previous chapters. Suppose that a body with a charge q_1 is at a distance R from a body with a charge q_2 and they are both stationary (figure 6-2). As in the case of gravitation, it is nec-

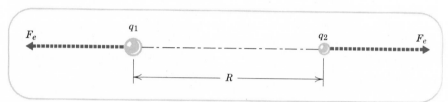

Figure 6-2 *Illustrating Coulomb's Law.*

essary to assume that the size of each body is small compared with their dis-tance apart. In the CGS system each body experiences an electrostatic force given by the equation

Coulomb's law

$$F_e = \frac{q_1 q_2}{R^2} \qquad (6\text{-}3)$$

R is measured in cm, q_1 and q_2 in statcoulombs, and F_e is then in dynes. The forces on the two bodies act along the line joining their centers. The two forces are equal in magnitude, but opposite in direction, obeying Newton's third law. If both charges are positive, or if both are negative, the forces are repulsive. If one of the charges is positive and the other is nega-tive, the forces are attractive.

The stipulation that the charges must be stationary is very important. If the charges are in motion there are additional forces, which produce the phenomena of magnetism. The study of the electrical effects of *stationary* charges is called **electrostatics.**

The mutual electrostatic potential energy of the two charges is given by another important formula:

The mutual electrostatic potential energy of two charges

$$\Phi_e = \frac{q_1 q_2}{R} \qquad (6\text{-}4)$$

If q_1 and q_2 are measured in statcoulombs and R in cm, the potential energy is measured in ergs. Notice that, as in the case of gravitational forces, the zero of potential energy corresponds to the situation when the two charges are infinitely far apart. When $R = \infty$, $\Phi_e = 0$.

The sign of the potential energy requires careful consideration. If either q_1 or q_2 is negative and the other is positive, the electrostatic force is attractive. The analogy with gravitation suggests that the potential energy should be negative. This is obviously so, because in equation 6-4 R is intrinsically positive and if either q_1 or q_2 is negative and the other is positive, the whole expression for Φ_e is negative. If q_1 and q_2 are both positive, or both negative, the mutual potential energy is a positive quantity. This is related to the fact that the force between the charges is then repulsive.

6-2 The Relative Magnitudes of Gravitational and Electrostatic Forces

Electrostatic forces are enormously large compared with gravitational forces. Let us consider this matter by comparing the electrostatic force with the gravitational force for a proton at a distance R from an electron. The electrostatic force is

$$F_e = \frac{e^2}{R^2} \qquad (6\text{-}5)$$

The gravitational force is

6-2 The
Relative
Magnitudes of
Gravitational
and
Electrostatic
Forces

$$F_G = \frac{G m_e m_p}{R^2} \qquad (6\text{-}6)$$

where m_e and m_p are the masses of the electron and proton respectively. The ratio is

$$\frac{F_e}{F_G} = \frac{e^2}{G m_e m_p} \qquad (6\text{-}7)$$

$$= \frac{(4.8 \times 10^{-10})^2}{(6.67 \times 10^{-8}) \times (9.11 \times 10^{-28}) \times (1.67 \times 10^{-24})}$$

$$= 2.3 \times 10^{39} \qquad (6\text{-}8)$$

This is an enormously large number! The electrostatic interaction is very much stronger than the gravitational interaction.

Since gravitational forces are so puny compared with electrostatic forces, it is interesting to consider why gravitation is so important in our everyday experience. The reason, of course, is that electricity comes in positive and negative forms, which can cancel one another out. On a large scale most matter is electrically neutral and produces no electrostatic forces, so that the gravitational forces are given a chance to reveal themselves. In a world composed entirely of negative electrons the gravitational forces would be completely overpowered by the electrostatic forces.

It seems strange that there should be two different kinds of interaction, that they should be similar in the sense that they both vary as $1/R^2$, and yet one is so very much stronger than the other. The ratio F_e/F_G depends only on the fundamental constants e, G, m_e, and m_p. Moreover, these constants are combined in such a way that the ratio is independent of the units of mass, length, and time. The electrostatic force is 2.3×10^{39} times stronger than the gravitational force, and this fact would not be influenced in any way if the standard meter were defined in terms of an entirely different platinum-iridium bar, the standard kilogram in terms of a different platinum-iridium cylinder, or if the rate of rotation of the earth changed overnight, thus altering the definition of the second of time. Many scientists believe that there must be a fundamental significance in a number such as F_e/F_G which is related to two of the fundamental interactions of nature, but which is independent of the arbitrariness of human decision.

The British astronomer and physicist, Sir Arthur Stanley Eddington, has suggested that F_e/F_G is directly related to certain properties of the universe as a whole. Modern theories of the universe strongly suggest that there is a limit to the observable universe. There is reason to believe that, however good our telescopes, we shall never be able to see beyond a distance R_u, which is called the **radius of the observable universe** and is approximately 10^{28} cm. The radius of an electron, r_e, is approximately 3×10^{-13} cm. Therefore

$$\frac{\text{Radius of universe}}{\text{Radius of electron}} = \frac{R_u}{r_e} = \frac{10^{28}}{3 \times 10^{-13}}$$

$$= 30 \times 10^{39} \text{ approximately} \qquad (6\text{-}9)$$

This number is also independent of the unit of length, is related to very fundamental properties of the universe, and is very large, yet its magnitude is not very different from the ratio F_e/F_G. There is a hint that, if we understood the nature of the universe better our theories would reveal that the factors that determine the ratio of the electrostatic to the gravitational interaction also determine the ratio of the radius of the universe to the radius of the electron.

Moreover, from the astronomical observations on that part of the universe which is nearest to us, we can make a rough estimate of the average density of matter in the universe and hence deduce the total number of fundamental particles in the universe. We find that, very approximately, the universe contains about 10^{79} protons, neutrons, and electrons. Note that

$$\left(\frac{F_e}{F_G}\right)^2 = 0.5 \times 10^{79} \tag{6-10}$$

Now there is a hint that the factors which determine F_e/F_G also determine the number of fundamental particles in the observable universe!

This is all very speculative and it may turn out to be far from the truth. On the other hand, it may be a very stimulating but tantalizing peep at future theories of the nature of the universe.

6-3 The Concept of a Field

So far we have been developing the idea that the universe is a collection of discrete particles moving through empty, featureless space. We have assumed that, in the ultimate analysis, the behavior of the universe must be described in terms of the motion of these particles, and that this motion can be explained and predicted if we know certain laws that tell us the acceleration given to a particle by its interactions with the other particles. We have introduced the first two of these laws, Newton's law of universal gravitation and Coulomb's law for electrostatic forces.

The next step is to consider a third type of force that exists between two *moving* charges and that is the origin of magnetism. Unfortunately the laws governing magnetic forces cannot be expressed as simply as the laws of gravitation and electrostatics. The next step in our program is therefore a difficult one. In the early nineteenth century, when the laws of magnetism were being discovered, the discovery was considerably aided by a new attitude to the nature of the universe embodied in the concept of electric and magnetic fields. Towards the end of the century, Clerk Maxwell was able to explain all the known phenomena of electromagnetism by a set of simple, elegant equations describing the behavior of the electric and magnetic fields. Simplicity and success are two powerful advocates for any scientific theory, so we must now turn our attention towards this new idea of a **field**.

Our previous attitude implies that two bodies at a great distance from one another, with only empty, featureless space between them, can never-

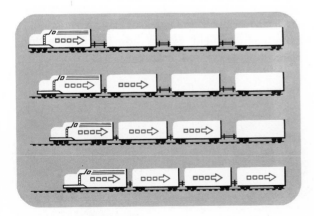

Figure 6-3 Transmission of an interaction along a line of railroad cars.

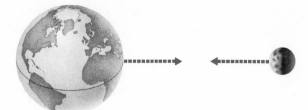

Figure 6-4 *The moon and the earth interact with one another even though the space between them is apparently empty.*

theless influence one another's motion. This is called **action-at-a-distance** and it worried many scientists and philosophers, who felt that two bodies could interact only if they were in "direct contact" with one another. To appreciate this point of view, imagine a railroad engine backing up to a line of railroad cars. The shock of impact is transmitted along the line as the first car hits the second, the second hits the third, and so on, until the disturbance eventually reaches the caboose (figure 6-3). However, in a situation such as the gravitational attraction between the earth and the moon, there are no visible, tangible bodies in between to provide the connecting links (figure 6-4). To overcome this difficulty, it was suggested that the whole of space is filled with an invisible material called the **ether,** having properties similar to a jelly. If you press your finger into a jelly, the strain is transmitted to all parts of the jelly and a pressure is exerted on a body embedded in the jelly some distance away (figure 6-5). In the same way, the earth was imagined to produce a strain in the ether which extended to the moon and transmitted a force to the moon (figure 6-6).

This idea of a material ether pervading the whole of space no longer appears to be tenable. Some of the most forceful arguments against it will be presented later in our discussion of the theory of relativity. Nevertheless, this line of thought is fruitful in the sense that it has taken our attention away from the material bodies and discrete particles and has raised the important question of whether space itself does not possess properties of interest to the physicist. Our original prescription for describing the instantaneous state of the universe was to define the position of each elementary particle by giving its Cartesian coordinates (See section 2-2). Now let us consider a possible alternative description in which physical properties are assigned to each point in space even if there is no particle located at that point. The procedure is to designate each point by giving the values of its x, y, and z coordinates and then to associate with this point certain other numbers that are a measure of such physical properties of the point as the gravitational field, the electric field, and the magnetic field. Exactly how this is done and what it means physically will be discussed throughout the rest of this chap-

Electricity

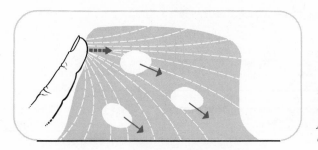

Figure 6-5 *Strain is transmitted through a jelly to objects embedded in the jelly.*

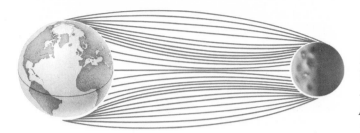

Figure 6-6 *Is there a jellylike ether between the earth and the moon? Is strain transmitted through this jelly to produce an attraction between the earth and the moon?*

ter. The important point to make at this stage is that this procedure is not only possible, but extremely useful. In particular, it clarifies and simplifies the complicated subject of electromagnetism. Later on it will help us to understand the nature of light and other electromagnetic radiations. It is an essential aspect of quantum mechanics and the general theory of relativity.

Since the above discussion has been somewhat abstract, a concrete analogy might prove helpful. Imagine a rapidly flowing trout stream containing several whirlpools that dimple its surface (figure 6-7). A complete description of the flow could be obtained by giving the velocity of the water at each point in the stream. Mathematically this would involve first setting up three Cartesian axes and then determining the Cartesian coordinates x, y, and z of the point P relative to these axes. The velocity v of a small sample of water located at P could then be specified and the procedure repeated for every point in the stream. The resulting description of the flow might be referred to as a **velocity field**.

A pictorial representation of this velocity field is obtained from the stream lines of figure 6-7. Each stream line represents the path followed by a small sample of water as it flows down the stream. (This statement is true only if the pattern of stream lines does not change with time. We are idealizing the flow of an actual trout stream.) At any point P the velocity v of the water is tangential to a stream line. In the vicinity of a whirlpool the water is moving around in circles. Each whirlpool produces a dimple on the sur-

Figure 6-7 *The velocity field of a flowing stream.*

face of the water. Are these dimples the analogs of elementary particles? A dimple is a highly localized entity that can move over the surface of the stream in the same way that an elementary particle moves through space. However, there is no material body located at the dimple. It is merely a point at which the stream lines behave in a peculiar manner. This suggests to us an extreme swing of the pendulum in our attitude toward describing the universe. Now the emphasis is placed on assigning properties to all the points in space so that space is filled with various fields. An elementary particle is then a point at which a field behaves in a peculiar way.

6-4 The Electric Field

In the vicinity of an arrangement of electric charges there exists an electric field which pervades the whole of space. Associated with any point P is a vector **E** which is called the **electric field vector,** or often simply the **electric field.** Its significance is that, if a small charge q is placed at the point P, then the electric force $\mathbf{F}_e$ acting on q is $q\mathbf{E}$.

Definition of the electric field

$$\mathbf{F}_e = q\mathbf{E} \tag{6-11}$$

The vector **E** varies from point to point, but it is a unique property of any particular point P, independently of whether or not there is actually a charge q at P. Since F_e is a force measured in dynes and q is measured in statcoulombs, the units of E are **dynes per statcoulomb.**

The electric field is therefore the electric force per unit charge. If q is a positive charge, $\mathbf{F}_e$ and **E** are in the same direction, but if q is a negative charge they point in opposite directions. This is taken into account in equation 6-11 if the scalar quantity q is given the appropriate sign.

To calculate the electric field due to a single point charge consider, as in figure 6-8, a point P at a distance R_1 from a charge q_1. Place a small

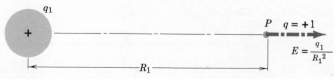

(a) The electrostatic force between charges q_1 and q a distance R_1 apart.

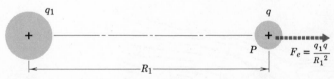

(b) The electric field at a distance R_1 from a charge q_1.

Figure 6-8 The electric field due to a single point charge.

positive charge q at P. The electrostatic force on q is

$$F_e = \frac{q_1 q}{R_1{}^2}$$

(6-12)

The electric field E at P is such that

$$F_e = qE$$

(6-13)

Therefore:

Electric field E at a distance R$_1$ from a point charge q$_1$

$$E = \frac{q_1}{R_1{}^2}$$

(6-14)

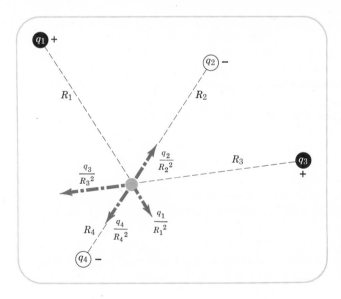

Figure 6-9 *The electrostatic field due to a distribution of stationary point charges. The resultant electric field at P is the vector sum of the various contributions represented by arrows.*

Note carefully that if q_1 is positive **E** points away from q_1, but if q_1 is negative **E** points toward q_1. The electric field produced by a positive charge points directly away from that charge, whereas the electric field produced by a negative charge points directly toward that charge.

Now suppose P is at a distance R_1 from a charge q_1, at a distance R_2 from a charge q_2, at a distance R_3 from a charge q_3, and so on. The electric field at P is the *vectorial* sum of a number of contributions similar to equation 6-14. This is illustrated in figure 6-9.

In the same way that the flow of the stream was represented by stream lines in figure 6-7, the electric field can be represented by **electric field lines.** The significance of a field line is that, at any point P, the tangent to the field line is in the same direction as **E.** Figure 6-10 shows the field lines

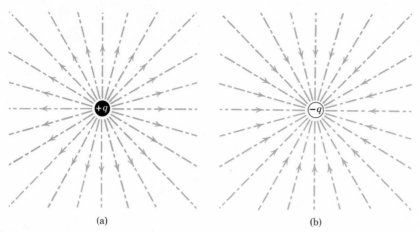

Figure 6-10 *The field lines for* (*a*) *a single point positive charge* (*b*) *a single point negative charge.*

for a single point charge. They are radial straight lines that diverge from a positive charge and converge on a negative charge.

Figure 6-11 shows the field lines in some more complicated situations involving two charges. Part *b* of this figure also illustrates how the electric field vector **E** is tangential to the electric field line at a point *P*. The field lines provide a very vivid pictorial representation of the field. One can imagine the two positive charges of figure 6-11*a* pushing one another apart, or the positive and negative charges of figure 6-11*b* pulling one another together. Moreover, when the lines are close together the field is strong, but when the lines are far apart the field is weak. The points marked *N* in parts *a* and *c* of the figure are points where the electric field is zero. Notice how the field lines veer away from these points. In part *a* the point *N* is midway between the two charges, since it is obvious that at this point the two charges make exactly equal but opposite contributions to the field.

Look at figure 6-11 and ask yourself the following question. Suppose the charges were fundamental particles such as protons or electrons. Is there really "something" at the point where the particle is located, or is it just a point from which all the field lines diverge or on which all the field lines converge? Should the emphasis be placed on the particles located at a few points in space or should the emphasis be on the field at all points in space? Present day physics talks about both particles and fields. Exactly how the future will resolve this issue is an open question.

6-5 The Electric Potential at a Point

There is an alternative way of describing an electric field in which the quantity given at each point *P* in space is not the vector **E** but a scalar quantity ϕ_e called the **electric potential** at the point. Its significance is that, if we place a small charge *q* at *P*, the sum of the mutual potential energies of *q* with all the other charges present is equal to $q\phi_e$. That is, the potential en-

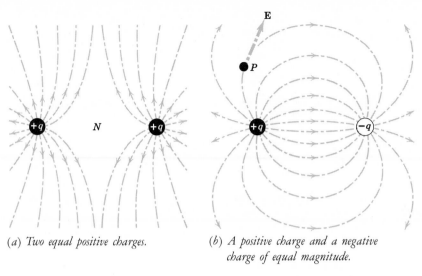

(a) Two equal positive charges.

(b) A positive charge and a negative charge of equal magnitude.

(c) A large positive charge and a small negative charge.

Figure 6-11 *Some typical field patterns.*

ergy Φ_e (a capital phi) of q at P is $q\phi_e$ (a small phi).

6-5 The
Electric
Potential at a
Point

> *The electric potential at a point*
>
> If a small charge q is placed at a point P where the electric potential is ϕ_e, the sum of its mutual potential energies with all the other charges present is
>
> $$\Phi_e = q\phi_e \tag{6-15}$$

If the energy Φ_e is measured in ergs and the charge q in statcoulombs, the units of ϕ_e are **statvolts.** As we shall explain more carefully later on, the statvolt is intimately related to the **volt,** which is a very well known practical unit applied in everyday life to batteries and the electricity supplied to our homes. In fact,

$$1 \text{ statvolt} = 300 \text{ volts} \qquad (6\text{-}16)$$

When a wall socket delivers approximately 100 volts, this means that the difference in potential between the two pins of the plug averages about $\frac{1}{3}$ statvolt. A 12-volt automobile storage battery has a potential difference of $\frac{12}{300} = \frac{1}{25}$ statvolt between its terminals.

We are now in a position to consider a very important practical unit of energy—the **electron volt.** The discharge tube of figure 6-12 consists of two metal plates, or *electrodes*, inside an evacuated glass tube. By means of some device such as a large number of storage batteries connected one after the other, electrons are added to one electrode to charge it negatively and electrons are removed from the other electrode to charge it positively. The potential ϕ_+ at the positive electrode is then greater than the potential

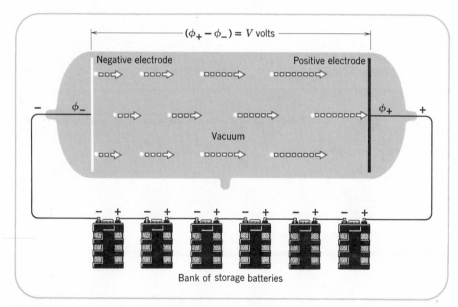

Figure 6-12 *A discharge tube accelerating electrons through a potential difference of V volts.*

ϕ_- at the negative electrode and we say that we have applied a potential difference (or voltage) V between the electrodes, where

$$V = (\phi_+ - \phi_-) \qquad (6\text{-}17)$$

Imagine an electron starting from rest in the vacuum in the immediate vicinity of the negative electrode. Since the electron carries a negative charge $-e$, it is repelled by the negative electrode and attracted toward the positive electrode. It therefore steadily accelerates toward the positive electrode and when it arrives there it has built up a speed v and a kinetic energy $\frac{1}{2}mv^2$. The law of conservation of energy requires that this gain in kinetic energy shall be achieved at the expense of a corresponding loss in potential energy. When it starts out at the negative electrode, the electron has a potential energy $-e\phi_-$. (Notice the negative sign which is needed because the charge on the electron is negative whereas the symbol e represents a positive number.) When it reaches the positive electrode, the potential energy of the electron is $-e\phi_+$. The loss of potential energy is the initial potential energy minus the final potential energy, or

Loss of potential energy

$$= -e\phi_- - (-e\phi_+)$$
$$= e(\phi_+ - \phi_-)$$
$$= eV \tag{6-18}$$

Equating this to the gain in kinetic energy, we obtain

$$\tfrac{1}{2}mv^2 = eV \tag{6-19}$$

In the early days of research in atomic and nuclear physics, many experiments were performed in which electrons or other charged particles were accelerated to higher energies in a discharge tube or some similar device. It was found very convenient to specify the energies of these particles just by quoting the voltage through which they had been accelerated *from rest*, and this became part of the language of physics. A particle with a charge of $+e$ or $-e$, which has been accelerated from rest through a potential difference of V volts, is said to have a kinetic energy of V **electron volts**. The electron volt is a unit of energy like the erg. One electron volt is the energy given to a particle with charge e when it is accelerated through a potential difference of 1 volt. To find its value in ergs, we must insert into the right-hand side of equation 6-19

$$e = 4.80 \times 10^{-10} \text{ statcoulomb} \tag{6-20}$$
$$\text{and} \quad V = 1 \text{ volt} \tag{6-21}$$
$$= \frac{1}{300} \text{ statvolt} \tag{6-22}$$

Then,

$$1 \text{ electron volt} = 4.80 \times 10^{-10} \times \frac{1}{300} \text{ erg}$$
$$= 1.60 \times 10^{-12} \text{ erg} \tag{6-23}$$

$$1 \text{ \textbf{electron volt}} = 1.60 \times 10^{-12} \text{ erg} \tag{6-24}$$

One electron volt is abbreviated to 1 eV, 10^3 eV is abbreviated to 1 KeV, 10^6 eV to 1 MeV and 10^9 eV to 1 GeV (or sometimes 1 BeV).

6-6 Electric Currents and Batteries

The practical applications of electricity are numerous and well known, but they hardly ever make use of the electrostatic forces between charged bod-

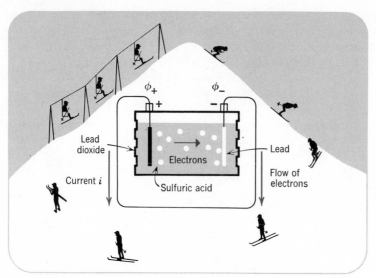

Figure 6-13 *A storage battery with a copper wire connected between its terminals compared with a ski lift and a ski run.*

ies. They rely instead on the effects produced by an electric current flowing through a metal wire that has no net electric charge. When an electrical appliance is plugged into a wall socket, electrons flow in through one pin of the plug, circulate through the wires inside the appliance and leave via the other pin of the plug. As many electrons leave as enter and there is never any net accumulation of charge inside the appliance. Actually, since the wall socket usually supplies 60 cycle alternating current, the electrons surge backward and forward sixty times a second. This is a nonessential complication and it will be simpler to discuss the direct current produced by a storage battery of the kind used to operate the ignition system of an automobile.

Figure 6-13 shows a storage battery with a copper wire connected between its terminals. The battery consists essentially of two metal plates immersed in dilute sulfuric acid. One of the plates is coated with sponge lead and the other is coated with lead dioxide. In section 6-1 we stated that when a glass rod is rubbed with a silk cloth the electrons have a preference for being in the silk rather than the glass and therefore some electrons migrate from the glass to the silk. Similarly, electrons prefer to be in lead rather than lead dioxide, and by a series of elaborate processes electrons are transferred from the lead dioxide plate through the sulfuric acid to the lead plate. The lead dioxide plate therefore has a deficiency of electrons, an excess of protons, and a positive charge. It is the positive terminal of the battery. The lead plate acquires an excess of electrons, is negatively charged, and is the negative terminal of the battery.

The electric potential ϕ_+ at the positive terminal is higher than the potential ϕ_- at the negative terminal. The battery operates in such a way that the potential difference V between its terminals is almost constant.

$$V = \phi_+ - \phi_- = 2 \text{ volts}$$
$$= 0.67 \times 10^{-2} \text{ statvolt, approximately} \quad (6\text{-}25)$$

Electricity

V decreases a little if a large current is drawn from the battery, and V also steadily decreases with time as the battery "runs down" because of chemical processes that contaminate its plates. However, these changes are never more than a few percent and we shall simplify the discussion by assuming that the current drawn from the battery is small and that the potential difference V between its terminals remains constant.

If a copper wire is connected between the terminals, electrons flow continuously from the negative terminal through the wire to the positive terminal and then through the battery back to the negative terminal. In copper, as in other metals, some of the electrons on the outside of the copper atom are very loosely held and are easily shaken free. These electrons do not remain attached to a particular copper atom, but wander freely throughout the whole of the copper wire. They are repelled by the negative terminal and attracted by the positive terminal and are therefore continually accelerating toward the positive terminal. This does not mean that the electrons move faster and faster the nearer they get to the positive terminal. They frequently collide with the copper atoms and, although an electron acquires an extra velocity toward the positive terminal in between collisions, this extra velocity is destroyed at each collision. However, in between collisions the electron does get a chance to move toward the positive terminal, and the net result is a slow drift of the electrons toward this terminal.

In the absence of the storage battery, with no current flowing, free electrons move through the copper with very high velocities of the order of magnitude of 10^8 cm/sec. (This is about 2 million miles per hour!) Since these velocities are randomly distributed in all directions, there is no net flow of electrons one way or another. When the storage battery produces a potential difference between the ends of the wire, the net result is equivalent to adding a small extra "drift velocity" v_d to each electron. This drift velocity points along the wire toward the positive terminal and is in the same direction and has the same magnitude for all the free electrons. The random chaotic motion therefore acquires a slight bias in this direction and more electrons move toward the positive terminal than away from it. However, v_d is always very small compared with the random velocity. A typical value of v_d might be 0.1 cm/sec, which is a billion times smaller than the random velocity of 10^8 cm/sec. One should therefore imagine an electron to be moving with a very high velocity, frequently colliding with the copper atoms and changing the direction of this velocity, but making slow progress in the direction of the positive terminal.

If a piece of wood were connected between the terminals, no current would flow. All the electrons in the wood are firmly held to their atoms and are not free to move from one terminal to the other. Wood is an **insulator** because it will not carry an electric current. Copper is a **conductor** of electricity. All metals are conductors.

When the electrons flowing through the copper wire arrive at the positive terminal they must be returned to the negative terminal, otherwise the charges on the two terminals would soon be neutralized and the current would cease to flow. The transfer of electrons through the sulfuric acid to the negative terminal against the repulsion of the negative charge on this terminal is a consequence of certain complicated chemical processes that occur inside the battery. In figure 6-13 a storage battery is compared with a ski lift and a ski run. The chemical processes inside the battery that lift the electrons from the positive terminal to the negative terminal are analogous to the ski lift that lifts the skiers from the bottom to the top of the mountain. The electron being pulled through the copper wire by the electrical attraction of the positive terminal is analogous to the skier gliding down the ski run by taking advantage of the gravitational attraction of the earth. The electron maintains a steady drift velocity v_d because of its numerous collisions with the copper atoms. Similarly, the skier can maintain a steady velocity down the run by making skillful use of the frictional forces exerted on him by the snow.

Quantitatively, the electric current i is defined as the net rate of flow of charge across any cross-section of the wire. Suppose that in t seconds a charge q flows past. Then

---

Electric current $=$ charge flowing past in one second

$$i = \frac{q}{t} \tag{6-26}$$

If q is in statcoulombs and t in seconds, the current i is measured in **statamperes.** The current is due to a flow of negatively charged electrons, but it is equivalent to a flow of positive charge in the opposite direction and it is traditional to assign the direction of the current on the assumption that positive charge is flowing. The direction of the current is therefore always opposite to the direction of flow of the free electrons.

It is interesting to consider the energy changes taking place when a copper wire is connected between the terminals of a storage battery (figure 6-13). The original source of the energy is the **chemical energy** of the materials constituting the battery. When current is drawn from the battery, the chemical reaction taking place is essentially

$$\begin{array}{ccccc}
\text{Pb} & + & \text{PbO}_2 & + & 2\text{H}_2\text{SO}_4 \\
\left[\begin{pmatrix}\text{One atom} \\ \text{of lead}\end{pmatrix}\right. & + & \begin{pmatrix}1 \text{ molecule of} \\ \text{lead dioxide}\end{pmatrix} & + & \left.\begin{pmatrix}2 \text{ molecules of} \\ \text{sulfuric acid}\end{pmatrix}\right]
\end{array}$$

$$\begin{array}{ccc}
\longrightarrow & 2\text{PbSO}_4 & + & 2\text{H}_2\text{O} \\
\text{change into} & \left[\begin{pmatrix}2 \text{ molecules of} \\ \text{lead sulfate}\end{pmatrix}\right. & + & \left.\begin{pmatrix}2 \text{ molecules} \\ \text{of water}\end{pmatrix}\right]
\end{array} \tag{6-27}$$

Any molecule appearing in this equation is a structure made from electrons, protons, and neutrons. The energy of the molecule is obtained by adding the kinetic energies of these particles to the mutual potential energies of the particles taken in pairs. The most important contribution to the potential energy is the electrostatic potential energy of the charged particles. The combined energy of the molecules on the left-hand side of equation 6-27 is greater than the combined energy of the molecules on the right-hand side. Therefore, when the chemical reaction takes place, energy is made available. Inside the battery this energy is used to lift electrons from the positive terminal to the negative terminal, where their electric potential energy is greater. The first step is therefore a transformation of chemical energy into the electrostatic potential energy of the charges which appear on the terminals of the battery.

Electric field lines originate on the positive charge at the positive terminal and pass through the copper wire to the negative charge at the negative terminal. There is therefore an electric field inside the copper that accelerates the free electrons and increases their kinetic energy. In the second step, then, the electrons flow from the negative to the positive terminal and their electrostatic potential energy is converted into kinetic energy. This is similar to the process we have already discussed in detail for the gas discharge tube of figure 6-12.

However, in the gas discharge tube the electrons can continue to accelerate without hindrance until they reach the positive electrode. This cannot happen in the copper wire because the electrons are continually colliding with the copper atoms. Any additional kinetic energy that an electron is able to pick up from the electric field between collisions is completely transferred to the copper atoms at the next collision. This increases the speed of vibration of each copper atom in the solid, which is equivalent to adding heat to the copper and raising its temperature. It is well known, of course, that a copper wire connected between the terminals of a storage battery warms up. The third step is therefore a conversion of the kinetic energy of the flowing electrons into heat in the copper wire.

The wire may become so hot that it glows like the filament of an electric light bulb, emitting light and heat radiation. There is then a fourth step, which the reader will understand better after he has read Chapter 8. There is a partial conversion of heat energy into the energy of electromagnetic waves traveling through empty space.

6-7 *Some Comments on Practical Units*

This book uses almost exclusively the CGS system of units, based on the centimeter, gram, and second. In this system the unit of charge is the statcoulomb, the unit of electric current is the statampere, and the unit of potential difference is the statvolt.

The corresponding practical units are the coulomb, the ampere, and the volt. They are associated with the MKS system of units, based on the meter, kilogram, and second. The practical units are so familiar because they are the ones used by electrical engineers for various historical reasons.

However, the conventional use of MKS units in electromagnetism introduces some unnecessary complications into the fundamental equations and tends to obscure the basic physical significance of these equations. We shall therefore continue to place the main emphasis on CGS units, but we will occasionally use practical units when it is unavoidable. It will be sufficient for the reader to know the conversion factors between the two systems.

Practical units of electricity

Charge: 1 **coulomb** $= 3 \times 10^9$ statcoulombs	(6-28)
Current: 1 **ampere** $= 1$ coulomb per second	(6-29)
$= 3 \times 10^9$ statcoulombs per second	
$= 3 \times 10^9$ statamperes	(6-30)
Potential difference: 1 **volt** $= \dfrac{1}{300}$ statvolt	(6-31)

Questions

A
1. If you were given a charged body, how would you decide whether its charge was positive or negative?
2. Does the mass of a body change when it is given a charge? If so, does it increase or decrease?

B
3. A charged body attracts small *uncharged* objects such as dust particles. Can you imagine why?
4. Are you more attracted to the idea of a universe consisting of discrete particles separated by empty space or to the idea of a continuous universe in which every point of space has physical properties? Try to justify your preference (a) on the basis of the physics you already know (b) by reference to your personal impression of the nature of the universe derived from direct observation of your surroundings.
5. Can you think of any other examples of fields not quoted in the text? In each case define the field carefully.
6. Can two field lines intersect at an angle?

C
7. Given N protons and N electrons, can you arrange them in some geometrical pattern such that the resultant electrostatic force on every single particle is zero? Think about this for N equal to 1, 2, and 3 and N very large. What conclusions are suggested about the structure of matter?
8. If a metal were composed of positively charged electrons (positrons) and negatively charged protons (antiprotons), would there be any way of discovering this by studying the phenomena described in this chapter?
9. Consider the following attempt to explain gravity in terms of electrical forces. "The earth has a net positive charge. Falling bodies contain equal numbers of protons and electrons but the negative charge on the electron is slightly larger than the positive charge on the proton." In particular, consider the following points. (a) Why does this theory not explain the fact that all bodies fall with the same acceleration? (b) Can you think of an assumption about the nature of the neutron which would make all bodies fall with the same acceleration? (c) Can the motion of the moon around the earth be explained by the theory? (d) Are there any difficulties in explaining the motions of all the bodies in the solar system on this theory? You may choose the charge on each body in the solar system to have any value that is needed to fit the facts. (e) What experiments might be performed to test the theory?

Problems

A
1. A charge of +5 statcoulombs is 2 cm away from a charge of +7 statcoulombs. What is the electrostatic force on each charge? What is their mutual electrostatic potential energy?

2. A charge of -3 statcoulombs is 1.5×10^2 cm away from a charge of $+9$ statcoulombs. What is the electrostatic force on each charge? What is their mutual electrostatic potential energy?

3. A mass of 10 gm carrying a charge of $+2$ statcoulomb is located at a point where the electric field is 50 dynes/statcoulomb. What force acts on it?

4. A mass of 2×10^3 gm carrying a charge of $+0.01$ statcoulomb is located at a point where the electric field is 3 dynes/statcoulomb. What is its acceleration?

5. What is the acceleration of a proton in an electric field of 10^3 dynes/statcoulomb?

6. Find the electric field at a point 10^{-8} cm from an electron.

7. What is the potential energy of an electron located at a point where $\phi_e = +1000$ statvolt?

B

8. An electron and a proton are stationary at a distance apart of 2×10^{-4} cm. What is the electrostatic force on (a) the proton (b) the electron? What is their mutual electrostatic potential energy?

9. An electron and a proton are instantaneously at rest at a distance apart of 5×10^{-9} cm. What is the acceleration of (a) the proton (b) the electron?

10. Make a rough estimate of the electrostatic potential energy of a hydrogen atom.

11. A charge of $+5$ statcoulomb is 100 cm from a charge of -9 statcoulomb. Find the electric field at the midpoint of the line joining the two charges.

12. Calculate the value of the vertical electric field that can support the weight of an electron.

13. If a current of 10^{-5} statamp flows through a wire, what is the net number of electrons passing any point in the wire per second?

14. A 12-volt storage battery promises to deliver 1 amp for 100 hours. If it does so, how many coulombs of charge will have passed between its terminals? How many electrons is this?

15. Without questioning the formulas of classical physics that have been presented so far, calculate the potential difference in volts through which an electron would have to fall from rest to acquire a velocity equal to the velocity of light (3×10^{10} cm/sec).

16. Two protons in a nucleus are 5×10^{-13} cm apart. Calculate their mutual electrostatic potential energy in electron volts.

17. Calculate the velocity of a 110 electron volt electron. Convert your answer into mph.

C

18. Referring to figure 6-11c, find an expression giving the position of the point N in terms of q_1, q_2, and their distance apart, R. What happens if the magnitude of q_2 is greater than the magnitude of q_1? Where is N if the two charges have equal magnitudes but opposite sign (as in figure 6-11b)?

19. A mass of 3 gm carrying a charge of $+5$ statcoulomb is in a uniform electric field, which has the same magnitude and direction everywhere. It starts from rest and 2 sec later has traveled a distance of 10 cm. Calculate the magnitude of the field.

Problems

20. An electron is initially at rest in a uniform electric field of 2.5 dynes/statcoulomb. What is its velocity 10^{-9} sec later?

21. A proton has an initial velocity of 10^8 cm/sec in exactly the opposite direction to a uniform electric field of 0.2 dynes/statcoulomb. How far does it travel before coming to rest?

22. Two 1 gm masses are at rest relative to one another. An equal number n of electrons is removed from each so that their gravitational attraction is exactly counterbalanced by their electrostatic repulsion? What is n?

23. What equal charge must be placed on the moon and the earth to counterbalance the gravitational force between them?

24. One ton (9.07×10^5 gm) of protons is taken from the earth, loaded on to a rocket and landed on the moon. At the same time the velocity of the moon is adjusted so that it remains in the same stable circular orbit. How many seconds difference does this make to the length of the month? Is it an increase or a decrease?

25. If the earth had its present mass, but were composed entirely of protons, what would be the escape velocity of an electron? (Use the formulas of classical physics. The velocity calculated in this way greatly exceeds the velocity of light and is not possible according to the theory of relativity.)

26. If the average kinetic energy of a molecule of an ideal gas is expressed in the form αT electron volts, what is the numerical value of α? What temperature is "equivalent" to 30 GeV?

27. If a small body carrying a charge of 10^{-6} coulomb moves freely through a potential difference of 500 volt, starting from rest, calculate its final kinetic energy in electron volts.

7 *Electromagnetism*

7-1 *Forces Between Moving Charges*

Figure 7-1 illustrates a simple experiment to demonstrate the forces between electric currents. The vertical wires W_1 and W_2 are freely hinged at the top and their lower ends dip into mercury pools. This leaves the wires free to move sideways, but also, since mercury is a metal and conducts electricity, it enables electric currents to be fed into the wires from the two storage batteries shown. When the storage batteries are connected so that the two currents flow in the same direction, the wires move toward one another. When the currents flow in opposite directions the wires move apart. Parallel currents moving in the same direction attract one another. Parallel currents moving in opposite directions repel one another. One way to memorize this is to notice that it is the other way round from electrostatic forces. Like charges repel one another and unlike charges attract one another.

In this experiment the wires always contain as many protons as elec-

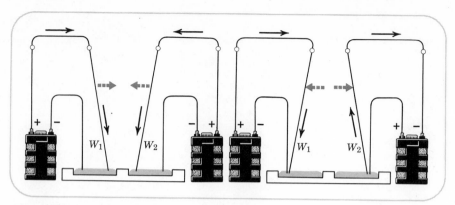

Figure *7-1 An experiment to demonstrate the forces between two parallel wires carrying electric currents.*

trons and do not accumulate any net charge. The forces between the wires are not the electrostatic forces discussed in Chapter 6, they are a consequence of the fact that electrons are moving inside the wires. When two charged particles are both in motion, they exert on one another a new kind of force that depends on their velocities and is zero if either velocity is zero. This new force is responsible for the magnetic properties of materials such as iron bar magnets and also for the magnetic effects of electric currents.

If we were to proceed as we have done previously, we would consider two charges q_1 and q_2 a distance R apart moving with velocities $\mathbf{v}_1$ and $\mathbf{v}_2$, as in figure 7-2a. We would then write down a law, similar to Newton's law of gravitation, or Coulomb's law of electrostatic forces, expressing the force in terms of the charges, their velocities and their distance apart. Unfortunately, it is not possible to formulate such a law in any simple form. Even a reasonable approximation to the law is very cumbersome. At this point, the concept of the magnetic field makes the discussion much easier.

To find the force exerted by q_1 on q_2 we proceed as follows. The moving charge q_1 produces at each point in space not only an electric field $\mathbf{E}$, but also a **magnetic field B** (figure 7-2b). These fields can be calculated in terms of q_1 and $\mathbf{v}_1$, and the position of the point relative to q_1. The values of $\mathbf{E}$ and $\mathbf{B}$ at the point P_2 where the second charge q_2 is located enable the force on q_2 to be calculated (figure 7-2c). The electric field $\mathbf{E}$ exerts a force $q_2\mathbf{E}$ that is independent of the motion of q_2 (see equation 6-11). In a manner that we shall explain below, the magnetic field $\mathbf{B}$ enables us to calculate that part $\mathbf{F}_m$ of the force which depends on the velocity of q_2. Looked at from this point of view, the magnetic field $\mathbf{B}$ is nothing more than a mathematical trick that helps us to calculate the forces between moving charges.

Electromagnetism The calculation of a magnetic force therefore proceeds in two stages. The first stage is to find a formula for the magnetic field $\mathbf{B}$ at a point in terms of the behavior of the moving charges that produce this field. This first stage is discussed in section 7-2. The second stage is to find a formula for the magnetic force $\mathbf{F}_m$ on a moving charge located at the point where the magnetic field is **B.** This second stage is discussed in section 7-3.

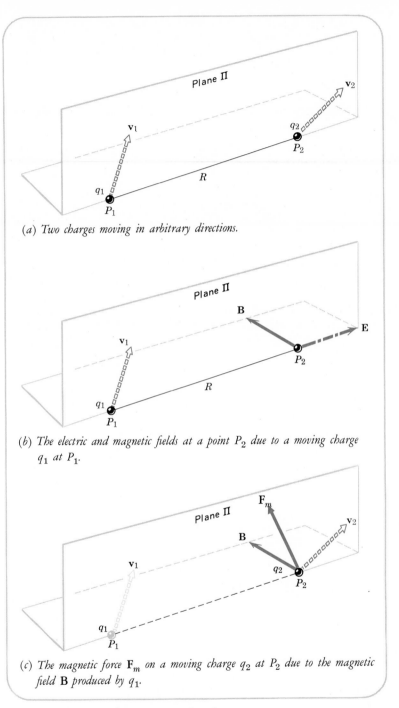

(a) Two charges moving in arbitrary directions.

(b) The electric and magnetic fields at a point P_2 due to a moving charge q_1 at P_1.

(c) The magnetic force $\mathbf{F}_m$ on a moving charge q_2 at P_2 due to the magnetic field $\mathbf{B}$ produced by q_1.

Figure 7-2 The forces between two moving charges.

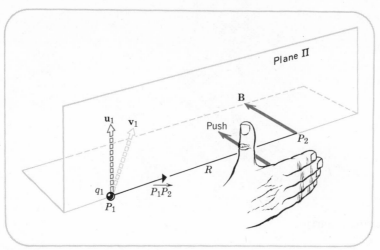

Figure 7-3 *The magnetic field* **B** *produced at a point* P_2 *by a moving charge* q_1 *at* P_1.

7-2 The Magnetic Field

Figure 7-3 is a modified form of part *b* of figure 7-2 and we shall use it to discuss the magnetic field **B** produced at the point P_2 by the moving charge q_1 at P_1. As explained in section 2-6 and figure 2-14, the velocity v_1 may be resolved into two rectangular components, one parallel to the line P_1P_2 and the other perpendicular to this line. The magnetic field at P_2 is produced entirely by the perpendicular component u_1 and the parallel component may be ignored. The magnitude of **B** is given by the following equation

---

Magnetic field produced by a moving charge

$$B = \frac{q_1 \, u_1}{R^2 \, c} \tag{7-1}$$

u_1 is the component of the velocity of the moving charge q_1 perpendicular to the line joining the charge to the point where B is measured.

If q_1 is in statcoulombs, R is in cm and u_1 is in cm/sec, then the units of B are called **gauss**.

Electromagnetism   The quantity c is a constant of nature. Its units are cm/sec, the units of a velocity, and its value is

$$c = 2.9979 \times 10^{10} \text{ cm/sec} \tag{7-2}$$

This is precisely the speed of light! It is one of the most important constants of physics. As we are now discovering, it enters in a fundamental way into the laws of electromagnetism. As we shall explain in a later chapter, it is the speed of light because light is an electromagnetic wave governed by the laws of electromagnetism. It plays a fundamental role in the theory of relativity because it is the maximum speed with which information can be transferred from one part of the universe to another.

The *direction* of **B** requires careful attention. **B** is perpendicular to both $\mathbf{u}_1$ and $\overrightarrow{P_1P_2}$. If $\mathbf{u}_1$ and $\overrightarrow{P_1P_2}$ are both in the plane Π of figure 7-3, then **B** is perpendicular to this plane. It is still necessary to decide whether **B** points outward toward the reader or inward away from the reader. In fact it points inward. The drawing of a hand in figure 7-3 illustrates the following general rule for obtaining the direction of **B.**

Right-hand rule number 1

Hold the right hand flat with the four fingers side by side in the same plane as the palm and with the thumb sticking out, but still in this plane. Point the thumb in the direction of $\mathbf{u}_1$. Point the fingers in the direction $\overrightarrow{P_1P_2}$, from the moving charge toward the point at which the field is being calculated. Then the magnetic field **B** at P_2 is perpendicular to the palm in the direction from the back of the hand toward the front, that is, the direction in which the hand would push.

The above rule has been formulated on the assumption that the moving charge q_1 is positive. If q_1 is negative the magnetic field **B** is in exactly the opposite direction.

Magnetic fields are frequently produced by electric currents flowing through wires. The magnetic field at any point is then the sum of a large number of contributions from all the electrons moving along the wire. The contribution from a single electron is determined by the laws given above, but the addition procedure often involves complicated mathematics. We shall therefore have to be content to quote the results in two important cases.

In the case of a current i in a long straight wire (figure 7-4), the magnetic field lines are circles centered on the wire. A **magnetic field line,** of course, is such that the magnetic field **B** at any point is tangential to the field line through that point. The magnitude of the magnetic field at a dis-

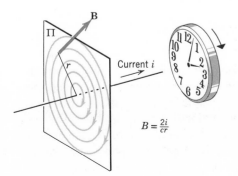

$$B = \frac{2i}{cr}$$

Figure 7-4 *The magnetic field produced by a current in a long straight wire. The magnetic field lines are shown in one plane Π perpendicular to the wire, but they exist everywhere in the whole space surrounding the wire.*

tance r from the wire is

$$B = \frac{2i}{cr} \tag{7-3}$$

The direction in which the field lines go round the wire is determined by the application of right-hand rule number 1 to the individual moving electrons, but it is simpler to remember a clock rule (figure 7-4).

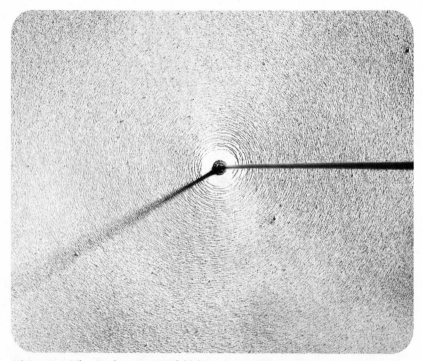

Figure 7-5 *The circular magnetic field lines surrounding a straight wire carrying a current, as revealed by the iron filings technique. (Photograph by Berenice Abbott.)*

Electromagnetism   Figure 7-5 was obtained by sprinkling iron filings on a sheet of paper perpendicular to a straight wire carrying a current. Iron filings are long and thin and align themselves in the direction of the magnetic field, in the same way that a compass needle points along the earth's magnetic field. The circles centered on the wire are easily seen in this photograph. This technique was known to Faraday and helped him to form his very vivid picture of a magnetic field as a pattern of lines in space.

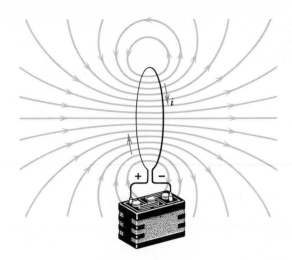

Figure 7-6 *Magnetic field lines due to a current in a circular loop of wire.*

Figure 7-6 shows the pattern of magnetic field lines in the important case of a current in a circular loop of wire. Although the lines are no longer perfect circles, it is still true that, if we stand at any point on the wire and look in the direction of the current, the lines go round the wire in a clockwise direction. The reader may already have noticed one important difference between electric fields and magnetic fields. Electric field lines sometimes start on a positive charge and end on a negative charge. Magnetic field lines never start or end, but always form closed loops. The field lines in figure 7-6 that appear to have a beginning and an end would also form closed loops if we could continue them beyond the edges of the figure.

7-3 The Magnetic Force on a Moving Charge

Once the magnetic field **B** at the point P_2 is known, we can turn to the question of the magnetic force $\mathbf{F}_m$ exerted by this field on a moving charge q_2 at P_2 (figure 7-7). The velocity $\mathbf{v}_2$ of q_2 should first be resolved into rec-

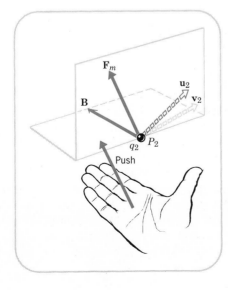

Figure 7-7 *The magnetic force $\mathbf{F}_m$ on a positive charge q_2 moving in a magnetic field* **B.**

tangular components parallel to the magnetic field **B** and perpendicular to **B**. Only the component u_2 perpendicular to **B** matters as far as the magnetic force is concerned. The magnitude of the force is given by the following equation.

Magnetic force on a charge moving in a magnetic field

$$F_m = q_2 \frac{u_2}{c} B \qquad\qquad (7\text{-}4)$$

u_2 is the component of the velocity of q_2 perpendicular to **B**.

If q_2 is in statcoulombs, B in gauss, u_2 and c in cm/sec, then F_m is a force measured in dynes. Notice that the velocity of light c enters the arena for a second time.

The *direction* of the force $\mathbf{F}_m$ is perpendicular to both **B** and u_2 and is given by another right-hand rule (figure 7-7).

Right-hand rule number 2

Hold the right hand flat as before. Point the thumb in the direction of u_2. Point the fingers in the direction of the magnetic field **B**. Then the magnetic force $\mathbf{F}_m$ is in the direction in which the hand would push.

This assumes that q_2 is a positive charge. If it is negative, the force is in exactly the opposite direction.

Let us reiterate one aspect of our discussion to clarify the significance of the electric field **E** and the magnetic field **B** at any point P in space. If a charge is placed at P, the electric field enables us to calculate that part of the force on it which is independent of its velocity, using equation 6-11. The magnetic field **B** enables us to calculate that part of the force that depends on the velocity of the charge, using equation 7-4 and right-hand rule number 2. The fields **E** and **B** at P are due to the surrounding charges and can be calculated from equations such as 6-14 and 7-1. In general, however, these calculations can be cumbersome.

To illustrate these considerations, we shall now show that a charged particle moving perpendicularly to a uniform magnetic field describes a circular path. (A magnetic field is said to be uniform throughout a certain region of space if **B** has the same magnitude and direction at all points in this region.) In figure 7-8 the velocity **v** is already perpendicular to **B** and so it *Electromagnetism* is the same thing as its component **u**. Application of right-hand rule number 2 at any point on the circle demonstrates that the magnetic force $\mathbf{F}_m$ is always directed toward the center of the circle. As explained in section 2-7, this is a necessary condition for uniform circular motion. Notice that, if you look in the direction of **B**, a positive charge goes round counterclockwise. Convince yourself that a negative charge would go clockwise.

If we are dealing with a fundamental particle, its charge is e. Suppose

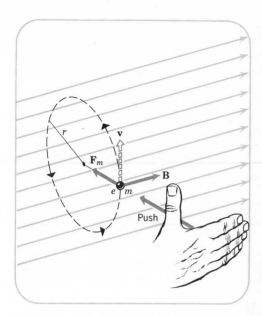

Figure 7-8 *The circular path of a charged particle in a uniform magnetic field.*

that its mass is m and the radius of the circle is r. Equation 7-4 now takes the form

$$F_m = \frac{evB}{c} \tag{7-5}$$

When a particle moves uniformly in a circle of radius r with velocity v, its acceleration is v^2/r (equation 2-20). If its mass is m, Newton's second law tells us that the force is the mass times the acceleration, or

$$F_m = \frac{mv^2}{r} \tag{7-6}$$

Combining equations 7-5 and 7-6,

$$\frac{evB}{c} = \frac{mv^2}{r} \tag{7-7}$$

which can be rearranged to give the radius

$$r = \frac{mvc}{eB} \tag{7-8}$$

The faster the particle moves, the larger is the radius. The larger the applied field B, the smaller is the radius. Heavy particles are more difficult to bend into circular paths than light particles.

Figure 7-9 is a photograph of the curved tracks of electrons moving in a magnetic field. The white tracks are trails of bubbles left behind when a charged particle passes through a superheated liquid in a device known as a bubble chamber.

Figure 7-9 *Curved tracks of electrons in a magnetic field that is perpendicular to the page and points toward the reader. At P a certain process suddenly produces two negatively charged electrons and a positron (which is identical with an electron except that it is positively charged). A white track is a trail of bubbles left behind after the passage of a charged particle. The electron e⁻ which moves off to the left follows a circular path in the magnetic field, but slows down as it loses energy in collisions with the atoms of the liquid through which it is passing. As its velocity decreases, the radius of its circular path decreases (equation 7-10) and so it spirals inward. The positron e⁺ which moves off to the right has the opposite charge and its path is curved in the opposite direction. The electron e⁻ in the center has a very high initial velocity and so the radius of its circular path is very large. (Courtesy of the Lawrence Radiation Laboratory of the University of California.)*

7-4 The Magnetic Force on a Current

Imagine a short segment of wire of length l carrying a current i in a magnetic field **B** (figure 7-10). Resolve **B** into two rectangular components parallel and perpendicular to the wire. Only the component B_n perpendicular to the wire exerts a force on the current. The magnitude of this force is given by the equation

$$F_m = \frac{ilB_n}{c} \tag{7-9}$$

If i is in statamperes, l in cm, B in gauss, and c in cm/sec, then the force F_m is in dynes.

The direction of the force is given by the following slight modification of right-hand rule number 2 (look carefully at the hand in figure 7-10).

Right-hand rule number 2 for a segment of current

Hold the right hand flat. Point the thumb in the direction of the positive current i. Point the fingers in the direction of B_n, the component of the magnetic field perpendicular to the wire. Then the magnetic force is in the direction in which the hand would push.

In preparation for our discussion of magnetic materials, let us now consider the forces between two parallel circular loops carrying currents that go round in the same direction. In figure 7-11a the magnetic field lines due to the current in loop I are shown. The field on a segment at the top of loop II tilts slightly upward. The force on this segment is perpendicular to both the field and the current. Application of right-hand rule number 2 shows that the force tilts slightly backward from an upward vertical direction. This force therefore has a horizontal component directed toward loop I. The situation is similar for all segments of loop II and the force on any segment has a component directed *toward* loop I. Loop II is therefore attracted toward loop I. The argument can be turned round to show that loop I is also attracted toward loop II. When the currents flow around the loops in the same direction, the loops attract one another.

The situation when the currents flow in opposite directions is shown in figure 7-11b. Now the force on a segment at the top of loop II is tilted slightly forward from the downward direction and has a horizontal component directed *away from* loop I. The two loops now repel one another.

> *Forces between two parallel circular current loops*
>
> Two parallel circular loops attract one another if their currents go round in the same direction, but repel one another if their currents go round in opposite directions.

The other situation we need to understand is a current in a loop at an angle to a uniform magnetic field. This magnetic field does not include the field produced by the loop itself but is an "externally applied" magnetic field due to the presence of electric currents elsewhere. For simplicity we shall consider a rectangular loop *ABCD* (figure 7-12a) carrying a current *i*. Applying right-hand rule number 2, the force on *AB* acts through the center

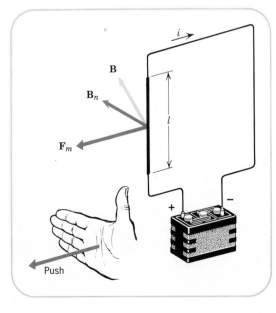

Figure 7-10 *The magnetic force on a length l of wire carrying a current i in a magnetic field whose component perpendicular to the wire is* $\mathbf{B}_n$.

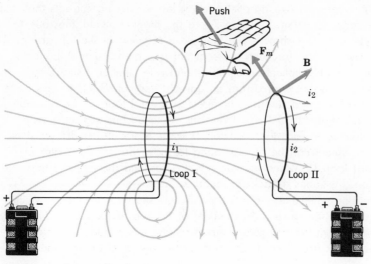

(*a*) *Currents in same direction.*

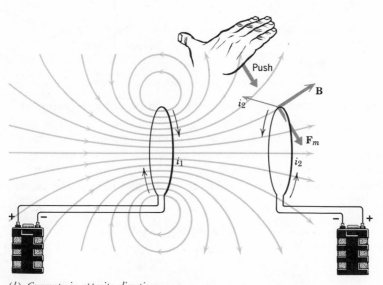

(*b*) *Currents in opposite directions.*

Figure 7-11 *The forces between currents in parallel circular loops.*

of *AB* and is vertically upward. The force on *CD* acts through the center of *CD* and is vertically downward. These two forces are equal and opposite and act along the same line, so they cancel one another. The forces on *BC* and *DA* are also equal and opposite, but, if we look at the loop from the top in the direction $\overrightarrow{BC}$ (figure 7-12*b*), these two forces are displaced sideways from one another and constitute a couple. This couple rotates the loop until its plane is perpendicular to the applied magnetic field. If we were to repeat this argument for various possible orientations of the loop relative to the magnetic field, we would come to the following general conclusion.

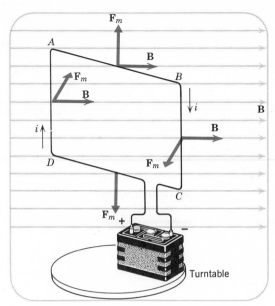

(*a*) *A rectangular loop free to rotate in a magnetic field, showing the directions of the forces on the four straight sides.*

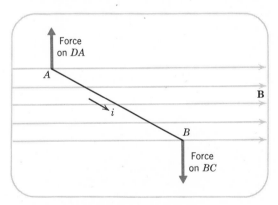

(*b*) *Top view, looking down on AB.*

Figure *7-12 Couple on a rectangular loop carrying a current in a magnetic field.*

Rotation of a current loop in a magnetic field

If a loop carrying a current is free to rotate in a magnetic field, it turns until its plane is perpendicular to the field and the current is in a clockwise direction from the point of view of an observer looking in the direction of the field.

7-5 Bar Magnets

To most people magnetism is not a property of electric currents, but of bar magnets. The following facts about the behavior of bar magnets are well known. If the magnet is freely suspended, it turns until it is pointing in an approximately north–south direction. This is the principle of the magnetic compass. The end of the bar pointing in a northerly direction is called the **north pole** of the magnet and the other end is called the **south pole.** If the magnet is suspended in any magnetic field, its north pole points in the direction of a field line. The earth produces a magnetic field with field lines running approximately, but not exactly, from south to north, and it is this magnetic field that lines up a magnetic compass. If the north pole of a magnet is brought near the south pole of another magnet, they attract one another. But if two north poles or two south poles are brought together, they repel one another. We shall now show how all these phenomena can be explained in terms of microscopic electric currents inside the material of the bar magnet.

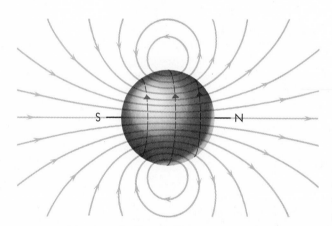

Figure 7-13 *The electron is a spinning cloud of negative charge which produces a magnetic field in the same way as a current loop.*

The properties of a bar magnet are a consequence of the fact that its electrons are spinning like tops. The electron may be visualized as a small spherical cloud of negative charge, which is spinning about an axis such as SN in figure 7-13. Any small portion of the electron describes a circular path about SN and is equivalent to a current in a circular loop of wire. The combined effect of the circular motions of all portions of the electron is a magnetic field which is similar to that produced by a current in a circular loop. We should remember, though, that the electron is composed of negative charge and is equivalent to positive charge spinning the other way round. With this in mind, the direction of the magnetic field lines for the spinning electron of figure 7-13 is seen to be consistent with the direction of the field lines for the current loop of figure 7-6.

Most materials show no net magnetic effects because, for every electron spinning about an axis in a certain direction, there is another electron spinning about a parallel axis but rotating the opposite way round, so that the two electron spins cancel one another. The important exceptions are the **ferromagnetic** materials, which are metals such as iron, cobalt, and nickel,

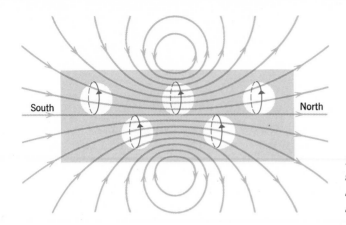

South   North

Figure 7-14 *The magnetic field of a bar magnet. Five spinning electrons are shown. In a bar magnet of normal size, there would be about 10^{24} spinning electrons.*

and the **ferrimagnetic** materials, which are nonmetallic compounds such as magnetite (the material used by the Chinese when they invented the magnetic compass). In these materials there is an excess of electrons all spinning about the same axis in the same direction and adding together to produce a large combined magnetic effect. Each spinning electron produces a magnetic field of the kind shown in figure 7-13. The combined effect of all the electrons is to produce the total field shown in figure 7-14, which is qualitatively similar to the field due to a single electron or the field due to a circular loop of wire carrying a current (figure 7-6). The line joining the south pole to the north pole is the axis of rotation of the electrons. If we look in the direction from the south pole toward the north pole, the electrons are spinning counterclockwise and are equivalent to loops of positive current going around clockwise. This means that *the magnetic field lines enter the south pole and leave the north pole.*

We shall now treat a bar magnet as a collection of circular loops of current that go around clockwise if we look from the south pole toward the north pole. We shall show that the properties of a bar magnet are explicable by analogy with the behavior of circular loops of current.

For example, we have already shown that, when such a loop is placed in a magnetic field, there is a couple acting on it which rotates it until the plane of the loop is perpendicular to the field and the current is clockwise when we look in the direction of the field. Applying these considerations to a bar magnet placed at an angle to an external magnetic field (figure 7-15), a couple acts on each spinning electron and the magnet is rotated until a vector pointing from the south pole toward the north pole is in the same direction as the magnetic field.

When the north pole of one magnet is brought up to the south pole of another magnet (figure 7-16a), the electrons in the two magnets are spinning in the same direction. The attraction of the north pole for the south pole is therefore a consequence of the fact that parallel loops with currents in the same direction attract one another (figure 7-11a). Conversely, when the north pole of one magnet is brought up to the north pole of another magnet (figure 7-16b), the electrons in the two magnets are spinning in opposite directions. Parallel loops with currents in opposite directions repel one another (figure 7-11b). A similar argument applies to adjacent south poles.

7-5 Bar Magnets

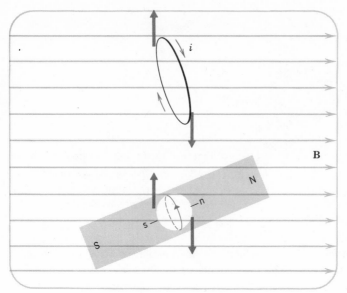

Figure 7-15 The alignment of a bar
magnet along an external magnetic field
is similar to the alignment of a current
loop. Looking along the field, the
current in the aligned loop must go
round clockwise. Looking along the field,
the electrons must be spinning
counterclockwise and therefore be
equivalent to positive current going
around clockwise. This means that, when
the bar magnet is lined up, looking
along the external field is the same as
looking from the south to the north pole.

7-6 *Electromagnetic Induction*

In figure 7-17 loop I is connected to a storage battery which drives a current round it in the direction shown. Loop II has no storage battery but is
connected instead to a galvanometer, which is an instrument for detecting
the flow of a current. The pointer of this galvanometer is normally at the
center of its scale, but deflects to the right if a positive current flows clockwise around loop II, and deflects to the left if a positive current flows counterclockwise around loop II.

When both loops are stationary there is no current in loop II. First consider what happens if loop I remains stationary but loop II moves toward it
with the velocity **v** shown in figure 7-17a (situation A.) A free electron in
the wire at the top of loop II acquires a steady velocity **v** toward the left
and therefore experiences a magnetic force in the magnetic field **B** produced
by loop I. To find the direction of this force, we must first replace **v** by its
component **u** perpendicular to **B** and apply right-hand rule number 2, as in
the figure. However, we must then realize that the electron is negatively
charged and so the force F_m is in the opposite direction to the push of the
hand. Similar forces act on the electrons in all parts of the loop and they
all drive the electrons around the loop in a clockwise direction, which is
equivalent to a positive current flowing in a counterclockwise direction. The
galvanometer is therefore observed to deflect toward the left.

Now suppose that loop II is stationary and loop I moves toward it with
a velocity −**v** (situation *B* of figure 7-17b). Under these circumstances the
free electrons in loop II have no steady velocity and experience no magnetic
force. If we conclude that there is therefore no current flowing in loop II,
we immediately come face to face with certain considerations that involve
Einstein's Theory of Relativity. Return for the moment to situation *A* in part *a*
of figure 7-17, and suppose that Observer I remains stationary with loop I
while Observer II moves with loop II. The description of situation *A* given
above is the point of view of Observer I, who sees loop I at rest relative to

himself and loop II moving toward him. However, even in situation A, Observer II sees loop II at rest relative to himself and loop I moving toward him, which is similar to situation B.

If there were a current round loop II in situation A, but not in situation B, we would have to conclude that there is some fundamental difference between the two situations. We would have to conclude that loop I and Observer I are *really* at rest in some fundamental sense, relative to an absolute space for example, whereas loop II and Observer II are *really* moving relative to this absolute space. This is exactly what the Theory of Relativity will not accept. It insists that the points of view of the two observers are equally valid.

In fact, an appeal to experiment reveals that in situation B there *is* a current in loop II, going round in the same direction and having the same magnitude as in situation A. Relativity is vindicated, but we must now look for an explanation of what is happening in situation B. The explanation is that, as loop I moves nearer to loop II, the strength of the magnetic field

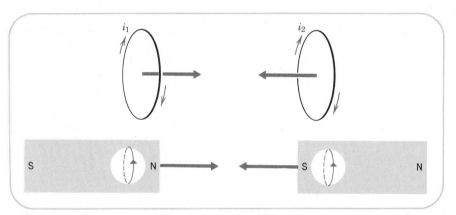

A north pole attracting a south pole is analogous to the attraction between parallel current loops in the same direction.

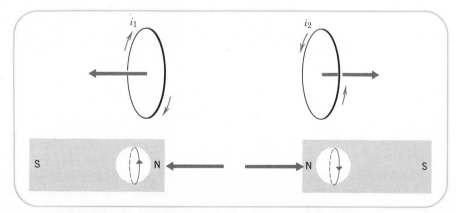

(*b*) *A north pole repelling a north pole is analogous to the repulsion between parallel current loops in opposite directions.*

Figure 7-16 *Forces between bar magnets interpreted in terms of the forces between spinning electrons.*

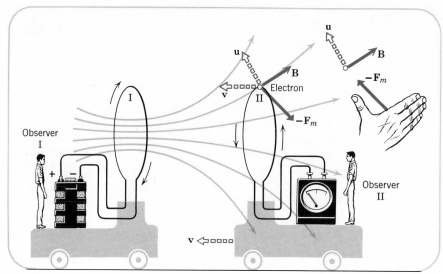

(*a*) *Situation A. The current induced in a circuit moving in a steady magnetic field.*

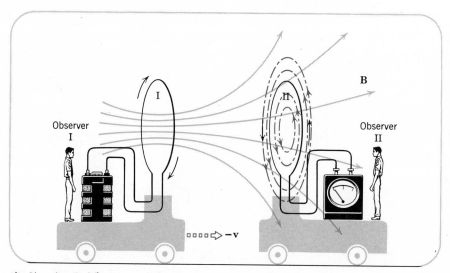

(*b*) *Situation B. The current induced in a stationary circuit by a varying magnetic field.*

Figure 7-17 *Two kinds of electromagnetic induction.*

produced by loop I in the vicinity of loop II increases. It is a fundamental law of electromagnetism that a magnetic field **B** that changes with time produces an electric field **E**. In the present case the electric field lines are circles, one of which coincides with loop II (examine figure 7-17*b* carefully). The electric field exerts an *electric* force e**E** on the free electrons in loop II and it is this electric force that drives them around the loop in situation B.

So, although the result is exactly the same in both cases, in situation *A* it is a magnetic force that induces the current whereas in situation *B* it is an

Electromagnetism

146

electric force. Notice, also, that Observer I sees only a magnetic field and no electric field, whereas Observer II sees, in addition to the magnetic field, circular lines of an electric field. This should dispel any illusions we might have that **E** and **B** are real "concrete" things existing in space. They are merely mathematical conveniences.

The production of a current in situations similar to those of figure 7-17 is the phenomenon of **electromagnetic induction,** discovered independently by Michael Faraday and Joseph Henry near the beginning of the nineteenth century. It leads to a new law of electromagnetism.

> **A magnetic field changing with time produces an electric field.**

This can be demonstrated very clearly by an electromagnetic induction experiment in which neither loop moves (figure 7-18). When there is no current in loop I, there is no magnetic field and no current in loop II. When there is a steady current in loop I, it produces a steady magnetic field and there is still no current in loop II. However, when the current in loop I is switched on, the magnetic field changes from zero to its steady value and

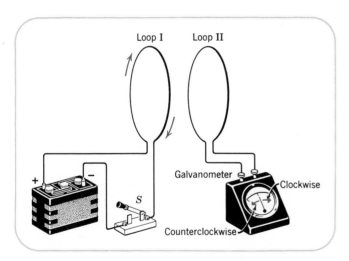

Figure 7-18 Electromagnetic induction with both loops stationary. When the current in loop I is changing, a current is induced in loop II. When the current in loop I settles down to a steady value, there is no current in loop II.

this increasing magnetic field produces counterclockwise electric field lines which drive a current counterclockwise around loop II. The galvanometer pointer is observed to deflect momentarily to the left, but soon returns to its mid-position as the magnetic field settles down to a steady value. If the current in loop I is switched off, the magnetic field falls to zero and this decreasing magnetic field produces *clockwise* electric field lines which drive a current *clockwise* around loop II. The galvanometer pointer is observed to deflect momentarily to the *right*.

7-7 *The Field*
Triumphant

7-7 *The Field Triumphant*

Electromagnetic theory reached its climax in the late nineteenth century when Clerk Maxwell was able to show that all these complicated electromagnetic laws can be summarized in a set of elegant equations describing the behavior of electric and magnetic fields in space. These equations do not relate the behavior at a point in space to charged bodies at a distance from this point, but to the behavior of the electric and magnetic fields at other points in the *immediate vicinity* of the first point. If a charge happens to be located at the point in question, then this is taken into account by extra terms in the equations. This is a complete swing of the pendulum away from our earlier action-at-a-distance approach to the universe, in which we assumed interactions between discrete particles at a distance from one another and did not worry about the space in between them. The present attitude might be crudely described as the "handing on" of interactions from a point in space to a neighboring point.

Before he could formulate his equations, Maxwell had to guess a new law of nature, which is a kind of inverse of electromagnetic induction, but which cannot be so readily demonstrated by simple experiments. Electromagnetic induction is based on the idea that a changing magnetic field produces an electric field. Maxwell's new law is that a changing electric field produces a magnetic field.

> An electric field changing with time produces a magnetic field.

Without advanced mathematics, it is not possible to state Maxwell's equations in a precise form. We shall have to be satisfied with an oversimplified qualitative statement hinting at the meaning of each equation. There are five equations. The first four are statements about the behavior of the electric and magnetic fields. The fifth equation gives physical significance to these fields by showing how they determine the force on a charged body.

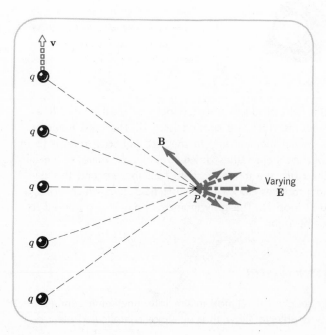

Figure 7-19 As the charge q moves past P, the electric field it produces at P varies with time and this gives rise to a magnetic field at P.

In order to appreciate the new attitude to physical phenomena underlying Maxwell's equations, let us consider the magnetic field produced by a moving charge (figure 7-19). In equation 7-1 we simply wrote down the expression for the field produced by the moving charge at a distance R, implicitly assuming a sort of action-at-a-distance. We shall now point out that the magnetic field at P may be looked upon as a consequence of a changing electric field at P. As the charge moves past P, its distance from P and its direction relative to P are continually changing. The electric field it produces at P is therefore continually changing in magnitude and direction. According to Maxwell's fourth equation, this produces a magnetic field in the vicinity of P, which is nothing more than the magnetic field of equation 7-1.

We have still not entered into the true spirit of Maxwell's equations because we have talked about the *electric* field at P as though it were produced by a charge *at a distance*. We ought really to describe the behavior of **E** and **B** at P only in terms of the behavior of **E** and **B** at points in the immediate vicinity of P. We should really proceed somewhat as follows. In the immediate vicinity of q there are diverging electric field lines originating on q (Maxwell's first equation). There are also circular magnetic field lines, first because a charge in motion is equivalent to an electric current and second because the electric field changes as the charge moves (Maxwell's fourth equation). Since the charge is moving, the electric and magnetic fields it produces in its immediate vicinity change with time. This produces electric and magnetic fields at points a little further removed from the charge, which in their turn produce electric and magnetic fields still further removed from the charge (Maxwell's third and fourth equations). In this way the disturbance travels out from the charge until it reaches P. In fact, it travels out as an **electromagnetic wave** with the speed of light, c, and takes a finite time to reach P.

This new concept that we have just encountered has revolutionary importance to our thinking about the nature of the universe. Electromagnetic interactions are not instantaneous, but travel from one particle to another with a finite speed, the speed of light. This was not noticeable in the classical experiments on electricity and magnetism that we have described so far. In these experiments on magnetic fields produced by currents and currents induced by changing magnetic fields, one part of the apparatus was rarely much further than 100 cm from any other part. An electromagnetic disturbance traveling at 3×10^{10} cm/sec covers 100 cm in only 3.33×10^{-9} sec! Such a short delay in time could never have been detected in the experiments. However, there was one phenomenon well known to nineteenth century physicists in which these considerations could not be ignored. Light is itself an electromagnetic wave which obeys Maxwell's equations and travels with a speed of 3×10^{10} cm/sec. We shall now turn our attention to the nature of waves and the character of electromagnetic radiation.

Questions

A

1. A dense beam of electrons all moving in the same direction has a tendency to spread out as it advances. Two overlapping beams of protons and electrons of equal density moving in opposite directions have a tendency to shrink. Explain this.

2. In which of the following cases is there no magnetic force between two electrons? (a) When they move directly toward one another. (b) When they move directly away from one another. (c) When they have the same velocity in a direction perpendicular to the line joining them. (d) When they have equal and opposite velocities in a direction perpendicular to the line joining them?

3. If an electron moves horizontally toward the north in a horizontal field pointing toward the west, in which direction is the magnetic force on it?

4. I am looking down on a proton that is moving horizontally in a uniform vertical magnetic field pointing up at me. Do I see it going round its circular orbit in a clockwise or counterclockwise direction?

5. If the earth can be considered to be a huge bar magnet, is the "north pole" of this magnet in the vicinity of the geographic north pole or the geographic south pole?

6. In the arrangement shown in the diagram, what is the direction of the force on the current-carrying wire?

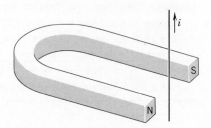

Question 7-6.

7. Use the right-hand rules to show that two parallel currents attract one another if they are flowing in the same direction, but repel if they are flowing in opposite directions.

8. When I look toward an electron, I see it moving in a counterclockwise circular orbit in a vertical plane. Specify the direction of the magnetic field responsible for this. (Hint: In this problem it is the direction of the magnetic field that is unknown initially. However, right-hand rule number 2 is based on the assumption that you are trying to find the direction of the force, given the directions of the velocity and the field.)

9. If the earth's magnetism is assumed to be due to a large circular loop of current in the interior of the earth, what is the plane of this loop and what is the direction of the current around it?

10. Oersted performed one of the first experiments to reveal a connection between magnetism and electric currents. He placed a straight horizontal wire carrying a current directly above a freely pivoted magnetic compass needle. The wire was originally parallel to the needle (see diagram). What did the compass needle do? Explain its behavior in terms of the forces acting on a microscopic current loop (a spinning electron) inside the needle.

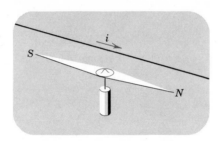

Question 7-10.

11. A circular loop of wire carrying a current hangs with its plane vertical and is free to rotate about a vertical axis (see diagram). What will it do when the north pole of a magnet approaches it from each of the positions shown?

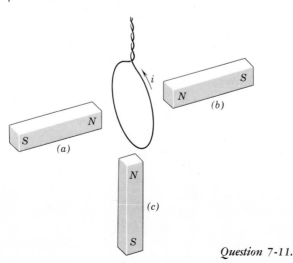

Question 7-11.

12. A student trying to find the directions of the forces between two mov-

ing charges always uses his left hand by mistake. Does he get the right answer?

C
13. A uniform horizontal magnetic field points toward the north. A horizontal copper rod is perpendicular to this field. If the rod is allowed to fall freely from rest, show that the two ends become oppositely charged. Which end is positive?

14. A horizontal copper rod is perpendicular to a uniform horizontal magnetic field. The rod is moved up and down in a vertical plane in an oscillatory fashion. Ignoring the friction of the air, does this operation require a continuous input of energy? If so, where does the energy reappear?

Question 7-14.

15. When an electric current flows in a long straight wire perpendicular to a magnetic field, a potential difference is produced across the wire (the Hall effect). Why?

16. Suppose that any one of the figures in this chapter were held up to a mirror. Apart from the reversal of the lettering, what would be wrong with the mirror image?

17. You are locked in a room without windows and told to determine approximately the direction of north. You are provided with the following and nothing else: (a) an ample supply of copper wire of all possible diameters, (b) a pair of wire cutters made of nonmagnetic steel, (c) a storage battery with its terminals marked + and −. Describe precisely what you would do.

Problems

A
1. Eight equal charges, each +0.2 statcoulomb, are spaced uniformly around a circle of radius 6 cm. If the circle rotates at 10^5 rpm, what is the magnetic field at the center of the circle?

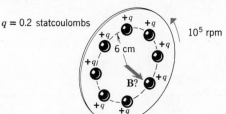

Electromagnetism

Problem 7-1.

2. A charge of +5 statcoulomb has a horizontal velocity of 4×10^7 cm/sec toward the north. What are the magnitude and direction of the

magnetic field that it produces at a point 2 cm due west of itself?

3. An electron has a velocity of 1.5×10^8 cm/sec vertically upward. What are the magnitude and direction of the magnetic field which it produces at a point 10^{-6} cm due north of itself?

4. A charge of -0.1 statcoulomb has a velocity of 2×10^6 cm/sec toward the east in a horizontal magnetic field of 300 gauss pointing due south. What are the magnitude and direction of the magnetic force on the charge?

5. A proton has a velocity of 1.5×10^9 cm/sec due north in a magnetic field of 5×10^5 gauss pointing vertically upward. What are the magnitude and direction of the magnetic force on it?

6. A long straight wire carries a current of 10^{10} statamp. What magnetic field does it produce at a distance of 1.5 cm?

7. A straight vertical wire 100 cm long carrying a current of 5×10^9 statamp flowing vertically upward is in a uniform horizontal magnetic field of 0.5 gauss pointing due north. What are the magnitude and direction of the magnetic force on it?

B

8. Two electrons are 10^{-7} cm apart. They have identical velocities of 2×10^8 cm/sec in a direction perpendicular to the line joining them. Find the magnitude of the magnetic force between them. What is the ratio of the magnetic force to the electric force?

9. A 2 MeV proton is moving vertically upward. What are the magnitude and direction of the magnetic field it produces at a point 3×10^{-7} cm southwest of itself?

10. What are the magnitude and direction of a magnetic field that could support the weight of an electron with a horizontal velocity of 1 cm/sec due east?

11. A proton has a velocity of 1.5×10^8 cm/sec perpendicular to a magnetic field of 15,000 gauss. What is the radius of its circular orbit?

12. A horizontal wire 10 cm long carries a current of 10 amp. Find the magnitude and direction of the smallest magnetic field that can support the weight of the wire, assuming that its mass is 2.5 gm.

C

13. A long straight wire carries a current of 5×10^{10} statamp. If the radius of the wire is 0.1 cm, what is the largest value of the magnetic field at any point outside the wire?

14. An electron has a velocity of 6.82×10^8 cm/sec parallel to a long straight wire carrying a current of 45.6 amp. If the perpendicular distance of the electron from the wire is 12.3 cm, calculate the force on the electron.

15. An electron moves in a circular orbit of radius 1.5×10^2 cm in a magnetic field of 0.2 gauss. Find (a) its velocity (b) its momentum (c) its energy in electron volts.

16. In a certain region of interstellar space, there are free electrons with kinetic energies of 10^{-3} eV moving in circular orbits of radius 2.5×10^6 cm. What is the magnitude of the magnetic field that causes this motion?

17. A fundamental particle of mass m moves in a circular orbit perpendicular to a magnetic field B. Find an expression for the time taken to complete one revolution. How does it depend on (a) the velocity v (b) the radius r (c) the mass m (d) the field B?

8 *Waves*

8-1 *The Nature of Waves*

In a wave something oscillates and it usually performs simple harmonic motion. Three typical examples of simple harmonic motion are illustrated in figure 8-1. Part (a) is the backward and forward swing of a pendulum, (b) is the up and down motion of a weight hanging on a spring, and (c) is the twisting and untwisting motion of a horizontal bar suspended by a wire. These motions are called periodic motions because they obviously repeat themselves periodically. The **period** T is defined as the time taken to complete one oscillation, that is the time to go from A to B and back again to A. The **frequency** is defined as the number of oscillations in one second and is denoted by ν (the Greek letter nu). Obviously, the frequency is the reciprocal of the period.

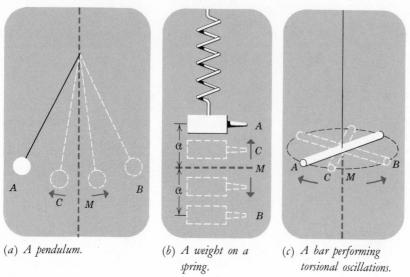

(a) A pendulum.

(b) A weight on a spring.

(c) A bar performing torsional oscillations.

Figure 8-1 *Three examples of simple harmonic motion.*

$$\text{Frequency} = \frac{1}{\text{Period}}$$

$$\nu = \frac{1}{T} \tag{8-1}$$

If one oscillation takes $\frac{1}{10}$ second, then 10 complete oscillations are performed in 1 second. Frequency is measured in *cycles per second,* written c/sec. The **amplitude** of the oscillations, α, is the distance between the central position and an extreme position. For example, in figure 8-1*b* the amplitude is the distance from the mid-position of the motion, *M*, to the highest position or the lowest position.

If we were to attach a pen to the weight and pull a piece of paper past it at a constant speed, the pen would trace out the curve shown in figure 8-2. This is a graph of the displacement from the mid-position against time. Such a curve is said to be **sinusoidal** or to describe a **sine-wave function.**

The most familiar examples of waves in everyday life are ocean waves or ripples on water. The most striking characteristic of a water wave is that it appears to be moving outward from its source. Figure 8-3 is a photograph of the ripples produced as a single water droplet falls into a tank of water. Everyone has seen this happen and can imagine the pattern of ripples moving outward from the point of impact. However, it is only the *shape of the surface* of the water that moves. A small drop of water at the surface merely bobs up and down as the wave passes over it.

To obtain a better appreciation of this last point, imagine a long horizontal spring with one end attached to the wall and the other end held in the hand and pulled taut. By moving the hand up and down rapidly two or three times, a wave can be made to travel along the spring as shown in

Waves

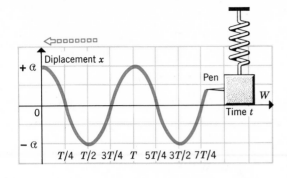

Figure 8-2 The variation of displacement with time for simple harmonic motion. The weight W is oscillating on the end of a spring. A pen attached to W traces a line on a piece of graph paper which is pulled past it from right to left at a steady rate.

figure 8-4. One of the coils of the spring P is painted gold so that its motion can be easily observed. Although the wave travels along the spring in a horizontal direction, P oscillates up and down on a vertical straight line and does not move in a horizontal direction. The motion of P is, in fact, simple harmonic. A wave is called a **transverse wave** if the oscillatory motion of any part of the system, such as P, is at right angles to the direction in which the wave is traveling.

The shape of the spring at a fixed instant of time is a *sine wave function* similar to figure 8-2. However, the difference between figures 8-2 and 8-4 should be clearly understood. If the vertical displacement of a particular coil, such as P, were plotted against time, a curve such as figure 8-2 would be obtained. Figure 8-4, on the other hand, shows the displacements of all the coils at a fixed instant of time, and is therefore a plot of displacement

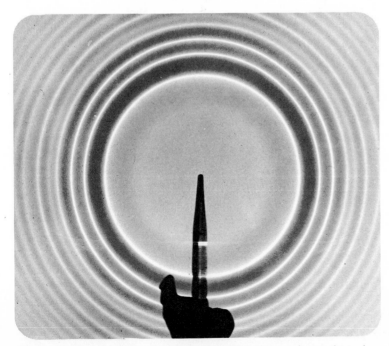

Figure 8-3 Circular ripples on a ripple tank. This photograph was obtained by allowing a single drop of water to fall into the tank. (Photograph by Berenice Abbott.)

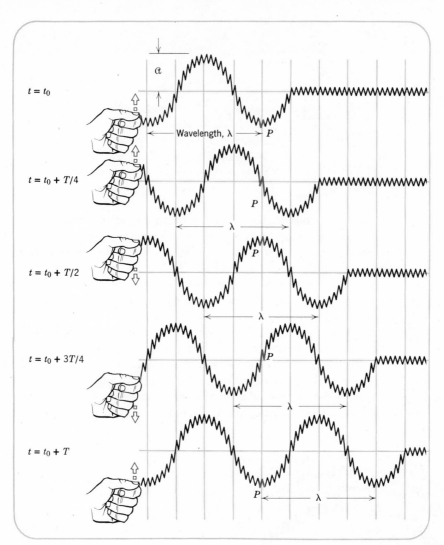

Figure 8-4 *A transverse wave on a long spring.*

against distance along the spring at constant t. It is a kind of "snapshot" of the shape of the spring at a fixed instant of time. In figure 8-2 the distance between adjacent peaks or adjacent valleys of the curve in the direction of the time axis represents the period T of the oscillation. In figure 8-4 the horizontal distance between adjacent peaks or adjacent valleys is a real distance in space. It is called the **wavelength** of the wave and is usually represented by the Greek letter lambda, λ.

Now let us consider the motion of the coil P as the wave travels along the spring. Careful study of figure 8-4 reveals that, while one wavelength is passing over P, the coil goes through one complete oscillation, which takes a time T. The wave therefore travels forward a distance λ in a time T. The **wave velocity** V must be

Waves

$$V = \frac{\lambda}{T} \tag{8-2}$$

The frequency of the wave is the number of oscillations that P makes in one second and is related to the period by equation 8-1. So

$$\nu = \frac{1}{T} \tag{8-3}$$

$$V = \nu\lambda \tag{8-4}$$

Wave velocity = Frequency × Wavelength (8-5)

This important equation is true for all types of waves.

8-2 Longitudinal Waves: Sound

Let us repeat the experiment with a long horizontal spring but now oscillate its free end backward and forward in a horizontal direction (figure 8-5). The spring remains straight, but a wave of alternating compressions and exten-

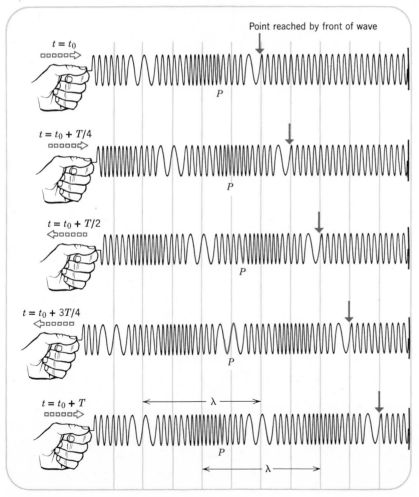

Figure 8-5 A longitudinal wave on a long spring.

sions travels along its length. The painted coil performs simple harmonic motion in a horizontal direction. The oscillatory motion of a part of the system is in the same direction that the wave is traveling and so this is a **longitudinal wave.** The wavelength is the distance from a point on the spring where the coils are closest together to the next nearest point where the coils are closest together. The reader should convince himself that equation 8-5 also applies to this case.

Sound is usually transmitted as a longitudinal wave in the air. Consider what happens when you speak to a friend. Air is expelled from your lungs through your windpipe, past two elastic membranes, known as *vocal cords,* inside the windpipe (figure 8-6). The flow of air past the vocal cords causes them to vibrate, and the size of the opening between them varies periodically. The air is consequently expelled in ''puffs'' that follow one another in rapid succession. The number of puffs per second determines the frequency (or *pitch*) of the sound emerging from your mouth.

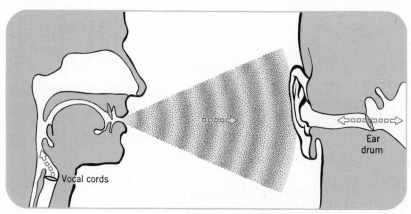

Figure 8-6 *Speech, sound, and hearing. The anatomy is simplified in order to illustrate the basic principles.*

These puffs of air do not travel all the way to your friend's ear. As a puff emerges it increases the pressure of the air in the immediate vicinity of the mouth. In between puffs the molecules of air near the mouth move further apart to relieve the excess pressure, but they collide with neighboring molecules and force them closer together so that a region of high pressure and density is created a little farther from the mouth. In fact the process continues and the region of high pressure moves outward from the mouth with the *speed of sound.* Each puff creates such a region of high pressure moving outward. The sound wave therefore consists of a train of high pressure regions following one another in rapid succession with the frequency of the sound and separated by regions of lower pressure (figure 8-6). The individual molecules do not travel with the high pressure regions, they merely oscillate backward and forward parallel to the direction in which the sound is traveling. The behavior of the molecules is similar to the behavior of the coils of the spring in figure 8-5. A region of high pressure and high density is analogous to a part of the spring where the coils are close together. A region of low pressure is analogous to a part of the spring where the coils are far apart.

When the high pressure regions impinge on a membrane known as the *ear drum* in your friend's ear, they cause it to oscillate backward and forward with the same frequency as your vocal cords. The mechanism of your friend's ear converts this oscillation into a train of pulses of electrical current traveling along one of his nerves to his brain. The frequency of arrival of the electrical pulses is exactly the same as the frequency of vibration of your vocal cords. He is thereby made aware of what you are saying.

8-3 *Electromagnetic Waves*

When an electric charge accelerates, it radiates an electromagnetic wave. In order of increasing frequency and decreasing wavelength, the important types of electromagnetic radiation are radio waves, infrared radiation (sometimes called heat radiation), visible light, ultraviolet light, x-rays, and γ-rays. Their production can always be ultimately related to the acceleration of electrons, protons, or other charged fundamental particles.

To illustrate the production and character of electromagnetic waves, let us consider the transmission and reception of a radio wave (figure 8-7). The transmitting aerial is a long metal rod or wire carrying an electric current oscillating backward and forward with a frequency which, in a typical case, might be a million times per second. In section 6-6 an electric current was ascribed to the motion of the free electrons in a metal. In the present instance each free electron performs a simple harmonic motion with a frequency of one million cycles per second. During its simple harmonic motion, the electron is accelerating and it therefore radiates an electromagnetic wave. The oscillating electron is continually changing its position and velocity and the electric and magnetic fields it produces in its immediate vicinity are continually changing. In accordance with Maxwell's equations and the considerations of section 7-7, these changing fields produce other fields in their vicinity and the disturbance travels outward with the speed of light.

When Maxwell's equations are applied to this situation and solved, it is found that, except in the immediate vicinity of the oscillating electron, the

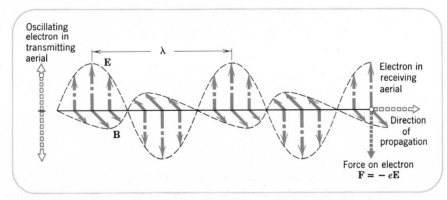

Figure 8-7 The electromagnetic wave produced by an oscillating electron. In the interests of clarity, the diagram ignores the fact that, as the wave travels away from the electron, its energy is spread over a greater area and the magnitudes of E and B decrease.

outgoing electromagnetic wave has the form shown in figure 8-7. In this diagram we are considering the electric and magnetic fields at a single instant of time at *points along a line* which, for convenience, is taken to be in a direction perpendicular to the motion of the electron. At each point the electric and magnetic fields are represented by two vectors. The electric field **E** is seen to be parallel to the motion of the electron, but its direction and magnitude vary along the line in such a way that the tips of the vectors trace out a sine wave function. The magnetic field **B** is perpendicular to the electric field and to the direction of travel and it traces out a similar sine wave function. In fact, in the CGS system, which we are using almost exclusively, the magnitude of the magnetic field at any point is exactly equal to the magnitude of the electric field.

$$E = B \qquad\qquad\qquad (8\text{-}6)$$

The diagram in figure 8-7 represents the electric and magnetic fields at points along the line *at a fixed instant of time*. The whole pattern should now be imagined to move along the line away from the electron with the speed of light. As various parts of the pattern pass over a fixed point, the vector representing the electric field oscillates between a maximum value in an upward direction and an equal value in a downward direction. Similarly, the magnetic field oscillates in and out of the plane of the page. This is similar to the behavior of the spring in figure 8-4. It is important to realize, however, that we are describing the electric and magnetic fields in empty space at *points* along a straight line and that nothing is moving in a direction perpendicular to this line. The significance of the picture is that, if an electron *were* present at any point in space, the oscillating electric field at that point would exert an oscillating force on it and the electron would be made to perform a simple harmonic motion similar to the motion of the electrons in the transmitting aerial. This is what actually happens to the free electrons in the receiving aerial. It is the mechanism whereby the behavior of the electrons in the transmitting aerial is reproduced by the electrons in the receiving aerial, so that communication becomes possible through the intervening empty space.

An electromagnetic wave is always a transverse wave, because the electric and magnetic vectors are always perpendicular to the direction in which the wave is traveling. The electric and magnetic vectors have the same magnitude and oscillate in step. They both assume a maximum value at the same time and are zero at the same time. Looking in the direction of propagation, when the electric vector **E** points upward, the magnetic vector **B** points to the right.

The constant c, which turns up in some of the fundamental laws of electromagnetism such as equations 7-1 and 7-4, is also present in Maxwell's equations. When these equations are manipulated mathematically to deduce the existence of electromagnetic waves, the wave velocity in a vacuum is found to be precisely this same constant c. In the first instance, therefore, the value of c can be determined by experiments that measure both the electric and magnetic forces between charges and use the fact that the ratio of these two forces depends on c. In the second instance, once it has been demonstrated that electromagnetic waves can be generated by oscillating electric currents, their velocity can be measured directly. This second step was first achieved by Hertz in 1885, twelve years after Maxwell had published the final form of his equations, and the resulting good agreement between the two entirely different methods of measuring c inspired confidence in those equations.

The really striking aspect of the situation, though, was that the velocity c of an electromagnetic wave was exactly the same as the velocity of *light*. The velocity of light had been deduced from astronomical observations by Roemer in 1675 and by Bradley in 1729 and had been measured in a terrestrial experiment by Fizeau in 1849. At the beginning of the nineteenth century the experiments of Young and Fresnel on interference and diffraction of light (to be described later in this chapter) had convinced scientists that light is a wave. One of Maxwell's outstanding achievements was the realization that light is an *electromagnetic* wave. Figure 8-7 is also a picture of a light wave. The only difference between light and radio waves is that the frequency of light is much greater and its wavelength is much shorter.

Figure 8-8 tabulates the various types of electromagnetic waves and shows the range of wavelengths associated with each type. The sizes of some familiar objects are included for comparison. There is no particular scale associated with this diagram. In fact it is necessary to vary the scale

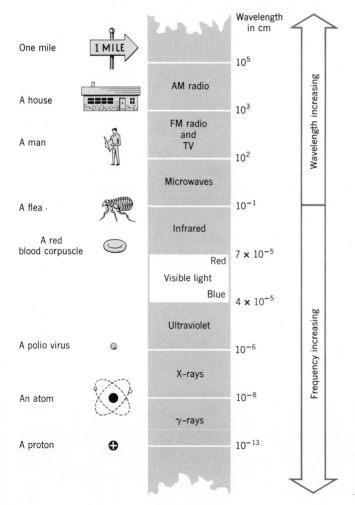

Figure 8-8 Types of electromagnetic waves.

from one part to another. Except in the case of visible light, the limits are only approximate and the ranges frequently overlap, so that radiation of a particular wavelength might be described as either of two overlapping types, depending on the method of its production.

Electromagnetic radiation visible to the human eye covers only a small range of wavelengths in the vicinity 5×10^{-5} cm. This is approximately the size of the smallest object that can be seen in an optical microscope. This is no coincidence, because there is a fundamental theory of the behavior of microscopes that reveals that the smallest object that can be seen is comparable in size with the wavelength of the electromagnetic radiation being used. The longest wavelength of **visible light** is red. As the wavelength is decreased, the color changes through orange, yellow, green, blue to violet, which has the shortest wavelength. An interesting question is why these particular wavelengths of electromagnetic radiation should have been chosen for human vision. Part of the answer is that these particular frequencies are strongly emitted by the sun but are not appreciably absorbed by the earth's atmosphere. There are radio waves that also satisfy this condition, but their very long wavelength makes them unsuitable for human vision because of the general principle that two objects cannot be distinguished from one another if their distance apart is much smaller than the wavelength of the radiation being used to view them.

Beyond the red end of the visible spectrum, where the wavelengths are becoming longer, there is first **infrared** radiation. It is sometimes called "heat radiation," because it is strongly emitted by a fire or central heating radiator and is readily absorbed by the human body to produce a sensation of warmth. A radiator emits no visible light and it cannot be seen in a dark room, but the heat radiation from it can easily be felt. At still longer wavelengths there are the **microwaves** used in radar, and then **radio waves** whose wavelengths range from a yard to a mile.

Beyond the blue end of the visible spectrum, where the wavelengths are becoming shorter, there is first **ultraviolet light.** This is strongly emitted by the sun, but almost completely absorbed by the earth's atmosphere. Then, at wavelengths comparable with the size of a molecule, there are **x-rays,** which can be generated by allowing high speed electrons in a discharge tube (figure 6-12) to bombard a metal target. The rapid deceleration of the electrons as they are stopped by the target atoms produces electromagnetic radiation because, as we have already emphasized, an accelerating charge always radiates an electromagnetic wave. Finally, at wavelengths comparable with the size of a proton or even less, there are the **γ-rays** which are emitted by radioactive nuclei.

8-4 *Energy and Momentum of Electromagnetic Radiation*

Since the radiation from the sun warms us up, electromagnetic waves clearly carry energy with them. They also carry momentum. Let us consider how this comes about.

Reverting to the problem of the magnetic forces between two moving charges (Chapter 7), suppose that the charge q_1 has a velocity v_1 perpendicular to the line joining the two charges, whereas q_2 has a velocity v_2 along this line (figure 8-9). First calculate the magnetic force exerted by q_1

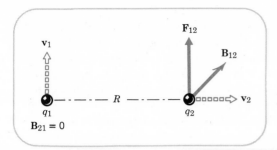

Figure 8-9 *Illustrating the failure of Newton's third law of motion for magnetic forces.*

on q_2. According to section 7-2, the magnetic field produced at q_2 by q_1 is

$$B_{12} = \frac{q_1 v_1}{R^2 c} \tag{8-7}$$

Right-hand rule number 1 tells us that $\mathbf{B}_{12}$ goes perpendicularly into the plane of the paper. According to section 7-3, the force exerted by this field on q_2 is

$$F_{12} = \frac{q_2 B_{12} v_2}{c} \tag{8-8}$$

Substituting the value of B_{12} from equation 8-7

$$F_{12} = \frac{q_1 q_2}{R^2} \frac{v_1 v_2}{c^2} \tag{8-9}$$

Right-hand rule number 2 tells us that this force acts upward toward the top of the page.

　　If this force, exerted by the first charge q_1 on the second charge q_2, is called the action, the reaction is the force exerted by the second charge q_2 on the first charge q_1. However, when we calculate the magnetic field $\mathbf{B}_{21}$ produced by q_2 at q_1, we must use the component of the velocity of q_2 perpendicular to the line joining the two charges. But q_2 is moving parallel to this line and its velocity has no component perpendicular to it. The magnetic field $\mathbf{B}_{21}$ is therefore zero and the magnetic force F_{21} is also zero! Since F_{12} is not zero, the two forces are not equal and opposite and Newton's third law of motion is violated.

8-4 Energy and Momentum of Electromagnetic Radiation

　　Force is equal to rate of change of momentum (section 4-1). The magnetic force exerted by q_1 on q_2 therefore gives q_2 an extra upward momentum, but there is no compensating change in the momentum of q_1. The law of conservation of momentum is apparently violated. For similar reasons there is also an apparent violation of the law of conservation of energy.

　　The way out of this dilemma is as follows. Suppose that the two charges are free to move subject to no external influence other than the forces that they exert on one another. Because they do exert forces on one another, they both accelerate, and an accelerating charge radiates electromagnetic waves. The electromagnetic radiation carries away the energy and momentum that was apparently lost in the previous analysis. We can reinstate the laws of conservation of energy and momentum if we make the following assumptions about the energy and momentum carried away by the radiation. Remembering the picture of an electromagnetic wave given in figure 8-7, suppose that the instantaneous electric and magnetic fields produced by the wave near a point P in space are **E** and **B**. The energy and momentum are being transported in the direction in which the wave is traveling, which is perpendicular to both **E** and **B.** Let S be the energy and P the momentum moving in *one second* across a surface of *unit area* near P perpendicular to the direction of propagation. Then, remembering that $E = B$, it can be shown that:

Transfer of energy and momentum by an electromagnetic wave

Energy crossing unit area per second

$$S = \frac{cEB}{4\pi} = \frac{cE^2}{4\pi} \tag{8-10}$$

Momentum crossing unit area per second

$$P = \frac{S}{c} \tag{8-11}$$

$$= \frac{EB}{4\pi} = \frac{E^2}{4\pi} \tag{8-12}$$

When electromagnetic radiation falls on a material body and is absorbed by it, energy is transferred to the body and warms it up. At the same time momentum is also transferred to the body and a force is exerted on it. This is often called the **pressure of light.** It can be verified experimentally by direct measurement of the force produced when light falls on a surface. In the apparatus illustrated in figure 8-10, the two mirrors are sus-

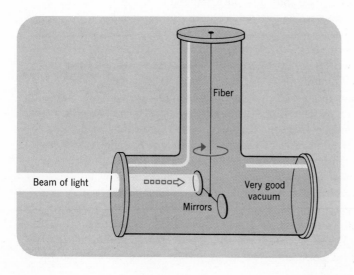

Beam of light

Fiber

Mirrors

Very good vacuum

Figure 8-10 *A demonstration of the pressure of light.*

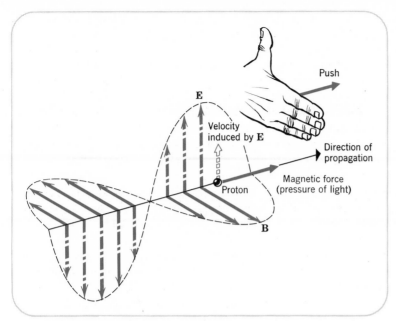

Figure 8-11 *The origin of the pressure of light.*

pended on a thin fiber inside a very good vacuum. If a beam of light falls on one of the mirrors, the pressure it exerts is seen to twist the suspended system.

Light intensities normally encountered on the surface of the earth are too weak to exert pressures of any consequence. However, the pressure of electromagnetic radiation is believed to be very important to the stability of many stars including the sun. The radiation streaming outward from the hot center of the star exerts an outward pressure on the material of the star. In the absence of this pressure the star would collapse because of the strong gravitational attraction between its inner layers and its outer layers. As often happens, a subtle effect in physics turns out to be essential to our existence. If light did not exert a pressure, the sun could not exist in its present form and life on earth could not exist.

The following simple example will give a good idea of why electromagnetic waves carry momentum. Figure 8-11 shows an electromagnetic wave incident on a proton and producing fields **E** and **B** in its vicinity. The field **E** exerts an upward force on the proton and gives it an upward velocity **v**. Once it has acquired this velocity, the magnetic field **B** exerts a magnetic force on the proton. Application of right-hand rule number 2 shows that this force is in the direction in which the wave is traveling. It is in fact the pressure of light. The proton acquires momentum in the direction of propagation and this must have come from the electromagnetic wave. The reader should verify that the force is always in the forward direction of propagation even when the electric field is downward and the magnetic field goes into the plane of the paper. If the charge in figure 8-11 were negative (an electron, say), it would be given a downward velocity by **E**, but **B** would still exert a forward force on this downward moving negative charge.

8-5 Interference

In the seventeenth and eighteenth centuries it was generally believed that light is a stream of particles. Newton tended to favor this point of view. It was not until the early nineteenth century that the opposing point of view, that light is a wave, received experimental support, mainly from the researches of Thomas Young and Augustin Jean Fresnel. They demonstrated that light exhibits the phenomena of interference and diffraction, which are very characteristic of waves but are difficult to explain in any other way.

In figure 8-12 S is the source of circular ripples on the surface of water. Each circle centered on S corresponds to the crest of a wave where the upward displacement of the water is greatest. These circles are sometimes called **wavefronts** and they move radially outward with the wave velocity. If a wavefront impinges on a very narrow gap in a barrier, one might expect a small portion of the wavefront to pass through the gap and to continue on its way undisturbed. What actually happens is quite different. On the far side of the gap the wave spreads out in all directions and new circular wavefronts centered on the *gap* are formed. This property of a wave which makes it spread out after passing through a narrow opening is called **diffraction** and is one of the most distinctive characteristics of the behavior of waves.

A good way to visualize what is happening is as follows. Imagine each *very small* portion of a wavefront to be a disturbance that acts as the source of an outgoing circular **wavelet.** The barriers of figure 8-12 block out everything except the wavelet from the small portion of the wavefront at the gap, and only this wavelet is seen on the other side of the gap. In the absence of the barriers, however, each small portion of the wavefront sends out its own circular wavelet and the wavelets from all parts of the wavefront add together to produce a new wavefront centered on the original source (figure 8-13). This is, in fact, the way in which the original wavefront is maintained on its outward motion.

Diffraction of light was discovered in the early seventeenth century by Francesco Maria Grimaldi, and the concept of wavelets was introduced later in the same century by Christiaan Huyghens, so they are often called **Huyghens' wavelets.** Nevertheless, it was not until the year 1801 that

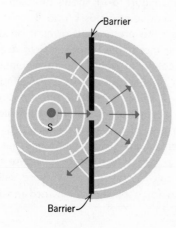

Waves

Figure 8-12 *Passage of a circular ripple through a very narrow opening.*

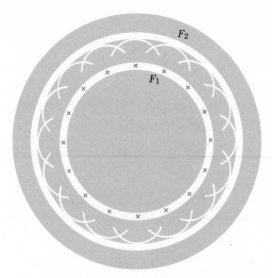

Figure 8-13 Huygens' wavelets adding together to form a new wavefront. F_1 is the original wavefront. Points such as those marked x on F_1 act as sources of Huygens' wavelets. These wavelets add together to form the new wavefront F_2, which has propagated ahead of F_1.

Thomas Young established the wave theory of light on a firm footing with a famous experiment on the **interference** of the light from two sources side by side.

To avoid nonessential complications, we shall not describe Young's original experiment, but a simplified version of it (figure 8-14). Light from a lamp is passed through a filter that transmits only one color, giving an electromagnetic wave of definite wavelength. The light then passes through a long, very narrow slit S_0 and emerges with circular wavefronts, which are viewed from above in the lower part of the figure. It then passes on to a second screen in which there are two *very narrow* parallel slits S_1 and S_2 very close together. Each slit acts as a source of its own circular wavefronts, but the waves from the two sources interfere with one another to produce on the viewing screen a set of equally spaced, vertical, bright strips separated by dark strips. These are called **interference fringes** and figure 8-15 is a photograph of a set of actual fringes.

A hint of how the interference of the waves can produce this effect is obtained by holding the book up to the level of the eye and viewing the right-hand side of the lower part of figure 8-14 at grazing incidence. One sees a bunch of alternately bright and dark beams radiating outward from a point M midway between the two slits.

The same kind of interference may be obtained with ripples on water if we dip two probes into the water side by side and oscillate them up and down in unison. Photographs of this effect in figure 8-16 clearly show the bright and dark radial beams. (We shall return shortly to the difference between the two photographs.) The explanation in this case is that, along the bright beams the crests of the waves from one source tend to coincide with the crests of the waves from the other source, and so the two waves augment one another to produce a crest twice as high. At the same time the troughs tend to coincide to produce a trough twice as deep. Along the dark beams, however, a crest from one source tends to coincide with a trough from the other source and they cancel one another, with the result that the surface of the water is neither raised nor lowered.

The explanation of Young's double slit experiment (figure 8-14) is simi-

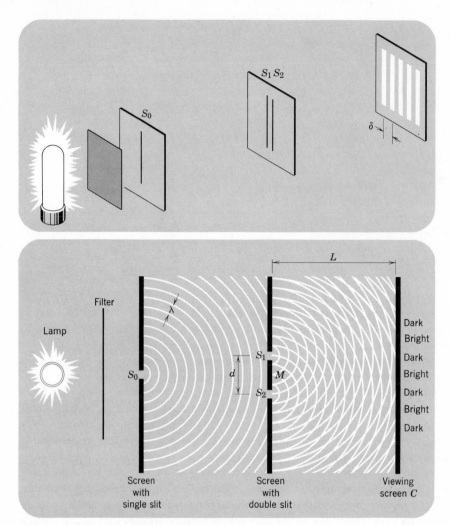

Figure 8-14 *Young's double slit experiment.*

lar except that light is an electromagnetic wave consisting of oscillating electric and magnetic fields. Along a bright beam, which produces a bright fringe, the oscillating electric fields **E** from the two slits are in the same direction and augment one another. The magnetic fields behave in the same way. Along a dark beam, which produces the dark region between two bright fringes, the electric fields from the two sources are in opposite directions and cancel one another. The magnetic fields also cancel.

To find the positions of the fringes it is necessary to solve the moderately complicated geometrical problem of where the crests of the waves

Waves

Figure 8-15 *Interference fringes from a double slit.* (*Photograph by Dr. Brian Thompson.*)

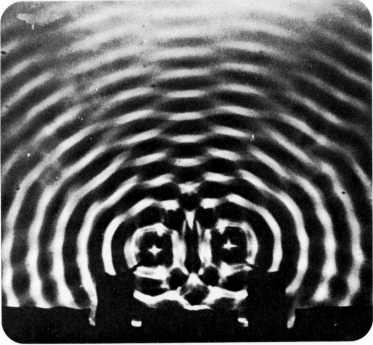

Figure 8-16 *Interference of ripples from two adjacent sources. Notice that when the wavelength is longer the beams spread further apart. (Reproduced with permission from Physics by the Physical Science Study Committee, D. C. Heath & Co.)*

from the two sources coincide. The result is that, if d is the distance between the two slits, L is the distance from these slits to the viewing screen and λ is the wavelength, the distance between neighboring bright fringes is

$$\delta = \frac{\lambda L}{d} \tag{8-13}$$

The proportionality to L is obvious if you again view figure 8-14 at grazing incidence. Since the bright beams diverge from M, the distance between the bright fringes obviously increases as the viewing screen is moved further away. The distance between the fringes also increases if the wavelength is increased. A comparison of the two parts of figure 8-16 illustrates this very well. The final, very important point is that the fringes are further apart if the two slits are closer together. We shall have need to remember this fact later in the book.

Interference fringes

The distance between neighboring bright fringes is

$$\delta = \frac{\lambda L}{d} \tag{8-14}$$

λ is the wavelength, L is the distance from the slits to the viewing screen, and d is the distance between the two slits.

If the wavelength is increased the fringes move further apart.

If the slits are placed *closer* together the fringes move further apart.

A simple rearrangement of equation 8-14 gives

$$\lambda = \frac{d\delta}{L} \tag{8-15}$$

All the quantities on the right-hand side can be easily measured, and the fringes therefore enable the wavelength of light to be measured even though it is only a few millionths of an inch.

8-6 Diffraction

Waves

The situation of figure 8-12, in which the wave spreads out in all directions on the far side of the gap, requires that the width of the gap shall be much smaller than the wavelength. If the width of the gap is comparable with the wavelength, or larger, the diffraction effects are more complicated. The reason is that the portion of the wavefront passing through the gap can be considered to be the source of a single wavelet only if this portion is very narrow. If a wide portion of the incident wavefront passes through the gap,

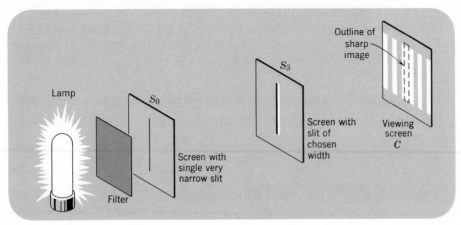

Figure 8-17 *Fresnel's experiment on diffraction by a single slit.*

then this portion must be divided up into very narrow parts each of which is the source of its own wavelet. The various wavelets then interfere with one another in a complicated way.

Diffraction of light was extensively investigated by Fresnel in a series of experiments such as the one illustrated in figure 8-17. This is similar to Young's experiment except that the middle screen does not have two narrow slits but a single wider slit S_3. If light were a stream of particles like bullets from a machine gun, these particles would either be stopped by the middle screen or would pass straight through the slit S_3 to form a sharp image of this slit on the viewing screen, as indicated by the broken colored line. In fact what is observed is a set of **diffraction fringes** spreading far beyond the sharp image. The central fringe is broad and very intense, whereas the side fringes are only half as wide and become fainter and fainter as one proceeds outward (figure 8-18.)

The graph above figure 8-18 shows how the intensity of illumination (brightness) of the viewing screen varies with distance from the center of the

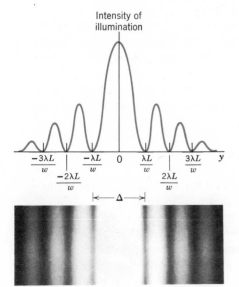

Figure 8-18 *The diffraction pattern produced by a single slit. For the sake of clarity, the graph has been distorted. The ratio of the intensity of a side fringe to the intensity of the central fringe is actually even smaller than it appears to be on the graph. (Photograph by Brian Thompson.)*

pattern. The algebraic expressions along the horizontal axis represent distances from the center. λ is the wavelength of the light, L is the distance from S_3 to the viewing screen, and w is the width of the slit S_3. Theory reveals that the distance between the two darkest points on either side of the central fringe is

$$\Delta = \frac{2\lambda L}{w} \tag{8-16}$$

Again we see that the pattern broadens out if the wavelength is increased. Also, the pattern is more spread out if the slit S_3 is narrower! This is the very reverse of what we would expect of a sharp image.

Diffraction by a single slit

The distance between the two darkest points on either side of the central fringe is

$$\Delta = \frac{2\lambda L}{w} \tag{8-17}$$

λ is the wavelength, L is the distance from the slit to the viewing screen, and w is the width of the slit.

If the wavelength is increased, the pattern broadens out.

If the slit is made *narrower*, the pattern broadens out.

If the opening in the middle screen is not a single slit but an aperture with some other shape, the diffraction pattern is usually more complex, but it can have considerable aesthetic appeal. Figure 8-19 is the diffraction pattern of a triangular aperture which may be regarded as a tapered slit that is wide at the bottom and narrow at the top. It illustrates admirably the fact that, as the slit becomes narrower, the diffraction fringes move further apart. Figure 8-20 is the diffraction pattern of a rectangular slit, with the length and width not very different. It is similar to the pattern for a very long, very narrow slit, but shows fringes in both the horizontal and vertical directions. The shape of the rectangular slit is shown in the bottom right-hand corner. In accordance with the general principle that the diffraction pattern is broader the narrower the slit, the long side of a rectangle in the diffraction pattern is parallel to the short side of the slit.

Returning to the single slit, imagine that the slit is gradually made narrower and narrower. The pattern continually spreads out until, when the width of the slit w is much less than the wavelength λ, the central bright fringe is very wide indeed. This means that the light passing through the very narrow slit spreads out in all directions, as in figure 8-12.

Now reverse the procedure and make the slit wider and wider. The pattern of fringes then shrinks. However, when the width of the slit is many times the wavelength, the theory becomes more complicated. The width of the central fringe is then approximately the width of the sharp image that would be formed if the light traveled along straight lines. The outer fringes

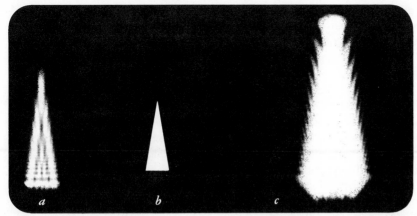

Figure 8-19 *Diffraction of light by an aperture with the shape of a long thin isosceles triangle. Part b is the image that would be formed if the light traveled strictly in straight lines. Part a, taken with a short exposure, shows the formation of crossed diffraction fringes inside the triangular image. Part c, taken with a longer exposure, shows the faint outer fringes. Considering the aperture to be a tapered slit, notice how the diffraction fringes move further apart as the slit becomes narrower at the top. (Photographs by Brian Thompson.)*

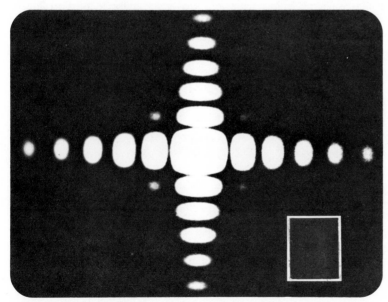

Figure 8-20 *The diffraction pattern of a rectangular aperture 0.8 cm × 0.7 cm. A magnified drawing of the aperture is shown in the bottom right hand corner of the photograph. Notice that the diffraction pattern is more spread out in a direction parallel to the narrower side of the aperture. (Photograph by Brian Thompson, from Wolf and Born, Principles of Optics, Pergamon Press, Ltd. 1959.)*

are very close together just outside the edge of this sharp image and are too faint to be seen except in a very narrow region near the edge. The fringes are in fact quite inconspicuous and the general appearance is close to what would be expected if light were a stream of particles traveling along straight lines.

We can now understand why in everyday life light does appear to travel along straight lines (the familiar light rays like a sunbeam) and why diffraction effects are rarely encountered. The wavelength of light is very small, about 5×10^{-5} cm, and the objects we normally look at are about a million times larger. We therefore normally encounter the situation in which the diffraction fringes are very closely spaced and confined to a narrow region near the edge of an image, so that we rarely notice them. It was this circumstance that made the particle theory of light so popular until Young and Fresnel performed their very careful experiments with very narrow slits.

Diffraction effects are important, though, for sound waves, which have wavelengths ranging from 1 inch to 20 yards. You can hear someone shouting round a corner even though you cannot see him. The long wavelength sound waves are diffracted through large angles as they impinge on the corner wall and the sound waves bend round the corner. The short wavelength light waves, however, effectively move along straight lines and cannot bend round the corner.

We can now see why an optical microscope cannot be used to obtain a clear image of an object whose size is much less than the wavelength of light. Illuminating such an object with light waves is similar to passing light through a slit of width less than the wavelength. The broad diffraction fringes formed spread out beyond the edges of the sharp image and the actual image appears blurred. Similarly, if we try to distinguish two small objects separated by a distance less than the wavelength, the two sets of diffraction fringes overlap and only a single blur is seen. In general, if we wish to obtain a clear picture of an object, we must illuminate it with electromagnetic radiation of wavelength less than the size of the object.

Questions

A

1. In the simple harmonic motions of figure 8-1, in which of the positions A, C, or M is the velocity (a) greatest, (b) zero?
2. An electromagnetic wave has a horizontal velocity directly toward you. At a certain instant the electric field at a point is vertically downward. What is the direction of the magnetic field?
3. Which of the following has the lowest frequency? (a) γ-rays, (b) blue light, (c) infrared light, (d) ultraviolet light.
4. Which of the following has the shortest wavelength? (a) radio waves, (b) x-rays, (c) red light, (d) ultraviolet light.
5. If the wavelength were comparable to the size of an apple, what type of electromagnetic radiation would it be?

B

6. In its mid-position the pendulum of figure 8-1a has kinetic energy. In its extreme position the pendulum has come to rest and has no kinetic energy. Where has the kinetic energy gone?

7. As the sun radiates away energy, does it also lose momentum?
8. In Young's double slit experiment, why must *d* be small and *L* large?
9. When photographing a small object through a microscope, is it better to use red or blue light? Assume that the film is equally sensitive to both colors.
10. If electromagnetic radiation could be used to study the structure of a proton, what wavelengths would be suitable? What type of radiation would this be?
11. A radio telescope operates on a frequency of 8×10^7 cycles per second. Assuming any advance in techniques permitted by the laws of physics, could it ever be used to take a photograph of two astronauts shaking hands on the moon?

C
12. In its mid-position the oscillating weight of figure 8-1*b* has kinetic energy. In its highest position the weight is at rest and has no kinetic energy. Where has the kinetic energy gone? (Be careful!)
13. The unit of length is now defined in terms of the wavelength of a certain spectral line and the unit of time in terms of the period of a different spectral line. Is this equivalent to *defining* the value of the velocity of light, *c*? What is the significance of an experimental determination of the value of *c*? Would it make any difference if the units of length and time were defined in terms of the same spectral line?
14. Is Newton's third law of motion applicable to the magnetic forces between two electrons moving in the same direction with different speeds v_1 and v_2 (a) if the line joining them is perpendicular to their velocities, (b) if the line joining them is not perpendicular to their velocities?
15. Two billiard balls collide inelastically. A small portion of each ball near the point of contact is warmed by the collision. Suppose that the red ball rapidly radiates away the heat before it has had time to spread to the rest of the ball. The white ball is less efficient at radiating heat and the heat spreads uniformly through the ball before it is radiated away. Can you think of a reason why the collision might appear to violate the law of conservation of momentum?
16. If Young's double slit experiment were performed with white light, containing all possible colors, what would the interference fringes look like?

Problems

A
1. What is the period of a sound wave with a frequency of 2.5×10^3 cycles per sec?
2. What is the frequency of a light wave with a period of 1.6×10^{-15} sec?

3. What is the wavelength of the light wave in the previous question?
4. The pendulum of a grandfather clock makes the floor vibrate with a period of $\frac{1}{2}$ sec. What would be the wavelength in air of the sound generated by this vibration? (Velocity of sound in air = 3.4×10^4 cm/sec.)
5. Approximately how long does it take an electromagnetic wave to travel (a) from this book to your eyes, (b) one city block, (c) to your radio re-

ceiver from a transmitting station ten miles away, (d) from New York to Los Angeles, (e) once round the equator, (f) from the sun to the earth?

6. In a Young's double slit experiment the light has a frequency of 6×10^{14} cycles per sec, the distance between the slits is 8×10^{-2} cm, and the distance from the slits to the screen is 1.2×10^2 cm. What is the distance between centers of adjacent fringes?

B

7. Assuming that the lowest audible note has a frequency of 16 cycles per sec, and the highest audible note has a frequency of 2×10^4 cycles per sec, find the corresponding wavelengths in air. In each case think of a familiar object that has a size comparable with the wavelength in question. (Take the velocity of sound in air to be 3.4×10^4 cm/sec.)

8. As you lie on the beach sunbathing, a typical energy flow for the sunlight might be 10^6 erg per cm^2 per sec. Calculate the corresponding electric field.

9. If the energy flow of the sunlight is 10^6 erg per cm^2 per sec, calculate the pressure exerted on a surface that is perpendicular to the sun's rays and that absorbs all the light falling on it. Compare it with atmospheric pressure.

10. In a Young's double slit experiment using blue light with a wavelength of 4.5×10^{-5} cm, the size of the room does not permit the distance between the slits and the screen to exceed 2.5×10^2 cm. What is the maximum permissible distance between the slits if the fringes must be separated by at least 0.1 cm?

11. In a Young's double slit experiment, the distance between centers of adjacent fringes is 0.01 cm. If the distance from the screen to the slits is doubled, the distance between the slits is halved and the wavelength of the light is changed from 6.5×10^{-5} cm to 4×10^{-5} cm, what is the new distance between fringes?

12. In a Young's double slit experiment, the distance between the slits is 0.062 cm and the distance from the slits to the screen is 472 cm. If the distance between centers of adjacent fringes is found to be 0.39 cm, what is the wavelength of the light?

13. Light with a wavelength of 5.5×10^{-5} cm passes through a slit 0.15 cm wide and forms a diffraction pattern on a screen 2.5×10^2 cm away. What is the width of the central fringe?

C

14. What would the magnitude of the electric field of a light wave have to be if it exerted a pressure equal to atmospheric pressure on a totally absorbing surface perpendicular to the direction of propagation?

15. Using a device known as a laser it is possible to produce very intense radiation with electric fields of the order of 10^5 dynes/statcoulomb. If this radiation falls perpendicularly on to a totally absorbing surface, calculate the pressure in atmospheres.

Waves

16. In a Young's double slit experiment, the filter passes red light and blue light, each with a sharply defined frequency. It is found that the fifth blue fringe coincides exactly with the fourth red fringe. (The central fringe is taken as the first fringe.) What conclusion can you draw? If you were given the distance between the slits and the distance from the slits to the screen, could you draw any further conclusions?

9 *Special Relativity*

9-1 The Question

When we were discussing electromagnetic induction in section 7-6, we came up against the question of whether there is an absolute space that is *really* at rest in the sense that an observer who is stationary in this space obtains a preferred view of the physical behavior of the universe. This is the basic question that led Albert Einstein to the Theory of Relativity.

If there can be such an observer at absolute rest, he is certainly not standing on the surface of the earth. An observer on the surface of the earth has a velocity of about 4×10^4 cm/sec (1000 mph) due to the rotation of the earth, a velocity of about 3×10^6 cm/sec (100,000 mph) due to the motion of the earth around the sun, and a velocity of about 2×10^6 cm/sec due to the motion of the sun relative to the nearby stars.

We might therefore ask the astronomers to investigate these motions in detail. However, an astronomer can only investigate the motion of his telescope relative to other objects and he usually chooses these objects to be the "fixed stars." We believe that the earth is rotating because the stars appear to perform a daily rotation in the sky. We believe that the earth revolves about the sun because the sun changes its apparent position relative to the stars throughout the year. Moreover, we believe that the sun is moving in a certain direction relative to the stars because, if we look in this direction, we notice that the stars, although they are moving in random directions relative to one another, appear to be moving as a body toward us. In the opposite direction the stars appear to be moving as a body away from us.

The astronomers might, in this way, decide on a **frame of reference** at rest relative to the stars. However, most of the stars in the sky belong to a group of stars known as the Local Galaxy. Far outside this local group of stars are other groups of stars known as extra-galactic nebulae. Relative to these objects the Local Galaxy is rotating at a rate that gives the sun an extra velocity of about 2.5×10^7 cm/sec. This is much greater than its velocity relative to nearby stars. The "fixed stars" are therefore not really fixed, and so the method must be extended to measure velocities relative to extragalactic objects. Clearly, in order to eliminate completely a possible motion of nearby matter, the observations should be extended to include all the matter in the observable universe. Unfortunately, our telescopes have still revealed to us only a fraction of what we believe is the whole observable universe, and the further we look the more difficult it is to obtain reliable observations. Moreover, modern theories of the nature of the universe, independently of whether they are right or wrong, strongly suggest that observations spanning these enormous distances must not be lightly interpreted in terms of laws that are valid over short distances in our immediate vicinity.

Nevertheless, it is conceivable that the astronomers might eventually provide us with a frame of reference at rest relative to the universe as a whole. The point at issue, though, is whether the laws of physics take a specially simple form for an observer who is stationary in this "astronomical rest frame." The fundamental laws of classical mechanics, for example, are mainly concerned with the *accelerations* that bodies induce in one another and these laws *are* independent of the velocity of the observer.

Figure 9-1 illustrates this last point and may also help to clarify the concept of a frame of reference. Suppose that a baseball player, standing in a car traveling at 20 mph, throws his ball vertically upward so that it has no horizontal velocity relative to the car. If he concentrates his attention on the ball, he sees it rising above him and falling again, but always remaining directly above his head (figure 9-1a). Meanwhile, out of the corner of his eye he sees the road, and all objects fixed on it, moving toward his rear at 20 mph. On the other hand, an observer standing still at the side of the road says that the ball initially had a horizontal velocity of 20 mph which it retained throughout its motion, and therefore it followed the curved path (a parabola) shown in figure 9-1b. If either observer wished to provide a mathematical description of the motion, he could set up as his **frame of reference** a set of Cartesian axes, as described in section 2-2. However, one set of axes would be rigidly attached to terra firma and the other set would be rigidly attached to the car and would move with a velocity of 20 mph relative to the first set.

The two observers have very different views of the motion of the ball and other moving bodies, but they agree in one important respect. They

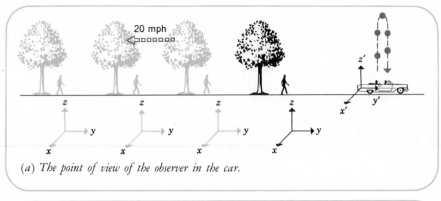

(a) *The point of view of the observer in the car.*

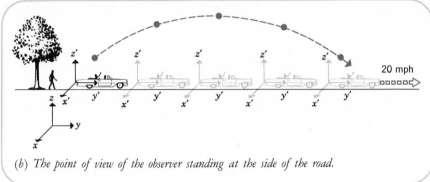

(b) *The point of view of the observer standing at the side of the road.*

Figure 9-1 *The motion of a ball seen from two different frames of reference.*

both concede the validity of Newton's laws of motion (Chapter 3) and they both agree that the acceleration of the ball is directed toward the center of the earth and has the value

$$g = \frac{GM_e}{R_e{}^2} \tag{9-1}$$

in accordance with Newton's universal law of gravitation (sections 3-3 and 3-5). The implicit realization that the fundamental laws of mechanics are independent of the velocity of the observer was one of the major achievements of Galileo and Newton in the seventeenth century.

As far as the laws of electromagnetism are concerned, we have already faced this issue in section 7-6. We discovered there that the phenomena of electromagnetic induction do not allow us to decide between two observers moving relative to one another. The two observers give slightly different theoretical descriptions of the phenomena, one of them deciding that there is an electric field present and the other deciding that there is not, but they are both agreed on the really important observational fact that an induced current flows around loop II of figure 7-17. (At this point the reader would be well advised to read section 7-6 again carefully.)

Perhaps, however, some other electromagnetic phenomenon might help us to resolve the issue. In particular, light is an electromagnetic wave and

Figure 9-2 *The velocity of sound is 750 mph relative to the air through which the sound is propagated.*

we can ask the crucial question, "With respect to what should the velocity of light be measured?" In the case of *sound* the answer to a similar question is well understood. Sound travels with a certain velocity *relative to the material through which it is traveling*. Suppose that the velocity of sound through the air is 750 mph. If I speak to you with a 50 mph gale blowing from behind me directly toward you, the sound emerging from my mouth travels at 750 mph relative to the air and so it travels from me to you with a speed of $750 + 50 = 800$ mph. On the other hand, if you speak to me against the direction of the wind, the sound travels from you to me with a speed of $750 - 50 = 700$ mph (figure 9-2.)

Perhaps the situation is the same for light. Perhaps, as was suggested in connection with the concept of a field, space is filled with a mysterious, invisible, imponderable, intangible, jellylike substance called the **ether.** Perhaps light travels through the ether in the same way that sound travels through the air and the velocity of light c is 2.9979×10^{10} cm/sec *relative to the ether.* If that were so, absolute rest could mean "at rest relative to the ether." There would be a special frame of reference stationary relative to the ether and we could recognize this frame because it would be the only one in which the velocity of light would be c. In frames moving relative to the ether, the velocity of light would be different. For example, we could chase after the light with a velocity $\frac{1}{4}$c relative to the ether and the light would then appear to be moving away from us with a velocity $\frac{3}{4}$c (figure 9-3). As we shall soon see, this is *not* what happens.

9-2 The Answer

At the end of the nineteenth century and the beginning of the twentieth century, these questions became important in physics and many experiments were performed to investigate whether any physical significance could be attached to the concept of an ether or a state of absolute rest. The experiment that is most often quoted in connection with the Theory of Relativity was first performed by Michelson and Morley in 1887. The principle of this experiment is illustrated in figure 9-4. Imagine that two flashes of light are sent out at right angles and are reflected back from two mirrors at equal

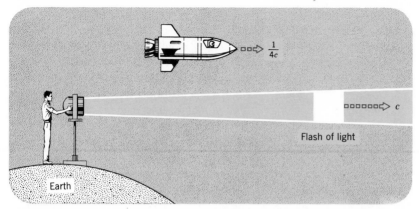

Figure 9-3 *According to prerelativistic ideas, if there were an ether, and if light always had a velocity c relative to this ether, a spaceship chasing the flash of light with a velocity $\frac{1}{4}c$ relative to the ether would see the light moving away with a velocity $\frac{3}{4}c$.*

distances from their source. They might be expected to arrive back simultaneously, but this would not be the case if there were an ether and if the apparatus were moving with a velocity **v** through this ether. As depicted in part *b* of figure 9-4, if we remain stationary relative to the ether and watch the apparatus move by, one flash of light travels along the zigzag path *ABC* and the other flash travels along the path *ADC*. These paths are not equal in length and the two flashes would then be expected to arrive back at different times.

The apparatus of Michelson and Morley was capable of detecting a difference in the time of arrival of about 10^{-17} sec! Nevertheless, when the experiment was performed no time difference was observed. The velocity of the apparatus relative to the ether could not have been greater than 10^6 cm/sec. But the velocity of the earth in its orbit around the sun is 3×10^6

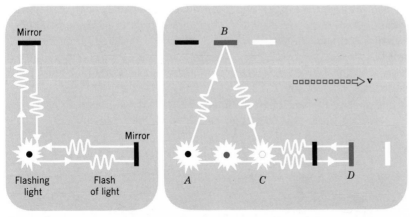

(*a*) *The apparatus is stationary relative to the ether.*

(*b*) *The apparatus has a velocity **v** through the ether.*

Figure 9-4 *The principle of the Michelson-Morley experiment, assuming that the ether exists.*

cm/sec. If, by chance, the earth had been almost at rest relative to the ether when the experiment was first performed, six months later the earth would have reversed its orbital velocity and would have had a velocity of about 6×10^6 cm/sec relative to the ether. This would have produced an easily detectable time difference, but when the experiment was in fact repeated six months later, there was still no detectable effect. It began to look as though the idea of a velocity relative to the ether had no physical meaning.

A theory as basic as the Theory of Relativity cannot be proved by one clever experiment or one clever argument. There must be an exhaustive interplay of theory, experiment, countertheory and counterexperiment, until all conceivable possibilities have been investigated and all ideas have been tested by observations. As is well known, the Theory of Relativity stimulates much more than the usual amount of interest, and so it has been subjected to particularly close scrutiny which, at the time of this writing, has lasted for about three quarters of a century. Nothing has emerged that would favor the concept of an ether or provide convincing evidence for the existence of a preferred state of absolute rest.

One aspect of Einstein's genius was that he saw clearly, as early as 1905, that this was the situation and what it implied. He postulated that all the basic laws of physics are independent of the velocity of the observer. Amongst the basic laws of physics are Maxwell's equations, and these imply that an electromagnetic wave travels with a velocity c. Any observer, whatever his velocity, must therefore see an electromagnetic wave traveling with a velocity c. The velocity of light is not to be related to an ether, but is always 2.9979×10^{10} cm/sec *relative to the observer who measures it.* It seems a simple and obvious solution to the problem, but its consequences are revolutionary.

Before proceeding, we must clarify one point. At present we are talking only about observers moving with constant velocities. To simplify matters, let us be content to say at this stage that this means constant velocity relative to the astronomical rest frame, that is, constant velocity relative to the whole universe. The original form of Einstein's theory was confined to observers moving with constant velocities and is called the **Special Theory of Relativity.** A later extension to accelerating observers is called the **General Theory of Relativity.** We shall discuss the general theory in the next chapter, where we shall discover how to recognize whether an observer is accelerating. We shall also see that the general theory is a means of incorporating gravitation into the theory of relativity.

In the special theory, when the observers are moving with constant velocities, Einstein's basic postulates can be stated in the following way.

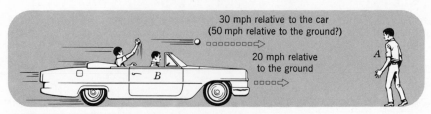

Figure 9-5 A problem in the addition of velocities.

9-3 The Boldness of the Answer

The hypothesis that the velocity of light is always the same relative to the observer is so simple, and provides so ready an explanation of the Michelson-Morley experiment and related experiments, that it is important to realize why it leads to such a revolution in our ideas. Let us first consider the "common sense" attitude to the following situation. A baseball player can throw a ball horizontally with a speed of 30 mph. Suppose that he stands up in a car that is traveling at 20 mph and throws the ball forward (figure 9-5). The ball is given a speed of 30 mph relative to the car, but the car itself is moving in the same direction at 20 mph. An observer standing still and watching this event "obviously" sees the ball fly past him at 30 + 20 = 50 mph.

Now consider an analogous situation involving light (figure 9-6). Spaceship B is traveling past spaceship A at the very high speed of 2×10^{10} cm/sec relative to A. B sends out flashes of light in the forward direction. The pilot of B measures the velocity of this light by setting up two detecting devices in the light beam at a known distance apart and measuring the time interval between the arrival of a flash at the two devices. He obtains the usual value of 3×10^{10} cm/sec since, according to Einstein's hypothesis, this is the velocity of light relative to any observer, independently of his po-

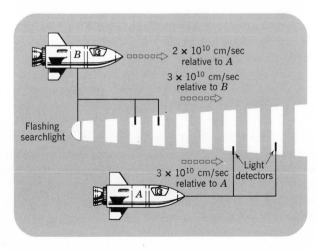

Figure 9-6 Both A and B measure a velocity of 3×10^{10} cm/sec for the light beam, even though B has a velocity of 2×10^{10} cm/sec relative to A.

sition or velocity. The pilot of A can also measure the velocity of the light flashes by setting up two similar devices in the light beam, although, of course, A's devices are moving with respect to B's devices with a velocity of 2×10^{10} cm/sec. The arguments that were applied to the baseball and the car suggest that the pilot of A should obtain a value of $3 \times 10^{10} + 2 \times 10^{10} = 5 \times 10^{10}$ cm/sec. However, this is contrary to Einstein's hypothesis, since *all* observers, measuring the velocity of *any* light beam, must obtain a value of 3×10^{10} cm/sec.

The natural reaction to this discussion is that it proves the absurdity of Einstein's hypothesis. Nevertheless, physicists now believe, with a high degree of confidence, that the pilot of A would obtain a value of 3×10^{10} cm/sec. It is common sense that is wrong. The outstanding achievement of Einstein's genius was the realization that the arguments applied to the car and the ball are based on certain subtle fallacies. We now believe that the velocity of the ball relative to the stationary observer would not be 50 mph but 49.999999999999933 mph. The correction introduced by the theory of relativity is so very small that it is clear why we are not aware of it in our everyday experience of cars and baseballs. The reason for this, as we shall see later, is that the velocities of the car and the ball are very small compared with the velocity of light. In the example of the two spaceships and the beam of light, the velocities are not small compared with c.

Reconsider the situation of figure 9-3 in terms of these new ideas. We must first abandon the idea that the light is traveling through an ether. The light flashes have a velocity c relative to the observer operating the search-light, not because he is stationary relative to the ether, or even because he is stationary relative to the source of the light, but because the light flashes must appear to *any* observer to have a velocity c. In particular, the observer in the spaceship also sees the light flashes moving away from him with a velocity c, not $\frac{3}{4}$c. In fact, however fast the spaceship chases after the light, the light flashes will always appear to the observer in the spaceship to be gaining on him with a velocity c! Chasing light is the ultimate frustration.

What Einstein clearly realized was that we cannot start to understand a concept such as velocity without devising some procedure for determining time at distant places. We must know where the body is at what time. If we look at it, we are relying on the light traveling from the body to our eyes and the journey takes a definite time. We do not see the body where it is now, but where it was at an earlier time when the light left it. To correct for this, it is necessary to know how fast the light traveled, and this immediately raises all the questions about the velocity of light with which we have already been preoccupied in this chapter.

This point can be emphasized by the following example. In the constellation Andromeda there is a celestial object called the Andromeda nebula which is visible to the unaided human eye. It is an extragalactic nebula, a group of stars that lies far away from our own local group of stars. It is so far away that light takes about 2 million years to travel from it to us. We see it, not as it is at the moment, but as it was 2 million years ago, at a time when man, if he existed at all, was very ape-like. We cannot completely exclude the possibility that one million years ago the Andromeda nebula exploded and is now dispersed throughout space. We should have to wait another million years to discover this.

Suppose that I step out into my garden on a clear moonlit night and first notice a moth hovering round the lamp near my door. I then glance at the moon, transfer my attention to the brightest star in the sky, Sirius, and finally look at the Andromeda nebula. From my point of view as a local observer, observing what happens at the particular spot on the planet Earth where I find myself, the time sequence of these events is unambiguous. It is: moth, moon, Sirius, Andromeda nebula. If, however, I ask what time the events occurred at the distant places where they originated, the sequence is reversed. Taking into account the time taken for the light to travel from the event to me, I must conclude that I saw light that was emitted by the Andromeda nebula 2 million years ago, light that left Sirius 8.7 years ago, sunlight that was reflected from the moon 1.28 seconds ago, and lamplight that was reflected from the moth one billionth of a second ago.

Even so, the situation has been oversimplified. The earth and the Andromeda nebula are moving toward one another with a relative velocity of about 3×10^7 cm/sec, which is about one thousandth of the velocity of light. Under these circumstances, should we assume that the light emitted by the nebula travels away from it with a velocity of 3×10^{10} cm/sec *relative to the nebula*, and that the earth moves forward to meet this light with a velocity of 3×10^7 cm/sec, so that the velocity of the light relative to the earth is $3 \times 10^{10} + 3 \times 10^7 = 3.003 \times 10^{10}$ cm/sec? Or should we assume that space is filled with an ether and that light travels with a velocity of 3×10^{10} cm/sec *relative to the ether*? In the latter case we should have to know the velocity of the earth relative to the ether before we could calculate the time taken for the light to travel from the nebula to the earth. The two assumptions about the behavior of light could easily lead to a difference of a thousand years or more in the two estimates of the time.

Now suppose that the skeleton of an early man has been discovered and, by means of refined dating techniques, the year of his death has been accurately determined. I wish to dramatize the situation by the statement, "If you look at a certain star in the Andromeda nebula tonight, the light that enters your eye left this star exactly one year after this prehistoric man died." However, this statement might be based on the assumption that the velocity of light is measured relative to the nebula. On the alternative assumption that the velocity of light should be measured relative to an ether, the statement might have to be modified to, "If you look at a certain star in the Andromeda nebula tonight, the light that enters your eye left the star nine hundred and ninety-nine years before this prehistoric man died." The point at issue is whether, at the instant the light left the star, the man had already died, or whether he had not yet even been born!

As we have already learned, Einstein's answer is that neither of the above approaches is correct. His postulate is that the velocity of light is c relative to the observer on earth. Suppose, however, that there is another observer in a third galaxy moving relative to our Local Galaxy. When he observes these events and assumes, as he must, that light has a velocity c relative to himself, will he agree with the observer on earth about the proper time sequence of the events? When Einstein's postulate is carried to its logical conclusions, it turns out that this other observer will not agree. In fact, different observers moving relative to one another form different ideas about times of events and even about distances in space. Moreover, it is basic to the underlying philosophy of the Theory of Relativity that the point of view of no one observer is to be preferred over the point of view of any other observer. They are compelled to agree to disagree.

All this leads to a complete revolution in our ideas about space and time. When first encountered it is very hard to swallow. All we can do is work out the consequences of Einstein's postulate and see if they are consistent with observations. Then we can work out the consequences of alternative postulates and show that they are inconsistent with observations. So far Einstein's theory has always been victorious, but the consequences are very strange and we must abandon some of our cherished prejudices.

9-4 *Prejudices and Observations—A Philosophic Entr'acte*

Consider how even our conventional ideas of space really depend on an elaborate comparison of theory with observations. Suppose that I enter the classroom to give a lecture on relativity and, by a supreme mental effort, I have abandoned all my prejudices concerning the nature of space. I am determined to rely only on direct observations and to accept only those theories that are consistent with all the observations. I am startled to discover that the students sitting in the front row are approximately ten times taller than the students sitting in the back row. I then formulate two hypotheses, between which I must decide by suitable experiments. Either students come in varying sizes and the university has conveniently provided smaller chairs in the rear for the smaller students; or else the space in the room has the very peculiar property that it shrinks a student as he moves to the rear. I therefore ask a student in the back row to move to the front row, and discover that, as he does so, he increases in size until he is comparable with the students already in the front row. My second hypothesis is confirmed!

However, as a scientist I have been trained not to accept a theory without subjecting it to every possible test. I therefore move to the rear of the room and discover that the students in the back row are then ten times taller than the students in the front row. My second hypothesis is shattered. In spite of my post-Renaissance upbringing, which has taught me not to look upon myself as the center of the universe, I am beginning to hypothesize that I possess some extraordinary power whereby students increase in stature the nearer they approach to me.

While I am still standing at the rear of the room, my colleague Professor X enters at the front and I solicit his help. He tells me that the students in the front row are obviously ten times taller than the students in the back row. I am reluctant to concede to my colleague the same powers that I possess myself, but even if I did we should presumably cancel one another out and all students would become the same size.

Professor X and I consult on these matters and come up with a very unorthodox theory. We conclude that our observations have been deceiving us. All students are really more or less the same size, but, due to some imperfection in our method of vision, they *appear* to grow smaller as they move away from us. We invent a device called a meter rule, with which we can measure the height of a student. We discover that, as a student holding a meter rule moves away from us, both student and rule shrink in the same way and maintain the same ratio to one another. We are even able to invent a system of geometry that provides a mathematical description of the situation. Upon later referring to the library, we are disappointed to find that we must concede prior publication to a certain Dr. Euclid.

Most of us perform this important piece of scientific research at a very early age, and for the rest of our lives accept its conclusions intuitively. Unfortunately, the upbringing of our children is such that they are confined to frames of reference whose relative velocities are small compared with the velocity of light. If we could take our babies and propel them through space with large variable velocities, we should presumably raise a generation to whom the results of the theory of relativity would be intuitively obvious.

9-5 Moving Clocks

One of the consequences of Einstein's theory is that a *moving clock runs slow*. To see how this comes about, imagine two observers A and B traveling through empty space in their spaceships with a relative velocity V (figure 9-7).

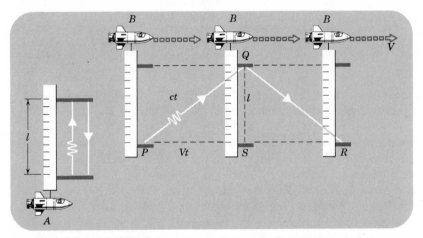

(a) *A decides that B's clock is slow.*

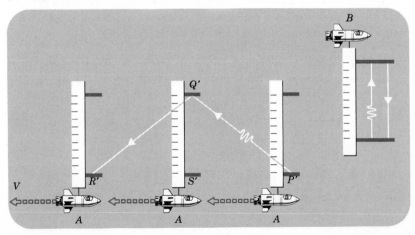

(b) *B decides that A's clock is slow.*

Figure 9-7 *Two observers A and B moving relative to one another compare their light clocks.*

Provide each with a specially designed "light clock" in which a pulse of light travels backward and forward between two mirrors a distance l apart. Every time the pulse arrives at the lower mirror it triggers some electronic equipment which produces the sound of a tick in a loudspeaker and also moves on a counter one unit.

Although this is a very unconventional clock, it can always be calibrated by measuring the time between ticks using a conventional clock. The result of the calibration experiment depends only upon the laws of physics and one of our fundamental postulates is that these laws are the same for A and B. Therefore, when A calibrates the light clock at rest in his spaceship against a conventional clock also at rest in his spaceship, he must get the same answer as B when he calibrates his light clock against his conventional clock. Any conclusions drawn from the behavior of the light clocks are therefore equally valid for any kind of conventional clock.

The disagreement between A and B commences when they look at the behavior of one another's clocks. Start with A's point of view (figure 9-7a). Looking at his own clock A says, in accordance with Einstein's postulate, that the light pulse has a velocity c relative to himself and therefore travels backward and forward a distance $2l$ in a time $2l/c$.

$$\text{Time between ticks of a stationary light clock} = \frac{2l}{c} \qquad (9\text{-}2)$$

However, when A looks at B's clock he sees the light pulse traveling along a zigzag path which is longer than $2l$. Since, according to Einstein's postulate, A sees B's light pulse also traveling with a velocity c, the time for the pulse to travel backward and forward is greater than for his own clock. The time between ticks of B's clock is greater and that means that B's clock is slow. If I have a clock that should tick once every second and I alter it to tick once every 2 seconds, then in 1 minute it will tick only 30 times and will register 30 seconds, or half a minute, so it is running slow.

It is quite simple to calculate the magnitude of the effect. If, according to A, the time between ticks of B's clock is $2t$, then the time for the light pulse to go from P to Q (figure 9-7a) is t and so the length of PQ is ct. Meanwhile the whole clock has moved a distance Vt, which is the length PS. Pythagoras' theorem applied to the right angled triangle PQS is

$$PQ^2 = QS^2 + PS^2 \qquad (9\text{-}3)$$
$$\text{or } c^2t^2 = l^2 + V^2t^2 \qquad (9\text{-}4)$$

Divide by c^2

$$t^2 = \frac{l^2}{c^2} + \frac{V^2}{c^2}t^2 \qquad (9\text{-}5)$$

Move the last term to the left-hand side

$$t^2 - \frac{V^2}{c^2}t^2 = \frac{l^2}{c^2} \qquad (9\text{-}6)$$

$$\text{or } t^2\left(1 - \frac{V^2}{c^2}\right) = \frac{l^2}{c^2} \qquad (9\text{-}7)$$

Take the square root

$$t\sqrt{1 - \frac{v^2}{c^2}} = \frac{l}{c} \tag{9-8}$$

$$\text{or } t = \frac{l}{c} \frac{1}{\sqrt{1 - \frac{v^2}{c^2}}} \tag{9-9}$$

Remember that the time for the pulse to travel backward and forward is twice this. So,

Time between ticks of a moving light clock

$$= \frac{2l}{c} \frac{1}{\sqrt{1 - \frac{v^2}{c^2}}} \tag{9-10}$$

If we compare this with equation 9-2, we obtain the general result that follows.

Slowing down of a moving clock

$$\frac{\text{Time between ticks of a moving clock}}{\text{Time between ticks of a stationary clock}} = \frac{1}{\sqrt{1 - \frac{v^2}{c^2}}} \tag{9-11}$$

It is not permissible to say that B's clock *is* slow compared with A's. This would introduce a basic difference between the two observers that is contrary to the principle of relativity. The correct statement is that, *according to A's observations*, B's clock is slow compared with his own. What is B's point of view? Figure 9-7b shows how the situation appears to B. We can apply to this diagram exactly the same arguments that we applied to obtain A's point of view. B says that the light pulse of his own clock travels up and down the same straight line whereas A's light pulse travels along a zigzag path and takes longer. Exactly the same argument as before leads to equation 9-11, but now the moving and stationary clocks are interchanged. B therefore says that A's clock is slow!

The disagreement between A and B is basic. If A and B were back on earth at rest relative to one another and A had a well-made wrist watch, whereas B had a poorly made one, they could easily come to an agreement. They could synchronize their watches so that they both read 12 noon when the sun was at its highest position in the sky. They might then discover that A's watch registered 1 P.M. when B's watch registered 12:55 P.M. A would then say that B's watch was slow whereas B might say that A's watch was fast, which is logically consistent. They could come to an agreement by noticing that A's watch again registered 12 noon when the sun was at its highest position the following day, and B could then adjust the rate of his watch to agree with A's from then on. When they are moving relative to one another, however, and they both have the best watch they can make,

A always says that B's watch is slow, whereas B insists that A's watch is slow. No adjustment of their watches can make them agree. They must agree to disagree.

Fortunately, very direct evidence for the slowing down of a moving clock is provided by an experiment with π-mesons. A π-meson or pion is a fundamental particle with a mass intermediate between the mass of an electron and the mass of a proton. It is created by a certain process and a little time later suddenly disappears by changing into other particles. Not all pions live for exactly the same time, but it is possible to define a quantity called the "half-life time," τ. Its significance is that 50% of the pions live for a time less than τ, whereas the other 50% live longer than τ. If the pions are at rest relative to the laboratory, or moving with velocities so small as to be negligible, their half-life time is found to have the value

$$\tau_0 = 1.8 \times 10^{-8} \text{ sec} \tag{9-12}$$

It is easy to obtain a beam of pions moving relative to the laboratory with a very high velocity V, comparable with c. Under these circumstances the half-life time is found to be

$$\tau = \frac{\tau_0}{\sqrt{1 - \dfrac{V^2}{c^2}}} \tag{9-13}$$

This is the form equation 9-11 takes in this case. A bunch of pions can be regarded as a type of clock. The time between ticks of the clock is the time for 50% of the pions to disappear. The laboratory is frame of reference A and, when the pions are at rest, they constitute a clock stationary with respect to observer A, so τ_0 is equivalent to the time between ticks of a stationary clock. However, when the pions move with velocity V relative to the laboratory, they constitute a moving clock. This clock would be at rest relative to an observer B moving with velocity V relative to A. τ is therefore equivalent to the time between ticks of a moving clock. Experiments of this kind with pions are in complete agreement with the relativistic formula. In a typical case 50 percent of the moving pions were observed to be still alive after time τ, whereas the number would have been only 2 percent if the lifetime had been τ_0 as for stationary pions. Similar results have now been obtained with other fundamental particles which can be created, live on the average for a certain time, and then disappear.

9-6 Past, Present, and Future

Two observers moving relative to one another also have interesting disagreements about whether two events occurred simultaneously and about which of two events occurred first. In figure 9-8a B has set up the following experiment on a long girder held below his spaceship and lined up parallel to the direction of the relative velocity. In the center of the girder at M is a small explosive charge. The flash of light from the explosion travels in both directions and enters the boxes P and Q at opposite ends of the girder at equal

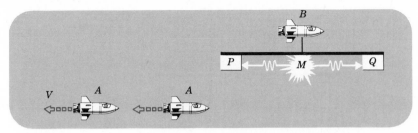

(a) *From B's point of view the explosions at P and Q occur simultaneously.*

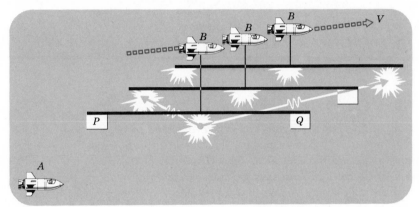

(b) *From A's point of view the explosion at P occurs before the explosion at Q. The direction of V is really parallel to the girder but, in order to improve the clarity, the three positions of the girder have been displaced sideways from one another.*

Figure 9-8 *A disagreement about simultaneity.*

distances from M. When the flash of light enters P or Q it operates a light sensitive device which immediately sets off another small explosion.

B's description of what happens is as follows. The light travels with a velocity c relative to himself and has equal distances to travel to P and Q. The light therefore arrives at these two boxes at the same time and the explosions at P and Q occur simultaneously.

A's point of view is illustrated in figure 9-8b. He says that the light is traveling with a velocity c relative to himself, but the apparatus is moving with velocity V. P moves forward to meet its light pulse, but Q moves away from its pulse. The light therefore reaches P first and the explosion at P occurs before the explosion at Q.

We thus have two well-defined events which B says occurred simultaneously, while A says that one occurred some time before the other. On the basis of our assumption concerning the constancy of the velocity of light, both observers' arguments are valid. Moreover, it is contrary to the principle of relativity to suggest that one of the observers is wrong and the other is right, because the latter has a preferred view of the universe. We are forced to conclude that there is no such thing as "absolute time" which is the same for all observers. Time is relative and is different for observers traveling with different velocities.

To illustrate the relativity of past and future, suppose that a box R is

9-6 Past, Present, and Future

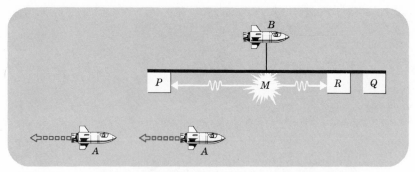

Figure 9-9 *A disagreement about past and future. Box R explodes just before Q. From B's point of view, P explodes at the same time as Q and therefore explodes after R. From A's point of view, P explodes much earlier than Q and therefore explodes before R.*

placed just to the left of Q and that it also explodes when the flash of light reaches it (figure 9-9). *B* says that the explosion at *R* occurs just before the explosion at *Q*, and that the explosion at *P* occurs at the same time as the explosion at *Q*. He therefore says that the explosion at *R* occurs *before* the explosion at *P*. However, *A* says that the explosion at *R* occurs just before the explosion at *Q*, but the explosion at *P* occurs much earlier than the explosion at *Q*. Therefore, according to *A*, if *R* is near enough to *Q* the explosion at *R* occurs *after* the explosion at *P*.

In view of these startling conclusions, it is relevant to ask if there is any limit to the peculiar behavior of time. There are, in fact, severe restrictions on what is possible. Certain possibilities are so unreasonable that not even the theory of relativity will permit them to occur. Look again at figure 9-8 and notice that the two explosions occur at *different places* in *B*'s frame of reference. There would be no confusion about events occurring at the *same place* in *B*'s frame. If, for example, *B* talks to *A* over the radio, *B*'s vocal chords are always in the same position relative to *B*, and so *A* hears the words in the correct sequence. He does not hear *B* talking backwards under any circumstances!

An extraterrestrial moving observer watching events on the surface of the earth can never see time running backwards. Neither can he observe events on earth before they happen to us. In fact, anything that happens on earth must send out light signals to the observer, and he is not aware of it until some time after we are, because of the time taken for the light to reach him. However, it is possible for this observer to be aware of an event at a *distant* place before we are, so that it is in his past when it is still to come in our future. Would it then be possible for this observer to send us a radio message about this event, enabling us to predict the future? The answer is no, because the radio message has to travel through space with the speed of light and is always delayed so long that the message arrives after we ourselves have become aware of the event. Even in the theory of relativity time passes in such a way that cause always precedes effect, but a cause can produce an immediate effect only in its immediate neighborhood. Effects produced at distant places have to travel there with a speed which cannot exceed the speed of light.

9-7 Contraction of a Moving Object

The two moving observers also disagree about distances between objects and lengths of objects. When they look into the matter, each decides that this is caused by the very peculiar behavior of the other observer's measuring rule. As long as a moving rule is held perpendicular to the direction of its velocity, it has the right length, but when it is held parallel to the direction of its velocity, it appears to shrink!

We now realize that we must be very cautious and so we shall devise a very careful experiment to investigate this matter. *B* decides to measure the length of *A*'s meter rule when it is parallel to their relative velocity in the following way (figure 9-10a). *B* holds out his clock so that *A*'s rule passes just to one side of it. A probe on the bottom of *B*'s clock touches a probe at one end of *A*'s rule and then passes on to touch a probe at the

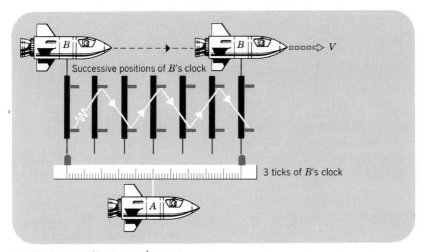

(*a*) *B measures A's meter rule.*

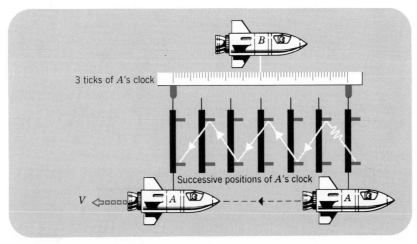

(*b*) *A measures B's meter rule.*

Figure 9-10 *A and B measure one another's rules.*

other end. An electronic device measures the number of ticks of the clock as the probe passes from one end to the other.

B's interpretation of what happens is as follows. From the number of ticks of his clock, he decides that it took a time t for the rule to pass by him. He knows that the rule has a velocity V and he therefore concludes that its length is Vt but, lo and behold, this turns out to be less than 100 cm. B decides that A's rule, which is moving relative to B, is too short.

A says that the explanation is obvious. B's clock is slow, as he has insisted all along, and therefore the time taken for B to move along the rule is really longer by the factor $1/\sqrt{1-(V^2/c^2)}$ and the length is really $Vt/\sqrt{1-(V^2/c^2)}$, which turns out to be exactly 100 cm.

Neither can be persuaded to change his mind. If they invert the procedure and A holds out a probe on his clock to measure the length of B's rule (figure 9-10b), exactly the same thing happens in reverse and A decides that B's rule is short. Again they must agree to disagree and formulate a new law about moving objects.

The contraction of a moving object

$$\frac{\text{Length of a moving object}}{\text{Length of the same object when stationary}} = \sqrt{1 - \frac{V^2}{c^2}} \quad (9\text{-}14)$$

Note carefully that it is only the length in the direction of the velocity that is contracted. The width perpendicular to the velocity is unaltered.

This contraction has interesting consequences for space travel if we aspire to reach objects outside the solar system. The nearest star is Alpha Centauri, which is 4.3 light years away. This means that light takes 4.3 years to travel from it to us. Since a rocket can never exceed the speed of light, it is clear that it would take the rocket at least 4.3 years to reach Alpha Centauri. This is certainly true from the point of view of an observer left behind on earth. However, from the point of view of the pilot of the rocket, the distance between the earth and Alpha Centauri is multiplied by the factor $\sqrt{1-(V^2/c^2)}$ and therefore appears shorter. If V is, say, 0.99c, then $\sqrt{1-(V^2/c^2)} = 0.14$ and the distance appears to be only 14% of its value as seen from the earth. The pilot therefore deduces that light takes only $0.14 \times 4.3 = 0.60$ years to travel from the earth to Alpha Centauri. Since he sees Alpha Centauri coming toward him with a velocity of 0.99c, he expects to get there in $0.60/0.99 = 0.606$ years, or just over seven months. By traveling at a velocity even closer to the velocity of light, the rocket might hope to arrive in an even shorter time.

In principle, therefore, it is possible to visit the stars without having to suffer a long tedious journey. In practice the power requirements to launch a rocket near the speed of light are quite prohibitive.

How does it happen that an observer on earth thinks that the rocket takes 4.3 years to reach Alpha Centauri, while the pilot thinks it takes only seven months? The answer is that, from the point of view of the observer on earth, a clock inside the rocket is slow. However, since the laws of physics are the same for the pilot as for the observer on earth, everything that happens inside the rocket keeps time with the rocket's clock. As seen by the observer on earth, inside the rocket electrons in atoms appear to be moving more slowly, molecules appear to be rotating more slowly, atoms in solids appear to be vibrating less rapidly, and chemical processes appear to proceed more slowly. In particular, the metabolic processes involved in the digestion of the pilot's food proceed more slowly and he needs to eat less often. At the rate of three meals a day, he needs to eat only $3 \times 365 \times 0.606 = 662$ meals, sufficient for 0.606 years. The bio-chemical processes responsible for the aging of the pilot also proceed more slowly and he ages by only 0.606 years. Unfortunately he is thinking and acting more slowly, so his lengthened life is no benefit to him.

A man's span of life is a real and important thing. Have we, then, violated the principle of relativity by producing a difference between the pilot and the observer left behind on earth? No, because it is only the observer on earth who is aware of the increased span of life of the pilot. If he confines his attention to what is happening inside the rocket, the pilot can think only the same number of thoughts and do only the same number of things as in a normal span of life. The only thing that is different for him is that the universe appears contracted in the direction of his motion. *Moreover, if the pilot looks back to earth, he comes to the conclusion that it is the observer on earth who is aging less rapidly.* We shall return to this question again in the next chapter, when we shall be in a position to say more about it.

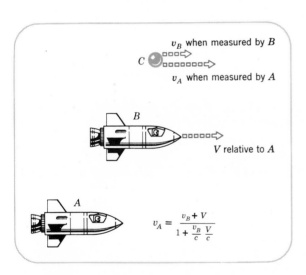

Figure 9-11 *The relativistic law of addition of velocities.*

9-8 *Addition of Velocities*

In figure 9-11 an object C is moving relative to both A and B in the same direction as their relative velocity V. A measures the velocity of C and finds it to be v_A. B also measures the velocity of C and finds it to be v_B. What is the relationship between v_A and v_B?

The answer of classical mechanics is simple and speciously convincing.

B is moving away from A by V centimeters in one second, while C is moving ahead of B by v_B centimeters in one second, so C moves away from A by $v_B + V$ centimeters in one second.

$$v_A = v_B + V \qquad\qquad\qquad (9\text{-}15)$$

However, this implicitly assumes that A and B are in complete agreement about measurements of length and time. We now know that they are not and a proper theoretical discussion is very complicated. We must be content to quote the result.

The relativistic law of addition of velocities

$$v_A = \frac{v_B + V}{1 + \dfrac{v_B}{c}\dfrac{V}{c}} \qquad\qquad\qquad (9\text{-}16)$$

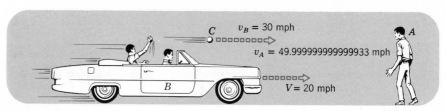

Figure 9-12 Application of the law of addition of velocities to the case of a baseball thrown from a car.

This formula can now be applied to the situation of section 9-3 in which a baseball is thrown from a car. Figure 9-12 is similar to figure 9-5, except that the observer standing by the roadside has been identified as A, the observer in the car as B and the ball as C. The relative velocity V of A and B is 20 mph. The velocity v_B of the ball C according to the observer B in the car is 30 mph. The velocity of light c in miles per hour is 6.7×10^8 mph. v_B/c and V/c are therefore both very small and the denominator of the right-hand side of equation 9-16 is only slightly larger than 1. It is in fact 1.00000000000000133. v_A is therefore only slightly less than the value $v_B + V = 50$ mph given by the classical formula. In fact $v_A = 49.999999999999933$ mph.

In general, relativistic formulas are almost identical with classical formulas when the velocities involved are small compared with the velocity of light. The small differences are never detectable in everyday life. Even for a rocket launched with the escape velocity of 24,800 mph (section 4-7) the factor $\sqrt{1 - (V^2/c^2)}$ appearing in equations 9-11 and 9-14 is 0.9999999993, which differs from 1 by less than 1 part in a billion. This means that a meter rule in the rocket would appear to an earth observer to have shrunk by about one tenth of one millionth of a centimeter! A clock in the rocket would appear to lose one second in about 50 years!

Now consider the other example discussed in section 9-3, involving the

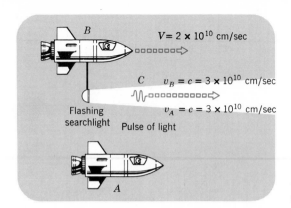

Figure 9-13 *Application of the law of addition of velocities to a light beam.*

speed of light relative to two spaceships. In figure 9-13, which is similar to figure 9-6, it is made clear that C is now a flash of light and B's estimate of the velocity of C is

$$v_B = c \tag{9-17}$$

Substituting this value of v_B into equation 9-16, A's estimate of the velocity of C is

$$v_A = \frac{c + V}{1 + \dfrac{cV}{c^2}} \tag{9-18}$$

$$= \frac{c + V}{1 + \dfrac{V}{c}} \tag{9-19}$$

Multiply numerator and denominator by c. Then

$$v_A = \frac{c(c + V)}{(c + V)} \tag{9-20}$$

$$\text{or } v_A = c \tag{9-21}$$

The law of addition of velocities automatically ensures that the velocity of light is c for both observers! So it must, because this was the initial assumption from which everything else was derived.

9-8 Addition of Velocities

Another important aspect of the law of addition of velocities is that it is not possible to exceed the speed of light by adding together two velocities less than the speed of light. Suppose, as in figure 9-14, that a large space station B is launched from earth with a velocity V = 0.99c. The space station then launches a small rocket C in the forward direction with a speed, relative to the station, of $v_B = 0.99c$.

According to the classical law of addition of velocities, the speed of the small rocket relative to the earth would be 1.98c. According to the theory of relativity the velocity of the rocket C, as measured by an observer A

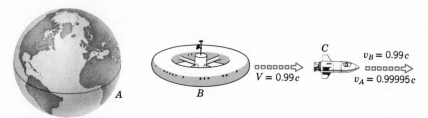

Figure 9-14 *An attempt to exceed the speed of light.*

on the earth, is

$$v_A = \frac{0.99c + 0.99c}{1 + \dfrac{(0.99c)(0.99c)}{c^2}} \qquad (9\text{-}22)$$

$$v_A = \frac{1.98c}{1 + 0.9801} \qquad (9\text{-}23)$$

$$v_A = 0.99995c \qquad (9\text{-}24)$$

Even though the classical sum of the velocities is larger than c, relativity predicts that the apparent velocity of the rocket from the point of view of the observer A on earth is still slightly less than c. In the next section it will become apparent that there is a very good reason why the velocity of a material body can never exceed the velocity of light.

9-9 *Relativistic Mechanics*

In the light of these new ideas of time and space and the way in which velocities add together, the whole of Newtonian mechanics must be reconsidered. The significance of such concepts as mass, force, momentum, and energy must be carefully examined and new definitions must be provided for these quantities. This reformulation of mechanics should be guided by the following principles. The fundamental laws should have the same form for all unaccelerated observers. Relativistic mechanics should become identical with classical mechanics for velocities very small compared with the velocity of light. Moreover, if possible, it would be desirable to arrange matters so that the laws of conservation of energy and momentum are still valid for all unaccelerated observers. It is not easy to justify this reformulation of mechanics without lengthy, cumbersome arguments or advanced mathematics. We shall therefore merely quote the results.

It is necessary to abandon the concept of mass as an invariable property of a body that remains constant under all circumstances. Instead, mass must be considered to vary with the velocity, v, of the body according to the formula:

Variation of mass with velocity

$$m = \frac{m_0}{\sqrt{1 - \dfrac{v^2}{c^2}}}$$

(9-25)

When the velocity v is zero the mass is equal to m_0, which is called the **rest mass** of the body. When v is very small compared with c, the mass m is very little different from the rest mass m_0. As the velocity is steadily increased, the mass m steadily increases, as shown in figure 9-15.

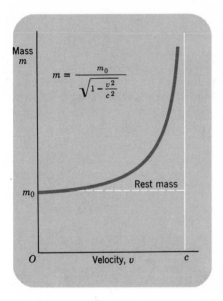

Figure 9-15 *The variation of mass with velocity.*

It is difficult to give a macroscopic body a velocity comparable with c, but it is now possible to produce fundamental particles, such as electrons or protons, with velocities comparable to c. Many radioactive nuclei emit electrons with velocities greater than $0.9c$. Moreover, a major part of current effort in experimental physics is devoted to accelerating charged fundamental particles up to velocities very near c. Many experiments have now been performed to measure the mass of such a particle as a function of its velocity. The results are in complete agreement with equation 9-25.

When v is only slightly less than c, v^2/c^2 is slightly less than 1 and $\sqrt{1 - (v^2/c^2)}$ is small. m is then very much larger than m_0. For example, if $v = 0.99c$, the reader can easily verify that $m = 7.07m_0$. As v approaches closer and closer to c, m increases without limit. As the velocity of the body approaches the velocity of light, its mass increases to infinity! This is brought out rather clearly in figure 9-15.

It is now easy to see that a material body cannot have a velocity greater than the velocity of light. If we try to accelerate the body, as its velocity approaches the velocity of light its mass becomes larger and larger and it becomes increasingly more difficult to accelerate it further. In fact, since the mass becomes infinite when $v = c$, we can never quite accelerate the body up to the velocity c. We can approach closer and closer to c, by making greater and greater efforts, but we can never quite get there.

The conclusion that a body cannot have a velocity greater than c is basic to the whole philosophy of the theory of relativity. Otherwise the body could carry a message to a distant place and, if the velocity of the body could be increased without limit, it would be possible to establish instantaneous communication with this distant place. If observers in different places were in instantaneous communication with one another, they would have no disagreements about simultaneity because they would all be simultaneously in touch with one another. They would then have no difficulty in adjusting their clocks to agree under all circumstances.

> **The velocity of a body can never exceed the velocity of light.**

Using the new definition of mass, the momentum of a particle can be defined as $p = mv$, exactly as it was in Newtonian mechanics.

> *Momentum*
>
> $$p = mv \qquad\qquad\qquad\qquad (9\text{-}26)$$
>
> $$p = \frac{m_0 v}{\sqrt{1 - \dfrac{v^2}{c^2}}} \qquad\qquad\qquad (9\text{-}27)$$

With this definition of momentum, the law of conservation of momentum is still true. In fact, if we consider the collision of two bodies and insist that the law of conservation of momentum shall be true *from the point of view of all unaccelerated observers*, then, after a little mathematical manipulation, we can show that mass must vary with velocity in accordance with equation 9-25.

In section 3-3 force was defined as mass multiplied by acceleration. In the theory of relativity it proves preferable to use the definition of section 4-1, that force is the rate of change of momentum with time. Suppose that, from the point of view of a particular observer, a body has a momentum p_1 at time t_1 and a momentum p_2 at a slightly later time t_2.

> *Force*
>
> Force = rate of change of momentum with time $\qquad\qquad (9\text{-}28)$
>
> $$F = \frac{(p_2 - p_1)}{(t_2 - t_1)} \quad \text{when } (t_2 - t_1) \text{ is sufficiently small} \qquad (9\text{-}29)$$

Since m is no longer constant, but varies with v, as the force accelerates the body and changes v, it also changes m. The rate of change of $\mathbf{p} = m\mathbf{v}$ results from changes in both m and $\mathbf{v}$. $\mathbf{F} = m\mathbf{a}$ is no longer true.

Finally, we come to one of the major revelations of Einstein's theory, the relationship between mass and energy. When light falls on a surface it warms it up and exerts a pressure on it. This can be explained by assuming that electromagnetic radiation carries energy and momentum (section 8-4). If an amount of energy S is incident on unit area per second, a quantity of momentum S/c is incident on the same unit area per second (equation 8-11). A total energy E of electromagnetic radiation is associated with a momentum E/c. (Here E represents energy and *not* the electric field.) The effect is the same as if the light were a mass m traveling with a velocity c and having a momentum mc given by

$$mc = \frac{E}{c} \qquad (9\text{-}30)$$

In this sense, therefore, an energy E of electromagnetic radiation is equivalent to a mass

$$m = \frac{E}{c^2} \qquad (9\text{-}31)$$

In the theory of relativity this equation is found to have a much deeper significance. Mass and energy can be considered as different manifestations of the same physical quantity. A quantity of energy E, whether it be kinetic energy, potential energy, chemical energy, nuclear energy, or any other form of energy, can always be considered to have a mass E/c^2. Conversely, the mass m of a material body can be considered to be equivalent to an amount of energy mc^2 and, under some circumstances, can even be converted into this amount of energy.

Mass and energy

$$E = mc^2 \qquad (9\text{-}32)$$

This is the most famous equation of modern physics. Its connection with atomic and hydrogen bombs and nuclear power is well known.

The full significance of the equation will become apparent as we apply it to various phenomena throughout the rest of this book. We shall see, for example, that a ray of light is bent in the gravitational field of the sun as though the light were a massive body. We shall see that the energy of electromagnetic radiation can be converted into the mass of fundamental particles. Conversely, fundamental particles can disappear and their mass reappear as the energy of electromagnetic radiation. The mass of a nucleus will be shown to be due, in part, to the energy of the particles inside it. At present, let us consider only the simple case of a body of rest mass m_0 moving with a velocity v. According to equation 9-32 its total energy is

$$E = mc^2 \tag{9-33}$$

but the value of m is given by equation 9-25 and so

> *Total energy of a body with rest mass m_0 and velocity v*
>
> $$E = \frac{m_0 c^2}{\sqrt{1 - \dfrac{v^2}{c^2}}} \tag{9-34}$$

If the velocity v is very small compared with c, it can be shown that there is an approximate expression for E.

$$E = m_0 c^2 + \tfrac{1}{2} m_0 v^2 \text{ approximately, when } v \text{ is much}$$
$$\text{smaller than } c \tag{9-35}$$

The second term $\tfrac{1}{2}m_0 v^2$ is the classical kinetic energy! The relativistic expression for the energy therefore becomes identical with the classical expression at very small velocities, except for the addition of the **rest mass energy,** $m_0 c^2$. Since m_0 is a constant, the rest mass energy remains constant and need not be taken into account in classical physics when the conversion of energy from one form into another is being considered. However, modern physics recognizes processes in which the rest mass is annihilated and an amount of energy $m_0 c^2$ is made available.

In classical physics it is taken for granted that mass cannot be created or destroyed and this is sometimes called the law of conservation of mass. Classical physics also has its law of conservation of energy. Now that mass and energy have been found to be interchangeable, these two laws must be combined into a single law of conservation of mass plus energy. It must be understood, of course, that in any equation expressing this law, the mass must be multiplied by c^2 in order to enter on the same footing as energy.

Finally, let us notice an interesting feature of equation 9-34. When $v = c$, m is infinite and E is infinite. If we try to accelerate a body, as its velocity approaches c its energy becomes very large. Its velocity cannot be made equal to c without giving it an infinite amount of energy.

Questions

A
1. Does the rotation of the earth on its axis give the apparatus a large enough velocity to affect the Michelson-Morley experiment?
2. What would happen to the theory of relativity if the velocity of light were infinitely large? Illustrate your answer by inserting $c = \infty$ in all the important formulas.

B

3. Show that the absence of the anticipated effect in the Michelson-Morley experiment could be explained if the velocity relative to the ether were in a vertical direction, perpendicular to both light beams? How could this possibility be eliminated?

4. Can you give a *completely conclusive* reason for believing that it is the earth that rotates rather than the stars? Try hard to criticize and demolish every argument that might be advanced.

5. Does the time between ticks of a moving light clock depend on whether it is aligned parallel or perpendicular to the direction of its velocity?

6. Does the slowing down of a moving clock depend on whether it is moving toward or away from the observer?

7. Do any of the phenomena in this chapter depend on the relative *positions* of the two observers, as distinct from their relative velocity?

C

8. As the baseball player in a car traveling at 20 mph throws his ball vertically upward relative to the car, a monkey hanging from the branch of a tree lets go and falls to the ground. Describe the motion of the ball from the point of view of a frame of reference attached to the falling monkey.

9. Terrestrial observations suggest that extragalactic nebulae are distributed uniformly in all directions. If an astronomical observatory were launched with a velocity nearly equal to the velocity of light, would the nebulae still appear to be distributed uniformly from the point of view of this observatory? Discuss this situation in relation to the precise meaning of the principle of relativity.

10. It is suggested that a light wave traveling from the south toward the north has a velocity that is less than that of a light wave traveling from the north toward the south. How would you attempt to test this?

11. A body is moving along the X-axis with a velocity that is not small compared to c. A force is applied to it parallel to the Y-axis. What happens to the component of its velocity parallel to the X-axis?

12. Is the number of atoms in exactly 1 gm of hydrogen gas at a very low pressure equal to the number of atoms in exactly 1 gm of solid hydrogen? Why or why not?

Problems

A

1. How many times does a cesium atom in an atomic clock vibrate between ticks of a light clock one meter long? (See section 2-3)

2. A spaceship is launched from earth with a velocity of $\frac{1}{2}c$. The ship then launches a small rocket in the forward direction with a velocity of $\frac{1}{2}c$ relative to itself. What is the velocity of the rocket from the point of view of an observer on the earth?

3. What is the mass of a proton when its velocity is $\frac{3}{5}c$?

4. A meter rule is moving parallel to its length. What is its apparent length when its mass is twice its rest mass?

5. What mass in grams is equivalent to an energy of 1 erg?

6. The "large calorie" used by dieticians is 1000 times the "small calorie"

defined by equation 5-38. What mass is equivalent to one large calorie? Comment.

B

7. A gale is blowing with a velocity of 4×10^3 cm/sec. How long will it take for the sound of a siren to travel to a point 5×10^4 cm away in the direction (a) from which the wind is blowing, (b) toward which the wind is blowing, (c) perpendicular to the velocity of the wind? The velocity of sound in air may be taken to be 3.3×10^4 cm/sec.

8. A wristwatch keeps perfect time on earth. If it is worn by the pilot of a spaceship leaving the earth with a constant velocity of 10^9 cm/sec, how many seconds does it appear to lose in one day from the point of view of an observer remaining behind on the earth?

9. A particle called a muon has a half-life time of 2.26×10^{-6} sec when at rest. What is its apparent half-life time when it has a velocity of $0.9c$?

10. A meter rule moves past with a velocity of 2.5×10^{10} cm/sec in the direction of its length. What is its apparent length?

11. A body has a rest mass of 5 gm. What is its mass when its velocity is 2×10^{10} cm/sec?

12. At what velocity is the mass of a particle three times its rest mass?

13. A spaceship is launched from earth with a velocity of 2×10^{10} cm/sec. A second spaceship is launched the following day and the pilot of the first ship sees it coming directly toward him with a velocity of 1.5×10^{10} cm/sec. What is the velocity of the second ship from the point of view of an observer on the earth?

14. Express the rest mass energy of an electron in ergs and in electron volts.

15. The cost of electrical energy supplied by a public utility company is ½ cent for 10^{13} ergs. At this rate, what is the cost of 1 gm of energy?

C

16. When at rest a box is a cube with a side of length 25 cm. What is its apparent volume when it moves with a velocity of 2×10^{10} cm/sec parallel to one of its sides?

17. With what constant velocity must a spaceship be launched toward Alpha Centauri in order to arrive there after one day by the pilot's reckoning?

18. The "edge of the observable universe" is believed to be about 10^{10} light years away. (Light takes 10^{10} years to travel from it to us.) If the pilot of a spaceship can expect to live another 50 years, with what constant velocity must he travel away from the earth in order to reach the edge of the observable universe before he dies?

19. According to an observer on the earth, a certain star is l light years from the earth. A spaceship is launched toward the star and takes l years to get there by the pilot's reckoning. What is the velocity of the spaceship?

20. A holds his meter rule parallel to the direction of his velocity relative to B. He mounts a pistol directly above the scratch at each end of the rule. Using a light signal originating at the mid-point of the rule, he arranges to fire the two pistols simultaneously in a direction perpendicular to the rule. B holds a wooden board parallel to A's rule very close to

the muzzles of the pistols, so that the bullet penetrates the board almost immediately after a pistol is fired. A says that the two bullets were simultaneously one meter apart and that the two bullet holes are therefore one meter apart. What result will B obtain if he measures the distance between the bullet holes? What is B's explanation of this result?

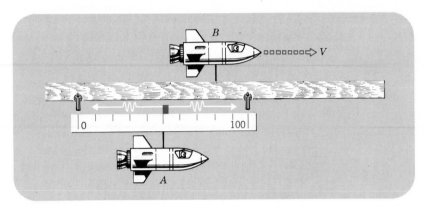

Problem 9-20.

21. A holds his meter rule parallel to the direction of his velocity relative to B. He has a paint brush sticking out at right angles to the rule directly opposite the scratch at each end. He rotates the rule about an axis through its center parallel to its length. B holds a graduated bar parallel to A's rule in such a position that the paint brushes make marks on it as they sweep past. What is the distance between the two marks according to B? How does B explain this observation?

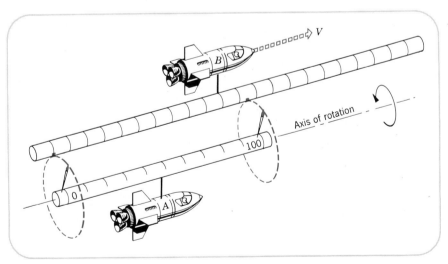

Problem 9-21.

22. A spaceship is launched from earth with a velocity of 1.5×10^{10} cm/sec. The ship then fires a small rocket back to earth with a velocity of 2.5×10^{10} cm/sec relative to itself. With what velocity does an earth observer see the rocket approaching him?

23. What is the momentum of an electron with a velocity of 1.8×10^{10} cm/sec?

24. A body is initially at rest. Half of its rest mass is destroyed and given as kinetic energy to the other half. What is the resulting velocity?

25. What is the distance between two electrons when their mutual electrostatic potential energy is equivalent to the sum of their rest masses when they are infinitely far apart?

26. A spaceship and its pilot have a total rest mass of 1000 kg. How much energy must be given to the ship during launching to enable it to reach Alpha Centauri in one day by the pilot's reckoning? (Hint: this problem can be solved without first calculating the velocity.)

27. A spaceship and its pilot have a total rest mass of 200 kg. To launch this ship into space, a mass M is annihilated and given as kinetic energy to the ship. The pilot expects to live another 50 years and in this time he hopes to reach the "edge of the universe," which is about 10^{10} light years away. How large must M be? Taking the average density of the rocks to be 3 gm/cm^3, what would be the height in miles of a hemispherical mountain with this mass?

10 *General Relativity*

10-1 *Accelerating Observers*

The special theory of relativity is restricted to non-accelerating observers and is mainly concerned with electromagnetic phenomena. The general theory of relativity considers accelerating observers and thereby becomes concerned with gravitational phenomena. The special theory is supported by an overwhelming mass of experimental observations. There are only a few observations supporting the general theory, which therefore rests on a much less secure foundation. Nevertheless, the issues raised by the general theory are very profound and well worth consideration, even though there may be some doubt whether the theory has provided the ultimate answers.

Although there is no experiment that can distinguish between an observer at rest and an observer with a constant velocity, it is very easy to decide whether an observer is accelerating. When a nonaccelerating observer looks at an isolated body, subject to no forces from other bodies, he sees it moving with a constant velocity in a straight line, which is Newton's first law of motion (section 3-1). When an accelerating observer looks at an isolated body, it appears to him to have an acceleration **a** and moreover all isolated bodies appear to have this same acceleration **a**. In reality, it is the observer himself who has an acceleration −**a** in the opposite direction. This method of detecting acceleration is very simple and unambiguous, because the nonaccelerating observer sees all isolated bodies moving along straight lines, whereas the accelerating observer sees most isolated bodies moving along curved paths.

Similar considerations may be applied to decide whether an observer is rotating. This is really the same question because, as was pointed out in section 2-7, uniform motion of a body in a circle inevitably implies an acceleration of the body toward the center of the circle. Imagine an observer on a turntable inside a closed room, watching the walls of the room turn around him. How can he decide whether he is rotating, or whether he is stationary and the room is rotating around him? Suppose that he places a smooth object on top of a desk that is so smooth that the frictional force it exerts on the object can be neglected. If the table is not rotating, when he releases the object it remains stationary in the position on the desk where he placed it (figure 10-1). However, if the turntable is rotating, the observer on the turntable sees the object move across the desk away from the axis of rotation, not in a straight line but curving to one side (figure 10-2). Another observer standing outside the turntable and not rotating would explain the situation as follows. Before the observer on the turntable releases his object it is moving in a circle. When the object is released it experiences no horizontal force because the top of the desk is frictionless. In accordance with Newton's first law of motion, the object therefore continues to move in a straight line tangential to the circle (figure 10-2b). Quite apart from the explanation, the test is a simple practical matter. The observer places a smooth object on a smooth desk. If it does not move, he is not rotating. If it moves across the desk along a curved path, he is rotating.

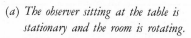

(a) *The observer sitting at the table is
stationary and the room is rotating.*

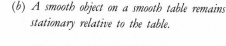

(b) *A smooth object on a smooth table remains
stationary relative to the table.*

Figure 10-1 *The observer is not rotating.*

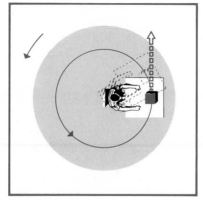

(a) The room is stationary and the observer on the turntable is rotating.

(b) When released, the smooth object on the smooth table moves in a straight line tangential to its original circular motion.

(c) Apparent motion of the smooth object in the observer's frame of of reference.

Figure 10-2 The observer is rotating.

Another method of deciding was pointed out by Newton. If a vessel of water is rapidly rotated the water is thrown outward and the surface becomes curved, with the level near the wall higher than the level in the center (figure 10-3). The observer on the turntable should therefore look at the surface of water in a vessel. If the surface is flat he is not rotating. If the surface is curved he is rotating. This phenomenon is related to a very simple, common sense attitude to the question of how the observer decides whether he is rotating. If he is rotating he becomes dizzy! Rotation causes the fluid in the ducts of the inner ear to be thrown outward, and the pressure of this fluid is responsible for the sensation of dizziness.

The earth is rotating on its axis and the stars therefore appear to move around the earth once every day. How can we know that it is not really the

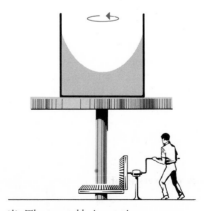

(a) The turntable is not rotating.

(b) The turntable is rotating.

Figure 10-3 When the vessel is rotating the surface of the water inside it is curved.

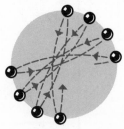

(a) *The pendulum oscillates in the plane of the paper and the earth rotates underneath it.*

(b) *As seen by an observer fixed to the earth, the plane of oscillation of the pendulum slowly rotates. (This is the viewpoint of an observer standing at the north pole looking up at the pendulum.)*

Figure 10-4 *The Foucault pendulum.*

earth that is stationary and the stars that move in a body around the earth? The rotation is so slow that the resulting curvature of the surface of water in a vessel is too small to detect. Neither can a desk top be made smooth enough to apply the test of figure 10-2. However, a device called a Foucault pendulum can be made to reveal the earth's rotation. It is a pendulum suspended in such a way that, as the point of suspension rotates with the earth, the couple exerted on the pendulum is negligibly small. Since there is nothing to twist it, the pendulum does not rotate, but continues to oscillate in a fixed direction. However, the earth rotates underneath the pendulum and the direction of oscillation of the pendulum is not fixed relative to the earth (figure 10-4). An observer standing on the earth sees the direction of oscillation of the pendulum slowly rotating, returning to its original position after approximately 24 hours. (The time taken for the earth to rotate once is not exactly 24 hours. There are complications due to the movement of the earth around the sun. Also, we are assuming that the pendulum is at the north pole. At any other latitude, the situation is more complicated and the time for one rotation is less than 24 hours, becoming zero at the equator.)

These are very real effects that we have just been discussing and the observations cannot be denied. Special relativity is based on the idea that no meaning can be attached to the concept of "motion of a body relative to space," but only to the relative motion of two observers or the relative motion of two objects. Do the considerations of the present section imply that we can nevertheless attach physical significance to the concepts of "acceleration relative to space" or "rotation relative to space"?

10-2 *Mach's Principle and Relativity Reinstated*

Within the limits of accuracy imposed by experimental difficulties, the direction of oscillation of a Foucault pendulum at the north pole is found to remain fixed *relative to the stars.* As an observer on earth watches the pendulum, he sees it rotating in exactly the same way that the stars in the sky above him are rotating. The "space" in which the pendulum does not rotate seems to be identical with the "astronomical rest frame" of section 9-1. When we devise a method of detecting rotation, what we are detecting is rotation *relative to the whole universe.*

Similarly, an observer moving with constant velocity is found to have no acceleration relative to the whole universe. An accelerating observer is accelerating relative to the universe, so that the stars on one side of him appear to be accelerating toward him, whereas the stars on the other side appear to be accelerating away from him. The significance of Newton's first law of motion is that an isolated body has a constant velocity relative to the universe.

The Austrian physicist and philosopher Ernst Mach interpreted all this as meaning that we cannot detect acceleration and rotation relative to "absolute space," but only relative to the matter in the universe. He suggested that there is a hitherto unrecognized interaction between a moving body and all the other matter in the universe, the most distant matter being the most important. This interaction depends on the relative acceleration of the body and the distant matter and is zero when the body is moving with a constant velocity relative to the distant matter.

The isolated body of Newton's first law is not really being left to itself. It experiences no strong force from nearby matter, but it is still subject to a Machian force from all the matter in the universe and this force constrains it to move with constant velocity (zero acceleration) relative to the universe.

Suppose that the body does experience a force F_s from nearby matter, such as a compressed spring (figure 10-5a). Then it acquires an acceleration a relative to the universe. Newton's second law tells us that, if m is the mass of the body,

$$F_s = ma \tag{10-1}$$

Can we adopt the underlying philosophy of relativity and describe the situation from the point of view of an observer sitting on the body and moving with it (figure 10-5b)? From the point of view of this observer, the body has no acceleration and so, if he can apply Newton's second law, he must conclude that there is no resultant force acting on the body. What cancels the force F_s exerted by the compressed spring?

Mach's answer is that the observer moving with the body sees the distant matter in the universe accelerating relative to himself with an average acceleration $-a$. This accelerating matter exerts a Machian force F_M on the body, which is proportioned to the acceleration and to a constant m characteristic of the body

$$F_M = -ma \tag{10-2}$$

Since, for this observer, the net force must be zero

$$F_s + F_M = 0 \tag{10-3}$$

$$\text{or } F_s - ma = 0 \tag{10-4}$$

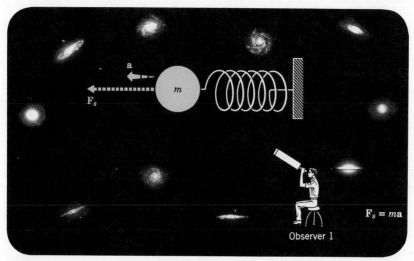

(*a*) *Point of view of an observer at rest relative to the universe.*

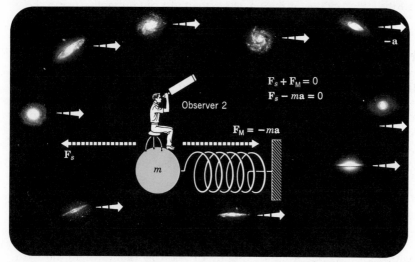

(*b*) *Point of view of an observer moving with m.*

Figure 10-5 *Mach's Principle and Newton's Second Law of Motion.*

This is identical with the form of Newton's second law deduced by the first observer (equation 10-1).

Notice that, if Mach's ideas are correct, we have the first real hint of the true physical significance of the inertial mass m. It has something to do with the Machian force on a body when it accelerates relative to the matter in the universe. Moreover, the remarkable fact that inertial mass is equal to gravitational mass (section 3-3) provides a strong hint that the Machian force is gravitational in nature.

The same ideas can be applied to rotation. The Foucault pendulum experiences forces from distant matter that constrain it to oscillate in a direction fixed relative to this matter. In the case of the rotating bucket, an observer rotating with the bucket sees the universe rotating around him, and the rotating distant matter exerts forces on the water that pile it up against the sides of the bucket (figure 10-6).

General
Relativity

214

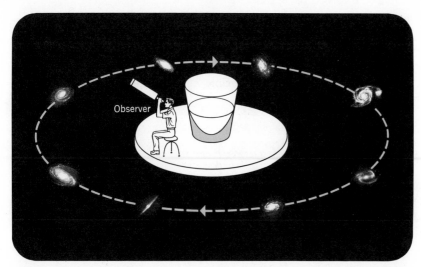

Figure 10-6 *An observer rotating with the bucket of water sees the universe rotating around him. The rotating universe exerts Machian forces on the water and piles it up against the side of the bucket.*

We shall not place Mach's principle in a colored frame because there is still much controversy amongst physicists about its validity. At the time of this writing, in the year 1971, there is no universally accepted, precise mathematical statement of Mach's principle. Einstein was very much influenced by Mach's ideas when he formulated the general theory of relativity, but the relationship between general relativity and Mach's principle is still somewhat obscure.

However, Mach's principle does show how we can still escape from the concept of absolute space and consider only the relative motion of different material objects. It encouraged Einstein to extend his principle of relativity to accelerating observers. One of the basic ideas underlying general relativity is that, when the laws of physics are properly understood, they have the same form for accelerating observers as for nonaccelerating observers.

> *The basic principle of general relativity*
>
> The laws of physics can be formulated in such a way that they are valid for any observer, however complicated his motion.

10-2 Mach's Principle and Relativity Reinstated

The laws of physics might seem to be different for rotating and nonrotating observers. When the rotating observer releases a body, it moves in a curved path. When the nonrotating observer releases a body, it moves in a straight line. However, the laws of physics should include the fact that the motion of a body is influenced by the motion of distant matter. The curved path of the body released by a rotating observer is not a consequence of rotation relative to "absolute space," but rather a consequence of the rotation of the stars relative to him. If the whole universe could be made to rotate with the observer, then a body released by him would move in a straight line. In fact, the concept of rotation of the whole universe relative to "absolute space" is meaningless, since no experiment could be performed to detect it. Only rotation relative to the matter in the universe is detectable.

Einstein's method of expressing the laws of physics in a form valid for all moving observers involves mathematical techniques too advanced to be described in detail here. Instead, we shall describe qualitatively two important aspects of his approach: the **principle of equivalence** and the need to use **non-Euclidean geometry.**

10-3 *The Principle of Equivalence*

Imagine an isolated observer inside a rocket in outer space, with no nearby matter to exert strong forces on him. Suppose that he cannot see outside his rocket and cannot directly measure his acceleration relative to the stars. He wishes to perform experiments inside his rocket to decide whether he is ac-

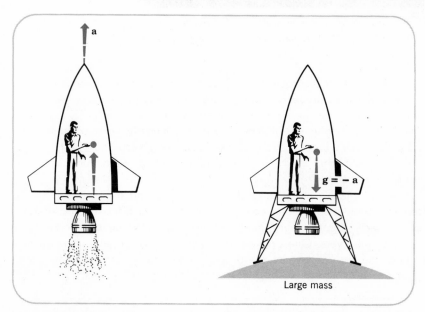

(a) *An isolated rocket with an acceleration* **a.**

(b) *A rocket stationary in a gravitational field.*

Figure 10-7 *Acceleration in empty space is equivalent to being stationary in a gravitational field.*

celerating. He holds up an object and releases it. Once released the object has no acceleration. If, however, the observer is in fact accelerating because his rocket engines are operating (figure 10-7a), the rocket accelerates relative to the released object, which therefore appears to the inside observer to fall in a direction opposite to the actual acceleration of the rocket. If the rocket is not accelerating, both the observer and the released object continue to move with the same velocity and so the object remains suspended in mid-air in the same position relative to the rocket where it was released.

Suppose then that the observer releases his object and observes that it falls to the floor with an acceleration $-a$. Can he assume that the real situation is that the rocket has an acceleration a? Only if he is absolutely cer-

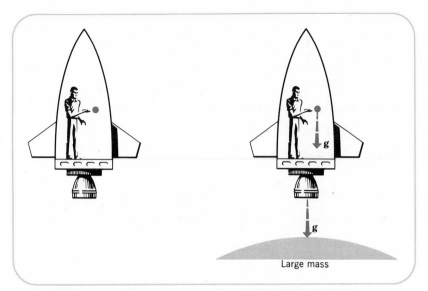

(a) *An isolated rocket with no acceleration.*

(b) *A rocket falling freely in a gravitational field.*

Figure 10-8 *Unaccelerated motion in empty space is equivalent to free fall in a gravitational field.*

tain that he is far away from the gravitational attraction of nearby bodies. If he is totally ignorant of what is going on outside his rocket, the true situation (figure 10-7b) might be that he is stationary in a gravitational field in which the object falls to the floor with a gravitational acceleration $\mathbf{g} = -\mathbf{a}$!

There is a similar dilemma if the observer releases the object and it remains suspended in mid-air. This might be because the rocket is traveling through empty space with a constant velocity, its engines inoperative (figure 10-8a). But it might equally well be that the rocket is falling freely in a gravitational field (figure 10-8b). In which case, when the object is released it continues to fall freely in exactly the same way as the rocket and remains in the same position relative to the rocket.

It is impossible to think of any experiment on the mechanics of moving bodies that will decide whether the observer is accelerating in empty space or not accelerating in the presence of a gravitational field. Einstein generalized this conclusion to any kind of experiment whatsoever that might be performed inside the rocket, including experiments in electromagnetism or experiments involving light. He thus obtained the **Principle of Equivalence,** which is one of the foundation stones of general relativity. It says in effect that acceleration in empty space is completely equivalent to unaccelerated motion in a gravitational field.

10-3 The Principle of Equivalence

Einstein's principle of equivalence

No experiment of any kind whatsoever performed inside a small, nonrotating, closed laboratory can distinguish between the effects of a gravitational field and the effects of an acceleration of the laboratory relative to the universe.

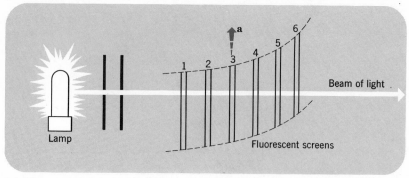

(a) The passage of the beam of light across the fluorescent screens, as seen by an observer who is not accelerating.

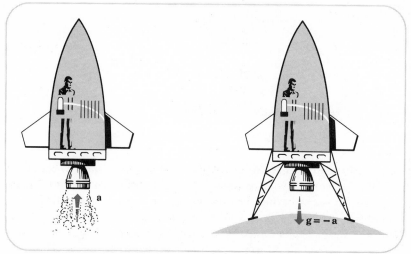

(b) The beam of light, as seen by an observer in an accelerating rocket.

(c) The rocket is stationary in a gravitational field.

Figure 10-9 Bending of a beam of light in (b) an accelerating rocket, (c) a gravitational field.

The principle of equivalence can be used to derive a very important result concerning the bending of a beam of light in a gravitational field. Imagine a laboratory inside an isolated rocket that has an acceleration **a**. The observer in the rocket produces a beam of light traveling in a direction perpendicular to **a** and observes its path by allowing it to pass through several semi-transparent fluorescent screens, on each of which it produces a fluorescent spot (figure 10-9). Another observer with a constant velocity, relative to whom the screens are accelerating, describes the experiment in the following way (figure 10-9a). The beam of light moves along a straight line. However, while the beam is traveling from screen 1 to screen 2, screen 2 acquires an extra upward velocity and moves upward relative to the beam, which therefore strikes screen 2 at a point lower down than the point at which it struck screen 1. As the beam passes across the screens their upward velocity relative to the beam becomes progressively greater, and so the vertical displacement between adjacent spots becomes increasingly larger.

The observer in the rocket, relative to whom the screens are stationary, therefore sees the beam of light bending downward as in figure 10-9b.

This is what happens when the rocket is not in a gravitational field, but has an acceleration **a.** According to the principle of equivalence, the phenomenon is exactly the same if the rocket has no acceleration, but is in the presence of a gravitational field producing a gravitational acceleration **g** = −**a** (figure 10-9c).

> **A beam of light is bent by a gravitational field.**

The light "falls downward" in the field, in a similar way to a material body. There is another way of looking at this. Suppose a flash of light is sent out, carrying a total energy E. This energy is equivalent to a mass $m = E/c^2$. It is not unreasonable to expect that this mass experiences a gravitational force in the same way as any other kind of mass.

The bending of light in a gravitational field has been observed. During an eclipse of the sun by the moon, it is possible to see stars very nearly in line with the edge of the sun. Light from these stars has passed very near the sun and has been bent in its gravitational field. The stars therefore appear to be displaced from their normal positions in a direction away from the sun (figure 10-10). The observed magnitude of this displacement agrees with the value predicted by the general theory within the observational accuracy of about 10 percent.

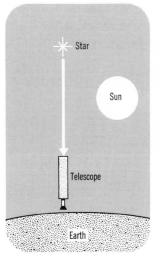

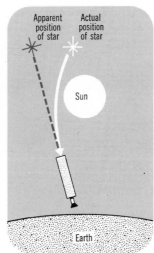

(a) *The light from the star passes through a weak region of the sun's gravitational field and is not appreciably bent.*

(b) *When the light from the star passes near the sun the bending is detectable.*

Figure 10-10 *Apparent displacement of a star very nearly in line with the edge of the sun.*

(a) The ray of light is bent by the
    gravitational attraction of the cow.

(b) The cow moves away and the posts no
    longer appear to be in line.

Figure 10-11 *Lining up the posts of a fence by sight. (In practice the bending
would be negligibly small, but it has been exaggerated to demonstrate the effect.)*

10-4 Non-Euclidean Geometry

Having shown that light is bent in a gravitational field and no longer travels
along a straight line, let us pause and consider what is meant by a straight
line. A good practical definition of a straight line, which might have been
accepted without question before the general theory, is that it is the path
followed by a beam of light in a vacuum! A farmer erecting a fence can
check that it is straight by sighting along the posts to see that they are all
"in line." Suppose, however, that the light is appreciably bent by the gravitational attraction of a nearby cow (figure 10-11a). Clearly the method
breaks down. When the cow walks away the fence no longer appears
straight (figure 10-11b).

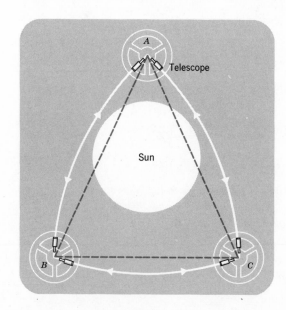

Figure 10-12 *An observational
proof that the sum of the three
angles of a triangle is greater than
180°.*

Unfortunately, astronomical observations, such as observations on the solar system, are made almost exclusively with telescopes and rely heavily on the behavior of light. Suppose that three astronomical observatories, A, B, and C, have been set up in space and are employing their rocket engines to hold themselves stationary relative to the sun (figure 10-12). Each observatory sights its telescope on the other two observatories in succession and measures the angle through which the telescope has to be swung. Because of the bending of light in the sun's gravitational field, it should be clear from the figure that the sum of the three angles measured by the three observatories is greater than 180°. The astronomers must now make a difficult choice. They can realize that light is bent in a gravitational field and try to think of a better definition of a straight line, which is not easy. Alternatively, they can decide, as a practical matter, to *define* a straight line as the path followed by a light beam. In which case they have proved from observations that the sum of the three angles of a triangle is greater than 180°, in flagrant contradiction to the precepts of Euclidean geometry (Mathematical Appendix B), which were presented to them as inviolable logic during the years of their early education. They must therefore seriously ask whether Euclidean geometry is necessary, or whether it is not possible to use other systems of geometry. This latter course was the one followed by Einstein.

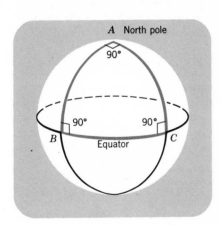

Figure 10-13 On the surface of the earth, it is possible to draw a "triangle" whose three angles add up to 270°.

10-4
Non-Euclidean Geometry

A somewhat similar situation is encountered on the surface of the earth. Imagine two pipe lines setting out from the north pole at right angles to one another and proceeding "straight" to the south pole. Straight, in this case, means following a line of longitude. A third pipe line follows the equator (figure 10-13). The first two lines cross the third line at right angles. A surveyor, who thinks that the earth is flat, measures the three angles of the triangle ABC and discovers that their sum is 270°. If the surveyor knows no way of drawing a line except on the surface of the earth, he cannot draw a "really straight" line. The best he can do is a line of longitude. The surveyor must invent a non-Euclidean geometry. Such a geometry has some peculiar features. If we confine ourselves to lines that must be drawn on the surface of the earth, the equator is a circle with its "center" at the north pole, N (figure 10-14). The "radius" of this circle is NP. If the distance around the equator is L, the length of the "radius" NP is $\frac{1}{4}L$. The ratio of the perimeter of the circle to its "radius" is not 2π but 4!

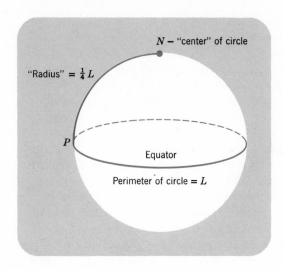

N – "center" of circle

"Radius" = $\frac{1}{4}L$

P

Equator

Perimeter of circle = L

Figure 10-14 *In the non-Euclidean geometry of the surface of a sphere, the ratio of the perimeter of the circle to its radius is not 2π but 4.*

The general theory assumes that the four dimensions of space and time show the same sort of peculiar behavior as the two dimensions on the surface of the earth. Space-time is curved! This curvature is produced by the gravitational effect of nearby matter. The curvature increases as the mass of the nearby matter is increased, and also increases if this matter is brought nearer.

10-5 *The Gravitational Red Shift*

A gravitational field distorts time as well as space. A stationary clock runs slow in a gravitational field. Imagine a clock at rest at a distance R from a mass M (figure 10-15a). It ticks more slowly than a similar clock at rest at a great distance from M.

Slowing down of a stationary clock in a gravitational field

$$\frac{\text{Time between ticks of a clock at a distance } R \text{ from } M}{\text{Time between ticks of a clock at an infinite distance from } M}$$

$$= \frac{1}{\sqrt{1 - \dfrac{2GM}{c^2R}}} \qquad (10\text{-}5)$$

Notice an important difference between the present situation and the behavior of *moving* clocks in the special theory (section 9-5 and equation 9-11). When two observers, A and B, are moving relative to one another, A decides that B's clock is slow, whereas B decides that A's clock is slow, and their disagreement cannot be resolved. In the present case everybody can agree that the clock in the gravitational field is slow compared with the clock outside the field. It is possible to agree that the clock outside the field is the "standard clock" and to make a suitable correction to the

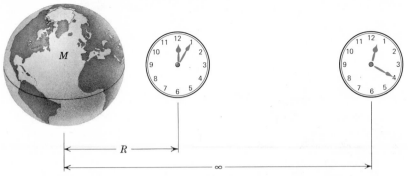

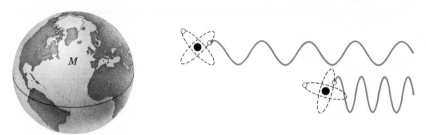

(a) *A stationary clock in a gravitational field runs slow.*

(b) *An atom in a gravitational field emits light of lower frequency and
longer wavelength.*

Figure 10-15 *The gravitational red shift.*

clock in the field to compensate for the effect of the field. The moving observers A and B cannot do this.

An atom is a kind of clock and it runs slow in a gravitational field. It
emits light of longer period, smaller frequency, and longer wavelength as
compared with an atom outside the field (figure 10-15*b*). When the wavelength of visible light emitted by the atom is lengthened, the light moves toward the red end of the visible spectrum.The effect is therefore called the
gravitational red shift. This red shift has been observed for light emitted by
atoms on the surface of the sun. It is particularly pronounced for white
dwarf stars. A white dwarf star is very small and very dense, so the quantity
M/R appearing in equation 10-5 is exceptionally large.

The gravitational red shift has also been observed in a terrestrial experiment making use of the earth's gravitational field. This experiment employed
γ-rays emitted by a nucleus and took advantage of a subtle effect, called
the Mossbäuer effect, which makes it possible to compare frequencies of
these γ-rays with extraordinary accuracy. A comparison was made between
the frequencies of the γ-rays emitted by two nuclei, one of which was on
the surface of the earth, while the other was at a height h above it. In this
case equation 10-5 takes a simpler form.

*10-5 The
Gravitational
Red Shift*

$$\frac{\text{Time between ticks of the lower clock}}{\text{Time between ticks of the upper clock}} = 1 + \frac{gh}{c^2} \qquad (10\text{-}6)$$

The nucleus at the lower height was found to emit γ-rays of smaller fre-

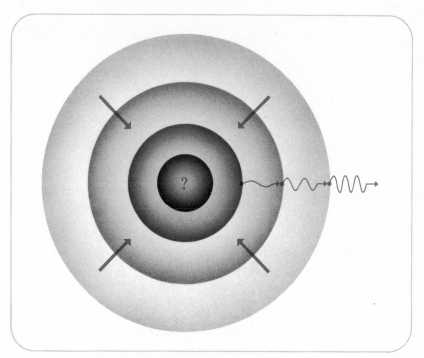

Figure 10-16 *Possible formation of a "black hole." As the star collapses the wavelength of the light emitted by an atom on its surface becomes longer and longer until eventually no light at all can be emitted.*

quency, and the magnitude of the effect was in good agreement with the theory.

Theoretical astrophysicists suspect that some massive stars cannot stop collapsing under their own weight. As the star shrinks, its radius R_s decreases and the factor $2GM/c^2R_s$ of equation 10-5 increases, eventually approaching the value unity. The right-hand side of equation 10-5 then approaches infinity as the denominator approaches zero. The period of the light emitted by an atom on the surface of the star becomes very long. The atoms become very sluggish in the strong gravitational field of the very dense, compact, small star. Therefore, as the collapse proceeds the star becomes redder and redder and eventually disappears from sight altogether when the surface atoms emit only infinitely long wavelengths (figure 10-16). Such invisible objects are called "collapsed stars" or "black holes." It is possible that the universe contains a large number of them.

10-6 *The Radar Test*

A recent test of general relativity applies the radar technique for tracking objects to the nearby planets and, in particular, to the planet Mercury whose orbit is nearest to the sun. The principle of radar is to send out a short pulse of radio waves which in the present case is reflected from the surface of Mercury and received back again after an appropriate time delay. The time lapse between transmission and reception determines the distance to the planet at the instant of reflection. The direction from which the echo returns provides the other information needed to locate the position of the planet completely at that instant. In this way the planet can be tracked over a period of several months and its orbit determined precisely.

However, the usual simple method of interpreting the data assumes that the radar pulse travels along a straight line with a constant velocity which is the same everywhere. According to the general theory of relativity, neither of these assumptions is valid in the strong gravitational field of the sun. We have already discussed the bending of a ray of light by the sun's gravitational field. There is another effect, more important in the present instance, which makes a light signal appear to travel more slowly the nearer it is to the sun.

The situation can be understood in principle with the help of an analogy in which the three-dimensional space around the sun is represented by a two-dimensional space on an imaginary surface. In the absence of the sun, this two-dimensional space would be a flat plain. The strong gravitational field of the sun curves the space in its vicinity and so the sun can be represented by a steep mountain in the center of the plain (figure 10-17). A radar pulse traveling across the plain far from the mountain is affected very little by it and has a velocity very close to the accepted velocity of light. However, a radar pulse passing near the sun is forced to climb the mountain and this slows it down and delays its arrival time.

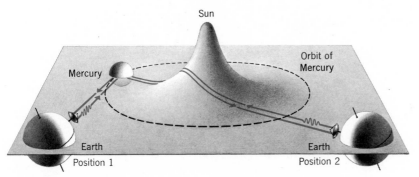

Figure 10-17 *The curved space near the sun is represented by a mountain in the center of a plain. A radar pulse passing near the sun is delayed by the need to climb the mountain.*

In terms of this analogy, the procedure in the radar experiment may be explained as follows. The orbit of Mercury is first carefully determined from a large number of observations confined to those occasions when the radar pulse traveling between the earth and Mercury does not pass near the sun (the earth is in position 1 of figure 10-17). For these observations, the effect of the sun's gravitational field is negligible because the pulses are always traveling on the flat plain far from the foot of the mountain. The final determination of the orbit is very accurate and has avoided the complications introduced by the curvature of space near the sun.

The next step is to observe the radar echoes when Mercury is just about to disappear behind the sun. The radar pulse is then obliged to pass near the sun and climb the mountain (the earth is in position 2 of figure

*10-6 The
Radar Test*

10-17). From the previous determination of the orbit, it is possible to predict accurately how long the radar pulse would take to travel to Mercury and back again if it traveled with the usual velocity over a flat plain. In fact, the pulse is observed to take a longer time than this because of the extra delay produced as it travels through the curved space near the sun. Moreover, the magnitude of the additional delay agrees with the value predicted by Einstein's theory to within the experimental accuracy.

10-7 *The Advance of the Perihelion of Mercury*

One other important observational test of the theory remains to be described. It is also concerned with the orbit of the planet Mercury, but it is a very subtle effect depending on some of the finer details of the theory. For this very reason, though, it provides a very sensitive test of the exact form of the theory. The orbit of Mercury is an ellipse, which might be described rather crudely as a slightly elongated circle. The sun is not at the center of

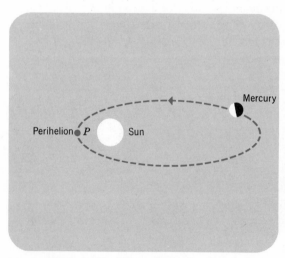

(a) *An elliptical orbit. This would be the orbit of Mercury in classical mechanics in the absence of perturbations by the other planets. (The elongation has been exaggerated. The actual orbit is almost a circle.)*

(b) *The ellipse is observed to be slowly rotating.*

Figure 10-18 *The advance of the perihelion of Mercury.*

this ellipse, but is displaced slightly to one side (figure 10-18a). The point P of the orbit at which the planet is nearest to the sun is called the perihelion. The perihelion is observed to be slowly revolving around the sun, and the orbit of Mercury is therefore as shown in figure 10-18b which, however, exaggerates the motion of the perihelion. This effect is called the advance of the perihelion of Mercury.

The advance of the perihelion is mainly a consequence of the gravitational attraction of the other planets in the solar system. However, when the influence of all the other planets was calculated with great accuracy, the result accounted for only 99.2 percent of the observed motion of the perihelion. There remained a small "residual" advance of the perihelion, corresponding to one complete revolution of the perihelion in about 3 million years, which defied all attempts to explain it using Newtonian mechanics. It was not possible, for example, to explain it by postulating the existence of an undiscovered planet. The general theory of relativity is able to give a completely satisfactory account of this residual advance and to derive its magnitude.

However, at the time of this writing there is still some controversy associated with the perihelion of Mercury. It has been suggested that a small part of the advance of the perihelion is a consequence of the fact that the rotating sun bulges out at its equator. If this effect were present, it would introduce a small distortion of the sun's gravitational field sufficient to influence the rate of advance of the perihelion and to destroy the good agreement between the observations and Einstein's theory.

10-8 *The Twin Paradox*

The following prediction of the theory of relativity has caused much controversial discussion among experts and nonexperts. A and B are identical twins. When they are 20 years old, B starts out on a journey to a distant star and back with a certain velocity near the speed of light, while A stays at home on earth (figure 10-19). B's clocks tell him that the journey takes 10 years, and on his return he feels and acts like a man of 30. However, upon landing he discovers that the earth clocks have advanced 60 years and his twin brother A is an old man of 80.

This phenomenon is clearly related to the slowing down of a moving clock. The earth-twin A, observing the space-twin B proceeding on his outward journey with a steady velocity, sees that B's clocks appear to be slow. In particular, the biochemical processes which cause B to grow older appear to A to be proceeding more slowly. On the other hand, if the space-twin B looks back at what is happening to the earth-twin A, he concludes that A's clocks are slow and that it is A who is aging less rapidly. According to the special theory, there is no reason to prefer A's point of view rather than B's. Yet we have stated that, when B returns to earth, there is a real difference in age between them, even when they are standing side by side stationary relative to one another.

This paradox might be expressed in the following way. The situation was originally described from the point of view of a frame of reference fixed relative to the earth (figure 10-19a). In this frame of reference B goes on a journey into space with a high velocity and on his return has aged less than A. However, since all motion is relative, would it not be equally valid to use a frame of reference fixed relative to B? In this frame of reference the earth and the earth-twin A make a journey into space with a high velocity (figure 10-19b). If the original postulate were true, then surely it would be A who would be younger than B when they met again.

The resolution of the paradox is that there is an important difference between the frames of reference of A and B. A's frame of reference is almost unaccelerated. The rotation of the earth and its motion around the sun produce accelerations that are too small to matter in this instance. B's frame of reference is an accelerating frame. While he is taking off, turning round,

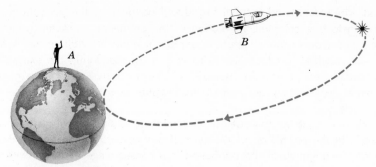

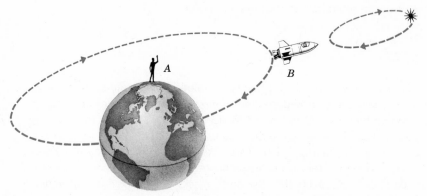

(a) *Twin B goes on a journey to a star, leaving twin A behind on earth. When he returns he finds he has aged less than A.*

(b) *From the point of view of a frame of reference fixed relative to B, the star comes on a journey to meet B. Meanwhile the earth and the earth-twin A go on a journey out into space. When the earth returns to B, will A have aged less than B?*

Figure 10-19 *The twin paradox.*

or landing, B's frame has a very large acceleration. According to the principle of equivalence an acceleration is equivalent to a gravitational field, and in a gravitational field clocks are slow.

The problem cannot therefore be satisfactorily discussed in terms of the special theory. It is necessary to apply the general theory. It can then be proved that the combined effects of B's velocities and accelerations are that, when he lands back on earth, his clocks have indeed registered a shorter period of time than A's clocks.

As a means of extending the span of human life this method has its limitations. Although B has aged 10 years compared with A's 60 years, he has eaten correspondingly fewer meals, performed fewer actions and had fewer thoughts. Nevertheless, it would be amusing to go on a year's vacation in space and return to discover what progress the earth will make over the next thousand years. (The time discrepancy can be increased by making the journey at a speed nearer to the speed of light.) Unfortunately, the necessary velocity cannot be achieved without the expenditure of an enormous amount of energy. The fuel required to propel one human being into space with this velocity would be sufficient to meet all the earth's power requirements for several centuries.

Questions

A 1. Imagine a freely falling elevator 1000 miles, or 1.61×10^8 cm, high! Why is it not possible to apply the principle of equivalence to the behavior of moving bodies inside this elevator? Use this example to demonstrate that the principle must be applied only to experiments performed inside a laboratory that is either very small or is in a uniform gravitational field.

B 2. Does an electric charge falling freely in a gravitational field radiate an electromagnetic wave? What reason is there for believing that it might? Why does the principle of equivalence suggest that it might not? (In the author's opinion, no completely satisfactory answer has yet been given to this question.)

C 3. A light clock similar to the one shown in figure 9-7 is placed in the sun's gravitational field and is aligned along a radius. Its two mirrors are at opposite ends of a meter rule. An observer at a great distance from the sun sends in a light pulse along a radius so that it is reflected in turn from the two mirrors of the light clock. What should the distant observer expect to be the time interval between the return of the two echoes? How might he interpret this experiment in terms of the effect of a gravitational field on the length of a meter rule and the velocity of light?

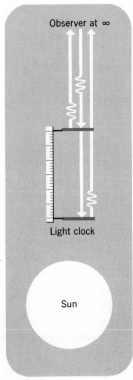

Observer at ∞

Light clock

Sun

Question 10-3.

4. Twin A does not stay at home, but both twins go on identical journeys in exactly opposite directions. How do the twins compare in age when they return? During the journey, what is each twin's opinion of the rate of the other twin's clock and his rate of aging?

Problems

A
1. In an experiment on the gravitational red shift using the Mossbäuer effect, one nucleus is at the top of the Empire State Building and the other is at street level. The height of the building is 3.17×10^4 cm. The γ-rays emitted by the lower nucleus have a frequency of 3.47×10^{18} cycles/sec. What is the difference in frequency between the γ-rays emitted by the two nuclei?

B
2. If the expectation of life at sea level is 70 years, how many seconds different is the expectation of life at the top of Mount Everest, which is 8.8×10^5 cm above sea level? Is it shorter or longer? Assume that the only relevant factor is the variation of the rate of a clock with gravitational potential.

3. The radius of a proton is about 0.8×10^{-13} cm. Compare the rate of a "clock" at the surface of a proton with a similar clock at infinity.

C
4. The dwarf companion of the star Sirius has a mass of 1.7×10^{33} gm. The lines in its spectrum show a gravitational red shift of 1 part in 10^4. Calculate its radius.

5. A satellite is placed in a stable circular orbit of radius R about a star of mass M. Compare the magnitudes of the two effects which slow down a clock inside the satellite. These two effects are (a) the effect of the velocity according to the special theory and (b) the effect of the gravitational potential according to the general theory.

6. Write down the form taken by equation 10-5 in the case when the clock is on the surface of a uniform spherical mass of radius r and density ρ. Show that it is possible for the clock to take an infinite time between ticks and find the condition for this. Find the necessary radius as a function of the density ρ. Calculate the radius (a) for $\rho = 5.5$ gm/cm^3, which is the density of the earth, (b) for $\rho = 2.3 \times 10^{14}$ gm/cm^3, which corresponds to all the fundamental particles present being in direct contact with their neighbors. Compare the answers with the radius of the earth and the radius of the sun.

7. The annual output of manufactured power by the whole of mankind is about 10^{27} erg. Check the statement made in section 10-8 about the power needed to send a man on a year's vacation which, by earth reckoning, would last for one thousand years of history. Take the mass of the man and his space capsule to be 500 kg.

8. A beam of light is projected horizontally near the surface of the earth. How far has it fallen after traveling a horizontal distance of 10 miles? First use Newtonian mechanics, treating the horizontal and vertical motions as entirely independent of one another. The relativistic answer is greater by a factor of 2.

11 $Quantum$ $Mechanics$

11-1 $Particles$ and $Waves$

At the beginning of this book, when we were introducing the concepts of classical mechanics, it was suggested that it might be possible to describe the universe in the following neat and tidy way. The universe consists of small, discrete particles moving along well-defined paths through empty space. The position of each particle must be known at each instant of time. However, the paths are not arbitrary, but are determined by the accelerations that the particles produce in one another. In principle, if the positions and velocities of all the particles are known at one single instant of time, it should be possible to calculate their positions and velocities at all other past and future instants of time. Before this can be done, though, it is necessary to know the laws governing the acceleration produced in one particle by another particle. One of the major objectives of classical physics was the discovery of these laws.

When we came to electromagnetic interactions we discovered that it is convenient to pay attention to empty space and to concentrate on the electric field **E** and the magnetic field **B**. The basic equations of electromagnetism, Maxwell's equations, are statements about the behavior of **E** and **B** in empty space. Moreover, electromagnetic forces are not transmitted instantaneously from one charged particle to another, but travel through the field with a finite velocity c. The consequent time delay before one part of the universe can influence another part is essential to the theory of relativity.

Maxwell's equations imply the existence of electromagnetic waves and it is clear that light is such an electromagnetic wave. Light, however, is something of which we have very direct experience. Our most powerful way of observing the universe is to look around us and make use of the light waves falling on our eyes. On a sunny day we have a feeling of warmth caused by the energy that electromagnetic radiation transfers from the sun to our bodies. It might well be argued that we are more intimately aware of the existence of electromagnetic waves than of the existence of electrons and protons.

To appreciate this point, suppose that a star explodes in the Andromeda nebula, producing what is called a supernova. A large amount of energy is released in a short time and a burst of electromagnetic radiation travels outward from the star. It takes about two million years to reach the earth. Soon after it starts out the explosion dies down and everything is quiet in the Andromeda nebula. Meanwhile, the earth is totally unaware of what has happened. Somewhere in the intervening space the energy is traveling in the form of an electromagnetic wave. It is difficult to avoid the impression that this electromagnetic wave is "something" that has just as much reality as an electron or a proton. It seems that the original picture of particles moving through empty space should be modified by the addition of electromagnetic waves traveling between the particles.

If we are to imagine the universe as consisting of both material particles and electromagnetic waves, let us compare and contrast their properties. Moving particles have energy and momentum, but so do electromagnetic waves. Particles have mass, but, according to Einstein's principle of the interchangeability of mass and energy, light also has mass. The path of a moving particle is bent in a gravitational field, but so is the path of a beam of light. In one important respect, however, particles and waves are very different. A particle is a highly localized entity, which is situated at a particular point in space at a particular instant. A wave is a diffuse entity which is spread over a whole region of space.

It took most of the nineteenth century to prove convincingly that matter is composed of atoms and that light is an electromagnetic wave. It was not until the very end of the century, in the year 1897, that J. J. Thomson discovered electrons in the "cathode rays" traveling from the negative to the positive electrode of a discharge tube (figure 6-12). In the early years of the twentieth century a series of experiments established without a shadow of doubt that these rays are really a stream of particles, that each particle is negatively charged, has a mass much smaller than an atom, and is an essential constituent of all atoms.

Then, almost before the experiments had provided the final vindication for the concept of charged particles communicating with one another via electromagnetic waves, the whole beautiful scheme fell apart. Certain experiments were performed in which light behaved, not like a wave, but like a stream of particles! It began to look as though the universe could again be considered to be a collection of particles. However, a few years later, other experiments indicated that electrons could sometimes behave like waves! Both light and material particles sometimes exhibit the characteristics of waves and at other times exhibit the characteristics of particles. This is called the **dual nature of light** and the **dual nature of matter.** These developments had the virtue of putting material particles and electromagnetic radiation on the same footing, but they introduced a new dilemma. How is it possible for any entity to behave sometimes like a localized particle and at other times like a diffuse wave?

In the present chapter we shall consider these new developments in more detail and we shall show how they led to **quantum mechanics** and to a complete revolution in our ideas of the nature of the physical universe. We shall present the decisive evidence which convinced physicists of the particle nature of light, and we shall show how it was immediately followed by decisive evidence for the wave nature of matter. Then we shall try to resolve the dilemma presented by this dual nature of light and matter and in doing so we shall introduce probability and uncertainty into the basic laws of physics!

11-2 The Birth of a Revolution

The first step in the direction of quantum mechanics was taken by Max Planck in the year 1900. He was trying to explain how the energy of the light emitted by a hot body is distributed amongst the various possible frequencies. Attempts to solve this problem on the basis of classical physics had run into serious difficulties and in particular the classical theories seemed to predict that an infinite amount of energy would be radiated at the highest frequencies.

Planck pointed out that the difficulties could be overcome by making the following rather startling assumption. When the hot body emits electromagnetic radiation of frequency ν, it does not emit it continuously but in bursts of well-defined energy ε. The quantity of energy in a burst is called a **quantum** of energy. It is proportional to the frequency of the radiation and the constant of proportionality is called Planck's constant.

Planck's quantum of energy

$$\varepsilon = h\nu \tag{11-1}$$

Planck's constant

$$h = 6.625 \times 10^{-27} \text{ erg sec} \tag{11-2}$$

Planck's idea was that the hot body cannot emit as small an amount of energy as it wishes. It must either emit a quantum of amount $h\nu$ or nothing. It can then go on to emit subsequent quanta, each of amount $h\nu$, but it cannot, for example, emit a total amount $\frac{3}{2}h\nu$. On a fine scale the emission of radiation is discontinuous.

Notice, however, how fine the scale is. For visible light of frequency about 5×10^{14} cycles/sec, the energy in a quantum is 3.31×10^{-12} erg. To acquire a kinetic energy of this magnitude, a grain of sand would have to fall through a height of only about 3×10^{-10} cm, which is much less than the diameter of a single atom! In our everyday experience of macroscopic bodies the energies involved are so large compared with Planck's quanta that we are not normally aware of the discontinuous nature of energy. This is an important point because, whenever we encounter a startling new idea, it is always relevant to ask why it was not apparent to the many generations of intelligent men who preceded us. It is only when we deal with very small entities of the size of atoms, or fundamental particles, that the quantum aspects of nature become apparent. If we ask how far a single proton must fall before it acquires enough kinetic energy to be able to create a quantum of visible light, the answer is interesting. A proton starting at rest a long way from the earth, at an effectively infinite distance, and falling freely all the way to the earth's surface, would still not have quite enough kinetic energy to create a quantum of *visible* light!

Planck's constant, h, is a very important fundamental constant of nature, which can be justifiably compared with the velocity of light, c. In the discussion of the theory of relativity we realized that some of our everyday prejudices about time, space, and motion are a consequence of the fact that we normally deal with velocities small compared with c. As soon as we encounter velocities comparable with c, unexpected new features appear. Moreover, c plays the role of an upper limit to all possible velocities. Similarly, as long as we are concerned with large objects we do not have to worry about the strange new ideas introduced by quantum mechanics. However, when we deal with very small objects such as atoms and fundamental particles, the unexpected again turns up. We have just encountered the first instance of this in the discontinuous nature of energy. The constant h then serves to set a *lower* limit on the smallness of such things as quantity of energy.

Planck was reluctant to go all the way and say that the hot body emits *particles* of light with energy $h\nu$. We now accept this as a permissible description of the situation and the particles of light are called **photons**. In 1905, the same year that he published his first paper on special relativity, Einstein pointed out that the concept of photons could explain certain puzzling aspects of a phenomenon known as the photoelectric effect. When electromagnetic radiation of a sufficiently high frequency falls on the surface of a metal, electrons are ejected from it (figure 11-1). This is the **photoelectric effect** and the ejected electrons are called **photoelectrons**.

The classical explanation is fairly obvious. The oscillating electric vector of the incident electromagnetic wave exerts an oscillating force on the free electrons inside the metal. These electrons are thereby set into oscillation with a steadily increasing amplitude until they are oscillating so violently that they are able to escape from the surface of the metal. However, this explanation is completely at variance with the facts. It suggests, for example, that increasing the intensity of the incident light, and thereby increasing the amplitude of the oscillating electric vector, should eject the photoelectrons with greater velocity. In fact the velocity of the photoelectrons is inde-

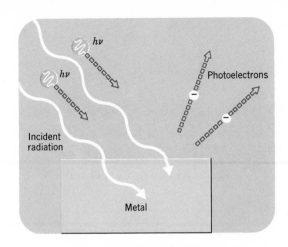

Figure 11-1 The photoelectric effect.

pendent of the intensity of the light and depends only on its frequency. The classical theory is also unable to explain the existence of a *threshold frequency* ν_c. The light is able to eject photoelectrons only if its frequency exceeds ν_c.

Einstein's explanation is simple. When a photon strikes the metal, it is completely absorbed and gives up all its energy, $h\nu$, to the photoelectron. A part W of this energy is needed to push the electron out of the metal, and the remaining part appears as the kinetic energy, $\frac{1}{2}mv^2$, of the escaped electron. Therefore,

$$h\nu = W + \tfrac{1}{2}mv^2 \tag{11-3}$$

If the electron is to escape at all, the frequency of the incident light must be at least large enough so that the energy of the photon, $h\nu$, exceeds the energy, W, needed to lift the electron out of the metal. The frequency ν must therefore exceed a threshold frequency ν_c given by

$$h\nu_c = W \tag{11-4}$$

Equation 11-3 may now be rewritten

$$h\nu = h\nu_c + \tfrac{1}{2}mv^2 \tag{11-5}$$

This equation obviously explains the fact that the velocity of the photoelectrons depends on the frequency of the incident light, but not on its amplitude. Experimentally, $\frac{1}{2}mv^2$ can be measured for various incident frequencies ν. The results are in complete accord with equation 11-5 and make it possible to deduce a value for Planck's constant, h, in exact agreement with the value needed to explain the radiation from a hot body.

The following analogy may prove helpful. A soccer ball of mass m at rest at the bottom of a pit of depth H is to be kicked out of the pit up a ramp (figure 11-2). The kick delivers a definite amount of energy, ε, which is analogous to the energy $h\nu$ given to the photoelectron by the photon. While climbing the ramp, the ball gains gravitational potential energy mgH, which is analogous to W. The velocity v of the ball as it emerges from the pit is

11-2 The Birth of a Revolution

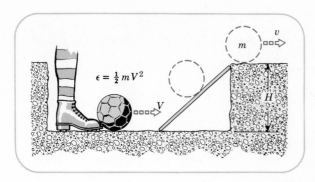

Figure 11-2 *An analogy to illustrate Einstein's equation for the photoelectric effect.*

therefore less than the velocity V with which it was kicked. The initial energy ($\varepsilon = \frac{1}{2}mV^2$) given to the ball by the kick is therefore used partially to lift the ball out of the pit and partially to provide it with kinetic energy when it emerges.

$$\varepsilon = mgH + \tfrac{1}{2}mv^2 \tag{11-6}$$

This is the analogue of

$$h\nu = W + \tfrac{1}{2}mv^2 \tag{11-7}$$

The ball will not escape at all unless it is kicked hard enough. The "threshold energy of kick" is

$$\varepsilon_c = mgH \tag{11-8}$$

The photoelectric effect enables us to come very close to "seeing" a particle of light. Modern detection techniques are so sensitive that it is possible to magnify the effect of each photoelectron so that it produces an audible click in a loudspeaker. With a sufficiently low intensity of illumination, it is then possible to "hear" the arrival of a single photon! The human eye is also extremely sensitive and can see a very faint light when it sends only about 100 photons into the eye every second.

Additional evidence for the particle nature of light is obtained from the x-rays produced when the fast-moving electrons in a discharge tube (figure 6-12) bombard a metal target. As explained in section 6-5, if V is the voltage across the discharge tube, the kinetic energy of an electron just before striking the positive electrode is Ve. When this electron is brought to rest, the maximum energy it can radiate as x-rays is Ve. If all this energy goes into a single photon, the maximum possible frequency, ν_m, of this photon is given by the equation

$$h\nu_m = Ve \tag{11-9}$$

In fact, x-rays produced this way are found to have frequencies between zero and an upper limit given precisely by this equation.

11-3 The Compton Effect and the Momentum of a Photon

The clinching argument for the existence of photons came from an experiment performed by Arthur Compton in 1922. When x-radiation passes through a solid, typically graphite, part of it is scattered in all directions. The scattered radiation has two components. One component has exactly the same frequency as the incident radiation. The other component, which is the one that interests us, has a frequency slightly smaller than that of the incident radiation (see figure 11-3a). Moreover, the decrease in frequency becomes greater as the angle of scattering, α, increases.

This effect is easily understood if we assume that it is a consequence of a collision between a photon of the incident x-rays and an electron in the solid. This collision may be discussed in exactly the same way as the collision of two billiard balls, except that we must use the relativistic formulas for the energy and momentum of the electron, which acquires a velocity comparable with c. The collision is depicted in part b of figure 11-3. Before the collision the electron is at rest for all practical purposes, but after the collision it has been knocked on with a large velocity and a large kinetic

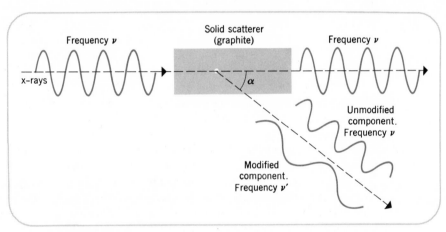

(a) The wave point of view.

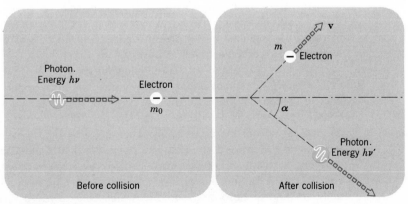

(b) The particle point of view.

Figure 11-3 The Compton effect.

energy. Since the collision must conserve energy, the kinetic energy of the electron can be gained only at the expense of the energy of the photon. Therefore, if the incident photon has a frequency ν and an energy $h\nu$, the rebounding photon must have a lower energy $h\nu'$ and consequently a lower frequency ν'. The increase in kinetic energy of the electron is $(h\nu - h\nu')$. The reduced frequency of the modified component is thus easily explained.

The complete theory of the Compton effect requires that, in addition to the law of conservation of energy, we should also apply the law of conservation of momentum. To do this we must have an algebraic expression for the momentum of a photon. As explained in section 8-4, a quantity of electromagnetic radiation with energy ε has momentum ε/c. A photon with energy $h\nu$ must therefore have momentum $h\nu/c$. Since $c = \nu\lambda$, where λ is the wavelength, this might also be expressed as h/λ.

Momentum of a photon

$$p = \frac{h\nu}{c} \tag{11-10}$$

or $\quad p = \dfrac{h}{\lambda}$ $\hfill (11\text{-}11)$

If this expression is used for the momentum of a photon, the two conservation laws can be applied to the collision and it is then possible to calculate the change in frequency of the x-rays, $(\nu - \nu')$, in terms of the angle of scattering α. The result is in completely satisfactory agreement with experiment.

Notice an important difference between the photoelectric effect and the Compton effect. In the photoelectric effect the photon completely disappears and *all* of its energy is given to the photoelectron. In the Compton effect there is still a photon after the collision, but its frequency is less than that of the incident photon and *part* of the energy of the incident photon has been given to the Compton electron.

The concept of photons with energy and momentum enables us to form a vivid picture of the pressure of light. It can now be visualized as due to a hail of photons striking the surface on which the pressure is exerted. It is therefore very similar to the pressure exerted by the atoms of a gas on the walls of their container (section 5-3).

We can also obtain a clearer understanding of the laws of conservation of energy and momentum when applied to electromagnetic phenomena. In section 8-4 we discussed the fact that the forces between two moving charges are not equal and opposite. The laws of conservation of energy and momentum do not therefore apply if we consider only the energy and momentum of the moving charges. We pointed out that the apparently lost energy and momentum is really carried away by electromagnetic radiation. We can now describe the situation as in figure 11-4. The two charges are continually emitting or absorbing photons, which carry away or bring up energy and momentum. The laws of conservation of energy and momentum are applicable if we include the photons as well as the charged particles.

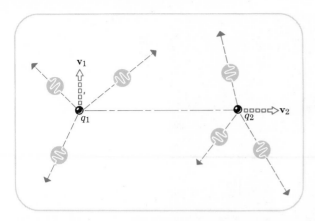

Figure 11-4 Two moving charges, producing accelerations in one another, radiate photons. The laws of conservation of energy and momentum cannot be applied unless the energy and momentum of these photons is included. (Compare figure 8-9.)

11-4 The Wave Nature of Matter

In 1922 the Compton effect provided conclusive evidence that electromagnetic radiation sometimes exhibits the properties of particles. Yet interference and diffraction experiments clearly indicate that electromagnetic radiation also exhibits the properties of a wave. In some sense it must combine the properties of both particles and waves. In 1924 Louis de Broglie made the bold suggestion that electrons and protons, which had been shown in many experiments to behave like particles, might also behave like waves. By 1927 Davisson and Germer in the United States, and G. P. Thomson in Scotland, had demonstrated interference and diffraction effects with electrons.

In these experiments the diffraction was not produced by slits, as in the experiments of Young and Fresnel with light (Chapter 8), but by crystalline solids. A crystal is an ordered array of atoms. When a wave falls on this array, each atom is the source of a scattered wavelet and the wavelets interfere constructively in some directions and destructively in other directions. The result is a complicated diffraction pattern, the details of which need not concern us here. Diffraction by crystals was first observed for x-rays and was successfully used both to measure the wavelengths of the x-rays and to study the spatial arrangement of the atoms in the crystals.

Part a of figure 11-5 shows a set of circular diffraction fringes obtained by passing x-rays through an aluminum foil. Part b shows a set of fringes produced by passing a beam of electrons through the same foil. The velocity of the electrons was adjusted to ensure that the wavelength associated with the electron beam was the same as the wavelength of the x-rays. The similarity of the two sets of fringes is striking evidence that electrons can be diffracted in exactly the same way as electromagnetic waves.

With modern improvements in technique, it is possible to observe diffraction of electrons by slits and straight edges. Figure 11-6 shows the interference fringes obtained when a beam of electrons passes through two parallel slits. It should be compared with the analogous fringes produced by light, as shown in figure 8-15. Figure 11-7a shows the fringes near the edge of the shadow when a beam of electrons passes near one of the straight edges of a small cubic crystal. Part b of the same figure shows similar diffraction fringes near the edge of a shadow produced by light. Subsequent investigations have revealed the existence of diffraction effects with other funda-

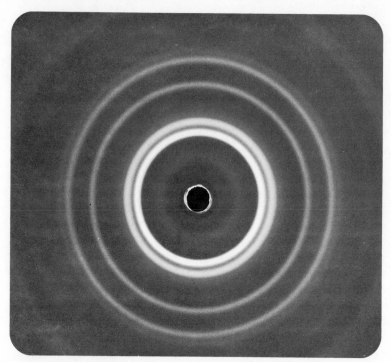

(a) The *diffraction pattern obtained by passing x-rays through an aluminum foil.*

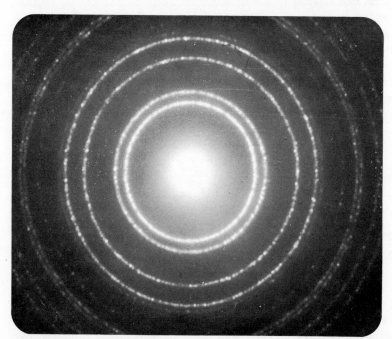

(b) The *diffraction pattern obtained by passing a beam of electrons through the same foil.*

Figure 11-5 *A direct comparison of electron diffraction and x-ray diffraction. The wavelengths are the same in both cases. (From the P.S.S.C. film Matter Waves.)*

Figure 11-6 *Interference fringes obtained by passing electrons through a double slit. (Photograph by G. Mollenstedt.)*

mental particles, such as protons and neutrons, and even with helium, neon, and argon atoms and hydrogen molecules.

The wave associated with a material particle is called a **de Broglie wave.** The quantity that oscillates in a wavelike fashion is normally denoted by the symbol ψ (the Greek letter psi), and is called the **wave function.** It is analogous to the vertical displacement of the surface of the water in the case of an ocean wave, or to the oscillating electric vector **E** in the case of an electromagnetic wave. The physical significance of ψ will be discussed in the next section.

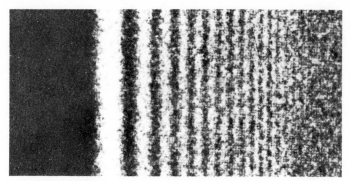

(a) *Diffraction fringes produced when a beam of electrons passes near the straight edge of a small cubic crystal. (Courtesy of Physics, P.S.S.C., D. C. Heath & Co., 1960.)*

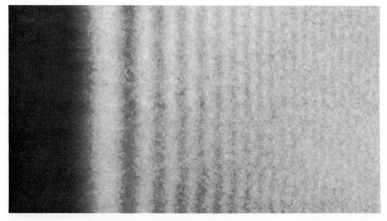

(b) *The diffraction pattern produced when light passes near a straight edge. (Photograph by Brian Thompson.)*

Figure 11-7 *Comparison of the diffraction patterns produced by electrons and by light passing near a straight edge.*

When we are talking about a particle, such as an electron, we ascribe to it a velocity $\mathbf{v}$, a mass m given by

$$m = \frac{m_0}{\sqrt{1 - \dfrac{v^2}{c^2}}} \, , \qquad (11\text{-}12)$$

a momentum $\mathbf{p}$ given by

$$\mathbf{p} = m\mathbf{v}, \qquad (11\text{-}13)$$

and a total energy ε given by

$$\varepsilon = mc^2 \qquad (11\text{-}14)$$

If the same particle can be described as a wave, the principal characteristics of this wave are its frequency ν and its wavelength λ. What is the connection between the quantities describing the particle and the quantities describing the wave?

de Broglie suggested that equations 11-1 and 11-11, which have been found to be valid for a photon, are applicable to all particles.

Formulas relating wave properties and particle properties

Total energy of a particle

$$\varepsilon = mc^2 = h\nu \qquad (11\text{-}15)$$

Momentum of a particle

$$p = mv = \frac{h}{\lambda} \qquad (11\text{-}16)$$

Equation 11-16 may be rewritten

$$\lambda = \frac{h}{mv} \qquad (11\text{-}17)$$

Interference and diffraction experiments enable the wavelength λ to be measured (compare the discussion associated with equation 8-15). The results are always in perfect agreement with equation 11-17.

Notice that the relationship between momentum and wavelength given in equation 11-16 for material particles is identical with equation 11-11, which applies to photons. However, equation 11-10, which is true for photons, cannot be applied to particles such as electrons. The reason for this is that the velocity of the wave associated with a photon is c and so

$$\text{For } photons \; \nu\lambda = c \qquad (11\text{-}18)$$

$$\text{and } \frac{1}{\lambda} = \frac{\nu}{c} \qquad (11\text{-}19)$$

For any particle whatsoever

$$p = \frac{h}{\lambda} \tag{11-20}$$

Only when the wave velocity is c is it permissible to make use of equation 11-19 and transform equation 11-20 into

$$p = \frac{h\nu}{c} \tag{11-21}$$

The wave velocity of a de Broglie wave is not c and so equation 11-21 is not valid for material particles.

There is another essential difference between a photon and a material particle. This becomes apparent when we consider the relativistic expression for the total energy of a material particle

$$\varepsilon = \frac{m_0 c^2}{\sqrt{1 - \frac{v^2}{c^2}}} \tag{11-22}$$

If we try to apply this to a photon, we must insert $v = c$ in the denominator, which then becomes zero and it appears that the energy might become infinite. This, of course, is the reason why a material particle can never attain the speed of light. The difficulty can be avoided for the photon if we assume that the photon rest mass is zero. Equation 11-22 for the energy then becomes $0/0$, which is indeterminate and may have any value whatsoever. It has, in fact, the finite value $h\nu$.

Conversely, if the rest mass of a particle is zero and its speed is less than c, then the numerator is zero and the denominator is not zero, so that the total energy is zero. A zero rest mass particle can possess energy only if it has the speed of light, because the numerator and denominator are then both zero.

> A photon is able to travel with the speed of light because its rest mass is zero. If a particle has rest mass zero, it must travel with the speed of light in order to have any energy.

11-5 The Role of Probability

What is the physical significance of the wave function, the quantity ψ that oscillates in a de Broglie wave? In what sense is it possible for an entity like an electron to combine the properties of a localized particle and a diffuse wave?

Consider first the situation in the case of light. Then the oscillating quantity is the electric vector **E**. If we produce a diffraction pattern with light, the brightness of the screen at any point of the pattern depends on the energy flow onto unit area of the screen at that point. According to section 8-4, this is equal to $cE^2/4\pi$. However, if we look on the light as a stream of particles, we say instead that n photons strike unit area of the screen per second and, since each has an energy $h\nu$, the energy flow is $nh\nu$ ergs per second per unit area. If both descriptions are permissible

$$nh\nu = \frac{cE^2}{4\pi} \qquad (11\text{-}23)$$

For light, therefore, the correlation between the wave point of view and the particle point of view is that the wave quantity E should be squared and E^2 is then a measure of the number of photons flowing across unit area per second.

Similar reasoning can be applied to electrons, or particles of any kind. Suppose that we allow a beam of electrons to pass through a single narrow slit and then allow them to fall onto a fluorescent screen, like the screen of a television tube. On the screen we shall see a diffraction pattern similar to the one shown in figure 8-18. The brightness of any part of the screen clearly depends on the number of electrons striking unit area there per second. From the wave point of view, however, the intensity of any part of the diffraction pattern depends on ψ^2. Is this, then, the answer? Is ψ^2 a measure of the number of electrons moving across unit area per second? Almost, but not quite, so let us proceed with the argument.

In front of the fluorescent screen, set up a row of counters as in figure 11-8. The graph behind the counters indicates the intensity ψ^2 of the de Broglie wave at various points across the screen. It is the same as the graph of figure 8-18. Its significance here is that it tells us the number of particles arriving at each counter. Each time an electron strikes a counter, its presence there is heralded by a click in a loud speaker and at the same time the number recorded on an instrument panel attached to this particular counter advances by one unit.

Let us first pass a beam of electrons through the slit and observe the number of particles recorded by each counter after 1,000,000 electrons have passed through. The number of electrons passing through a counter is related to the value of ψ^2 in the vicinity of this counter as shown by the graph to the right of the figure. Most of the electrons arrive inside the broad central fringe and the counter C at the center might register 500,000 arrivals. (The figures to be quoted have the right order of magnitude, but no precise significance, since they depend, amongst other things, on the width of a counter.) No electrons arrive at the center of the first dark fringe where $\psi^2 = 0$. The counter D_1 opposite this point will, however, receive a few particles, say 10, since it has a finite width and extends into regions where ψ^2 is slightly larger than 0. The counter B_1 opposite the center of the first outer bright fringe might count 20,000 particles. This is much smaller than the number counted by C, because the value of ψ^2 at B_1 is only about one twentieth of its value at C. Similarly, the counter B_2 opposite the center of the second outer bright fringe might count 8,000 particles.

Now comes a nasty question. What would happen if we were to send a *single* electron through the slit on its own? Would it spread itself out over the whole screen and trigger all the counters? Would it strike a definite

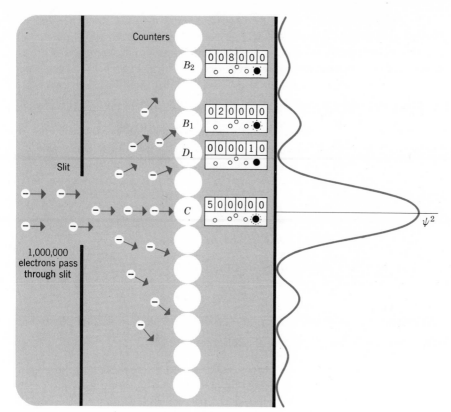

Figure 11-8 *The physical significance of ψ.*

point on the screen? If so, which point would it strike? The very popular central point?

The answer, which is supported by an impressive amount of experimental evidence, is as follows. Only one counter will respond, as might be expected if the electron is a particle that can be in only one place at a time. However, it is quite impossible to predict in advance which counter this will be. The best that can be done is to assess the **probability** that the electron will strike a particular counter. This probability is proportional to the square of the wave function, ψ^2, in the vicinity of the counter.

To illustrate the precise meaning of probability in this instance, suppose that we proceed as follows. Pass a single electron through the slit, on its own. It is quite impossible to predict which counter this electron will strike. Suppose that, in fact, it strikes B_1. Now pass through a second electron. It might strike C. A third electron might strike B_2 and a fourth electron again strike C. Repeat this experiment 1,000,000 times, passing through 1,000,000 electrons one at a time. The numbers registered by each counter will be the same as in the original experiment in which the 1,000,000 electrons were passed through together in a beam. (There will really be small differences, but these are secondary and will be ignored for the sake of exposition.) In 500,000 of the experiments the electron strikes C. In 10 of the experiments it strikes D_1. Notice, though, that it is impossible to predict which 10 experiments these will be. The result of any single experiment is unpredictable. The net effect of a large number of experiments *is* predictable.

11-5 The Role of Probability

245

In 500,000 out of 1,000,000 experiments the electron strikes C. The probability that the electron strikes C is defined as 500,000/1,000,000 or 0.5. Clearly a probability of 1.0 means complete certainty, whereas a probability of 0 means that the event never occurs. The probability that the electron strikes B_1 is 20,000/1,000,000 or 0.02. The probability that it strikes B_2 is 8,000/1,000,000 or 0.008. The probability that it strikes D_1 is only 10/1,000,000 or 0.00001.

The number of particles striking a counter is proportional to ψ^2. The probability that a single electron strikes a particular counter is consequently proportional to the value of ψ^2 in the immediate vicinity of this counter. The significance of the wave function ψ for a single particle may therefore be expressed as follows:

The physical significance of the wave function ψ

If a particle is described by a wave function that has the value ψ at a particular point, then ψ^2 is a measure of the probability that the particle will be found in the vicinity of that point.

The de Broglie wave is a wave of probability!

These considerations are readily extended to photons. If **E** is the electric vector of the electromagnetic wave, E^2 at a point is a measure of the probability that a photon will be found in the vicinity of the point. In a certain sense, a light wave is a wave of probability!

11-6 *The Philosophical Implications*

According to the contemporary interpretation of quantum mechanics, it is quite impossible to predict in advance which counter a particular electron will strike. This is *not* believed to be a consequence of the presence of unknown factors that guide the electron along its path by means that we have not yet discovered. The idea is rather that the future behavior of an electron is not completely determined by its past history. Several possibilities are available for its future behavior and one of these is chosen *purely by chance*, for no reason that can ever be determined. This applies to all aspects of the behavior of an electron, not only to its diffraction by a slit. It also applies to all moving bodies although, as we shall see, its consequences are more important for bodies of small mass, such as fundamental particles, atoms and molecules.

This indeterminism of quantum mechanics is in sharp contrast to the determinism of classical mechanics, as discussed in section 3-1. There it was suggested that the past, present, and future behavior of all the particles in the universe is completely determined. In principle, by applying the laws of classical physics to these particles, I could predict their future behavior and discover that I am predestined to be killed by a car as I cross the road in search of my lunch. Moreover, it would be beyond my power to do anything to prevent this. Quantum mechanics suggests a very different state of affairs. Whatever the past behavior of the particles in the universe, several

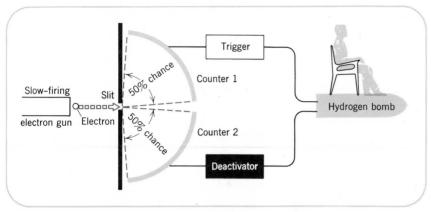

Figure 11-9 An experiment in philosophy.

possibilities are available for their future behavior. The possibility that is actually realized is purely a matter of chance. The universe may not have predetermined my death in an automobile accident, but it may well be "playing roulette" to decide whether I shall be killed by the automobile or not.

The behavior of electrons is very remote from our everyday experience of human bodies and automobiles. Perhaps these considerations about probability and chance are important to the physicist in his calculations of the behavior of small particles, but they have no relevance to the macroscopic bodies of normal experience. The following experiment is designed to demonstrate that the indeterminism of the behavior of electrons may have serious macroscopic consequences.

The electron gun at the extreme left of figure 11-9 ejects single electrons at a rate slow enough to ensure that only one electron is in the apparatus at any time. Consider the very first electron ejected after the gun is switched on. This electron passes through a very narrow slit which produces a broad diffraction pattern. If it should happen to be diffracted to the left, it strikes Counter 1. The arrival of the electron at Counter 1 results in a pulse of electric current, which travels to the component labeled "Trigger." This in its turn explodes a hydrogen bomb. If the electron should happen to be diffracted to the right, it strikes Counter 2, which delivers its pulse of current to the component labeled "Deactivator." This component operates a mechanism inside the hydrogen bomb which renders it incapable of exploding at any future time.

The student sits on the hydrogen bomb contemplating the philosophical implications of quantum mechanics. The outcome of the experiment is a matter of great consequence to him. Yet, according to the contemporary point of view, there is no way of predicting whether the first electron will be diffracted to the right or to the left. The best team of scientists available, having at their disposal all that they might require in the way of technical or computational facilities, could still not predict the outcome. Moreover, rightly or wrongly, the current belief is that it will never be possible to predict the outcome of such an experiment, however much our understanding of the physical nature of the universe progresses. The element of chance is fundamental to the behavior of nature. All that can be said is that the chances are 1 in 2 that the student survives.

11-6 The
Philosophical
Implications

In spite of the importance of this element of chance, the behavior of the universe is not completely chaotic. Although we cannot completely predict the outcome of an experiment, we can certainly "lay odds" on the various possibilities. Some consequences are much more likely than others. If we had to bet on the outcome of the experiment of figure 11-8, we would be well advised to place our money on central counter C. The probability that the electron hits this counter is 1 in 2, whereas the probability that it hits D_1 is only 10 in 1,000,000. Moreover, if we pass through a large number of electrons, either together or one at a time, we can be quite certain that approximately one half of them will hit C. If the experiment of figure 11-9 were repeated with 1,000,000 students, we could be quite certain that approximately 500,000 of them would survive. The larger the number of experiments performed, the more nearly would the survival rate approach 50%. The laws of probability operate in such a way that, even when it is not possible to make any prediction about a single event, it is possible to predict with some accuracy the outcome of a large number of events.

The macroscopic bodies of everyday experience contain an enormous number of electrons, protons, and neutrons. Although it is not possible to predict the behavior of a single one of these particles, it is possible to predict with good accuracy the combined behavior of the large number of particles. The indeterminacy of quantum mechanics is therefore not obvious in our everyday experience.

Questions

A

1. Which of the following types of radiation has a photon with (a) the least, (b) the greatest momentum? (i) Radio waves, (ii) ultraviolet light, (iii) red light, (iv) violet light, and (v) x-rays.
2. Which of the following equations are not applicable to all particles? (a) $\frac{1}{2}mv^2 = h\nu$ (b) $\varepsilon = mc^2$ (c) $\varepsilon = h\nu$ (d) $p = h/\lambda$ (e) $p = h\nu/c$ (f) $p = mv$.

B

3. In the Compton experiment, the radiation scattered *without* change in frequency can be considered to be the result of a collision between an incident photon and the whole block of scattering material. Explain carefully how such a collision can change the direction of the momentum of the photon without producing any appreciable change in its energy.
4. How would the Compton effect differ if protons were involved rather than electrons?
5. The classic example of determinism in Newtonian mechanics is the accuracy with which it is possible to predict the time of a future eclipse of the sun. Imagine that you have at your disposal all technical facilities that the human race might be expected to develop in the foreseeable future. Devise a method of perturbing the moon's orbit in order to change the times of future eclipses. Arrange for the exact result of the procedure to depend on an unpredictable factor, as in the experiment of figure 11-9.

C

6. Invent a precise definition of a particle. Suggest experiments to discover if photons of light satisfy this definition.

7. Invent a precise definition of a wave. Suggest experiments to discover if light satisfies this definition.

8. Consider a process in which a photon of visible light strikes an isolated electron at rest in otherwise empty space. The photon disappears and its energy is given as kinetic energy to the electron. Show that this process cannot simultaneously conserve both energy and momentum. Why is this objection not valid for the photoelectric effect?

9. A single electron passes through a narrow slit and is diffracted to one side. The angle of diffraction is determined as accurately as possible by placing a dense array of very small counters on the other side of the slit and observing which counter responds to the arrival of the electron. Is there any limit to the accuracy with which this can be done? Use this example to discuss the precise meaning of the uncertainty principle.

10. For the situation described in the previous question, is it permissible to apply the law of conservation of momentum to the passage of the electron through the first slit? (This is a very profound question. Think about it carefully. Discuss it exhaustively.)

Problems

A

1. If a photon has an energy of 2×10^{-14} erg, what is its frequency?

2. What is the energy of a photon of a radio wave with a wavelength of 2×10^4 cm?

3. What is the wavelength of a photon if its energy is 1 erg?

4. A photon has an energy of 10^{-20} erg. What type of electromagnetic radiation is it?

5. What is the minimum voltage that must be applied across an x-ray tube to produce x-rays with a frequency of 10^{20} cycles/sec?

6. What is the momentum of a photon with a frequency of 4.5×10^{17} cycles/sec?

7. If a photon has a momentum of 3×10^{-21} gm cm/sec, what is its wavelength?

8. What is the momentum of a 1 MeV photon?

9. In a Compton scattering process the incident radiation had a frequency of 1.000×10^{19} cycles/sec and the scattered radiation had a frequency of 0.990×10^{19} cycles/sec. What energy was given to the scattered electron? Express the answer in ergs and in electron volts.

B

10. A grain of sand has a mass m of 10^{-5} gram. Through what height must it fall from rest in order to have a kinetic energy equal to the energy of a quantum of visible light with a frequency of 5×10^{14} cycles/sec?

11. A beam of infrared radiation has a wavelength of 10^{-4} cm. What is the energy of one of its photons in electron volts?

12. What is the wavelength of a 1 MeV photon?

13. What is the equivalent mass of a photon of red light with a wavelength of 6.75×10^{-5} cm?

14. What is the frequency of a photon if its energy is equal to the rest mass energy of an electron? What type of electromagnetic radiation is it?

15. What is the minimum voltage that must be applied across an x-ray tube to produce photons each of which has an energy equal to the rest mass energy of an electron?

16. What is the minimum voltage that must be applied across an x-ray tube to produce photons with a wavelength equal to the radius of a proton $(0.8 \times 10^{-13}$ cm)?

17. To meet the requirements of the United States Golf Association, a golf ball must have a mass of 45.6 gm and acquire a velocity of 7600 cm/sec when tested on a certain machine. What is its de Broglie wavelength under these circumstances?

18. What is the de Broglie wavelength of an electron with a velocity of 2×10^6 cm/sec?

19. What is the de Broglie wavelength of a 5 eV electron?

20. Find the frequency and wavelength of the wave associated with (a) a 100 eV photon (b) a 100 eV electron.

21. A proton is at rest a long way away from the earth. It falls toward the earth under the influence of the earth's gravitational attraction. When it reaches the earth, all its kinetic energy is converted into the energy of a single quantum of electromagnetic radiation. What is the frequency of this quantum?

22. Calculate the period of a photon whose energy is equal to the mutual *gravitational* potential energy of two electrons 10^{-13} cm apart. Comment on the order of magnitude of the answer.

23. What is the de Broglie wavelength of an electron with a velocity of $(3/5)c$?

24. What is the de Broglie wavelength of a 1 GeV electron?

25. In a Compton scattering process the incident radiation had a frequency of 1.200×10^{20} cycles/sec and the scattered electron acquired a velocity of 1.5×10^{10} cm/sec. What was the frequency of the scattered photon?

26. A beam of blue light with a wavelength of 4.5×10^{-5} cm shines perpendicularly on the surface of a polished metal mirror and is totally reflected back in the direction from which it came. If 10^{20} photons strike the mirror in one second, what is the force exerted on the mirror by the pressure of light?

27. Make a rough estimate of the de Broglie wavelength of a baseball hit for a home run.

28. Find the order of magnitude of the de Broglie wavelength of most of the atoms in helium gas at room temperature.

Quantum
Mechanics

12 *More Quantum Mechanics*

12-1 Heisenberg's Uncertainty Principle

In 1927 Heisenberg tried to resolve the problem of exactly what is meant by saying that the electron is a particle. He decided that the essential characteristic of a particle is that, at a fixed instant of time it must have a definite location at a definite point in space and must have a well-defined velocity. He then asked how it might be possible to determine experimentally this definite position and definite velocity. He came to the startling conclusion that, in the case of a small particle like an electron, it is *not* possible to measure precisely both its position and its velocity. In any conceivable experiment there is always an uncertainty in the measured value of the position, or the velocity, or both. Moreover, this is not because of imperfections in the design or construction of the apparatus used in the experiment. With the best possible apparatus that could ever be constructed, the uncertainty would still be present. It is an unavoidable consequence of the way in which nature behaves.

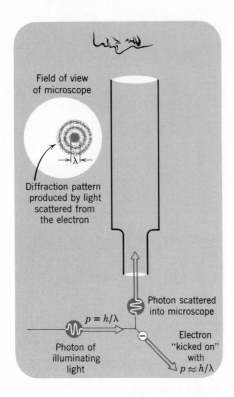

Field of view
of microscope

Diffraction pattern
produced by light
scattered from
the electron

λ

Photon scattered
into microscope

$p = h/\lambda$

Photon of
illuminating
light

Electron
"kicked on"
with
$p \approx h/\lambda$

Figure 12-1 If we try to look at an electron through a microscope, using illuminating light of wavelength λ, the electron is seen as a blur of width λ and the electron is "kicked" and given an extra momentum of about h/λ.

If, for example, we try to follow the path of an electron by looking at it through a microscope (figure 12-1), we must illuminate it with light of wavelength λ. However, as we have pointed out earlier on several occasions and particularly in section 8-6, a microscope does not give a sharp image, but a blurred diffraction pattern spread out over a distance of about the wavelength λ. There is therefore still an uncertainty of about λ in the position of the electron.

Moreover, the electron must scatter the illuminating light into the microscope. A photon of the incident light collides with the electron and rebounds into the microscope (figure 12-1). This is the Compton effect and it results in the electron being given a "kick" by the photon and receiving an extra momentum comparable in size with the momentum of the photon, h/λ. We may now proceed, if we wish, to see where the electron is at a slightly later time in order to attempt to measure its velocity and momentum, but we cannot possibly hope to measure the values these quantities had before we gave the electron its kick. Merely by trying to make measurements on the electron, we have tampered with it, we have given it an extra momentum of about h/λ and thereby we have produced an uncertainty of about h/λ in what its momentum was before we interfered. Very careful study of the situation reveals that these uncertainties are unavoidable. There is no clever trick we can employ, or no subtle way of applying the theory of the Compton effect to improve the measurements.

There is nothing to prevent us from inventing a "microscope" using electromagnetic radiation of smaller wavelength λ, thereby reducing the spread of the diffraction pattern and determining the position of the electron more precisely. Unfortunately, the uncertainty in the momentum, h/λ, is inversely proportional to the wavelength and so it would be made much worse. Conversely, we could use very long wavelengths to give the electron a very gentle kick and produce very little uncertainty in its momentum, but the diffraction pattern would then be very broad and there would be a large uncertainty in the position of the electron.

In general, whatever procedure we think up for simultaneously measuring the position and momentum of any particle, we find that we can measure either of these quantities as accurately as we wish, but only at the expense of a large uncertainty in the other quantity. The uncertainty in position multiplied by the uncertainty in momentum is always approximately equal to Planck's constant h.

Heisenberg's uncertainty principle for position and momentum

(Uncertainty in position)

$$\times \text{ (Uncertainty in momentum)} \approx h \quad (12\text{-}1)$$

The symbol $\approx$ means "has a value not very different from." We have to be a little vague because we have not gone into the laborious mathematical analysis that gives a precise significance to the uncertainty of a quantity.

If the velocity of the electron is small compared with c, we can forget about the variation of mass with velocity and put the mass equal to the constant rest mass, m_0. Then, since momentum $p = mv$, we can assume that

$$\text{Uncertainty in momentum} = m_0 \text{ (Uncertainty in velocity)} \quad (12\text{-}2)$$

Heisenberg's uncertainty principle is then

(Uncertainty in position)

$$\times m_0 \text{ (Uncertainty in velocity)} \approx h \quad (12\text{-}3)$$

This can be rearranged to give

$$\text{Uncertainty in velocity} \approx \frac{h}{m_0 \text{ (Uncertainty in position)}} \quad (12\text{-}4)$$

This statement is not true when the magnitude of the velocity is comparable with c. Equation 12-1 is always true.

Since m_0 is in the denominator of equation 12-4, the uncertainty in velocity is large for small, light particles but becomes very small for large, massive particles. That is why the uncertainty principle is not conspicuous in everyday life when we are dealing with large objects.

12-2 Golf Balls and Electrons

The uncertainty principle can be illustrated by a game of miniature golf in which the player is attempting to putt the ball through a gap in a wall so that it falls into a hole on the other side (figure 12-2). As it passes through the gap the uncertainty in the position of the ball cannot possibly exceed

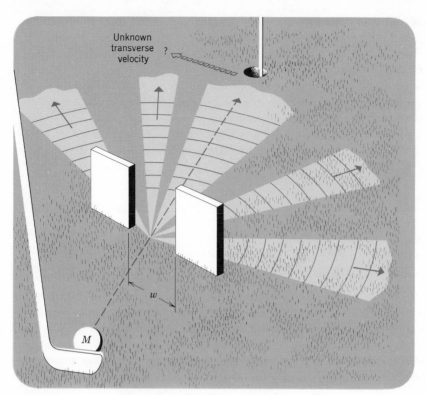

Unknown
transverse
velocity ?

w

M

Figure 12-2 *The uncertainty principle applied to miniature golf.*

the width of the gap, w. There must therefore be an uncertainty of at least
h/w in the component of the momentum in a direction parallel to the width
of the gap. There is no guarantee that the ball will pass straight through,
because that implies complete certainty that the ball acquires no transverse
component of velocity or momentum. The behavior of the ball is determined
by the laws of probability and we cannot predict precisely what it will do
after passing through the gap, but the most likely thing is that it will acquire
a transverse momentum of about h/w and a transverse velocity h/Mw,
where M is the mass of the ball.

Why, then, are we confident that, with sufficient skill, we can be certain of putting the ball into the hole? The answer is that, in the expression
h/Mw for the uncertainty in the transverse velocity, a very small quantity h
appears in the numerator whereas a very large quantity M appears in the
denominator. The uncertainty in the transverse velocity is therefore negligibly
small.

Another way of looking at this is that a moving golf ball can be represented by a de Broglie wave. After passing through the gap (a single slit as
in figure 11-8) the de Broglie wave spreads out into a diffraction pattern
(figure 12-2). The significance of this pattern is that it gives the probability
that the ball will be deflected in a given direction, but it says nothing at all
about what will happen in a single particular instance, the actual putt. However, the de Broglie wavelength is

More
Quantum
Mechanics

$$\lambda = \frac{h}{p}$$ (12-5)

or $\lambda = \dfrac{h}{Mv}$ $\hspace{4cm}$ (12-6)

where v is the velocity that the putter gives to the ball. In a practical example, M is 45.6 gm and v might be about 20 cm/sec, in which case λ would have an extremely small value of about 10^{-29} cm. For so small a wavelength, the diffraction fringes would be very close together. The reason we can rely on golf balls to travel along straight lines is the same as the reason why we can usually rely on a ray of light to travel along a straight line (section 8-6).

However, when a small object like an electron passes through a narrow slit, the de Broglie wavelength is no longer small and there is a broad diffraction pattern. The electron is likely to be deflected through a large angle and acquire a large transverse momentum. The behavior of a single electron is quite unpredictable, as we have already emphasized. It may acquire a large or small transverse momentum to the right or to the left. It really is an *uncertainty* in transverse momentum that is involved.

All that can be said is that the most likely thing to happen is that the electron will be deflected inside the broad central fringe, and the classical theory of diffraction then shows that its transverse momentum is likely to be somewhere between h/w to the right and h/w to the left, where w is the slit width. The uncertainty in transverse momentum is then about h/w. Since passing the electron through the slit fixes its position with an uncertainty equal to the slit width w,

(Uncertainty in position)

$$\times \text{ (Uncertainty in momentum)} \approx w \times \frac{h}{w} \quad \text{(12-7)}$$

$$\approx h \quad \text{(12-8)}$$

This demonstrates the validity of Heisenberg's uncertainty principle for this particular situation.

Notice that the uncertainty in position as the electron passes through the slit can be made very small by making the slit very narrow. However, as we very carefully pointed out in section 8-6, when the slit is made narrower the diffraction pattern spreads out and there is a greater uncertainty in the transverse momentum. A gain in fixing the position is achieved only at the expense of a loss in the accuracy of determination of the momentum.

12-3 Wave Packets

One way of making a wave look something like a particle is to consider a **wave-packet**. This is a wave confined to a region of space of length l, as shown in figure 12-3a. Since the square of the wave amplitude at a point is related to the probability of finding the particle at that point, there is a high probability of finding the particle anywhere inside the region occupied by the wave-packet, but no probability of finding it outside. Consequently,

$\hspace{2cm}$ Uncertainty in position $= l$ $\hspace{3cm}$ (12-9)

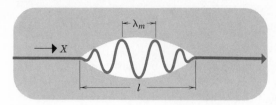

(a) This wave-packet can be synthesized by adding together a large number of infinitely long waves similar to the ones shown below.

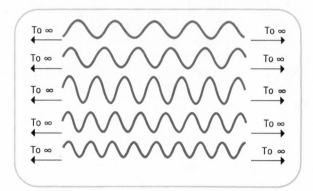

(b) Some of the Fourier components of the wave-packet.

Figure 12-3 A wave-packet and its Fourier components.

It might seem that we should take the wavelength to be the length λ_m shown on the figure, and assign to the momentum the precise value

$$p_m = \frac{h}{\lambda_m} \tag{12-10}$$

However, the strict mathematical formulation of quantum mechanics allows us to apply a relationship of this kind only to an *infinitely long* wave. Fortunately, the wave-packet of finite length can be replaced by a large number of infinitely long waves of various wavelengths, similar to the ones shown in part *b* of figure 12-3. These are called the *Fourier* components of the wave-packet. If their wavelengths and amplitudes are properly chosen, when they are all added together they completely cancel one another outside the wave-package, but reproduce the wave of figure 12-3a inside the wave-packet.

*More Quantum Mechanics*The mathematics reveals that most of the Fourier components have wavelengths inside a narrow range straddling λ_m. Each component has its own particular wavelength λ, corresponding to a particle momentum h/λ. The resulting range of values of p turns out to have a width of about h/l. The wave-packet does not have a unique momentum, but a whole range of values of momentum, so

Uncertainty in momentum $\approx h/l$ (12-11)

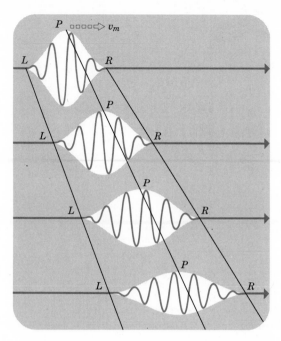

Figure 12-4 *Distortion of a de Broglie wave-packet as it travels through space.*

Combining equations 12-9 and 12-11,

(Uncertainty in position)

$$\times \text{ (Uncertainty in momentum)} \approx l\frac{h}{l} \quad (12\text{-}12)$$

$$\approx h \quad (12\text{-}13)$$

Heisenberg's uncertainty principle is seen to be an intrinsic aspect of the nature of such a wave-packet.

An interesting property of a de Broglie wave-packet is that, as it moves forward, the various Fourier components get "out of step" and the wave-packet broadens as it proceeds on its way. As shown in figure 12-4, the left-hand edge L travels more slowly than the right-hand edge R. This is related to the uncertainty in the velocity of the particle. When the particle starts out, we are not quite certain where it is, but we are also not quite certain what its velocity is. The uncertainty in where it will be at a later time therefore becomes greater as time goes on. However, the peak P of the wave-packet travels with a velocity v_m, which is a kind of "mean velocity" of the particle and has the value

12-3 Wave Packets

$$v_m = \frac{h}{m\lambda_m} \quad (12\text{-}14)$$

This is the de Broglie relationship for an infinitely long wave with wavelength λ_m.

These ideas about wave-packets can be used to discuss another form of the Heisenberg uncertainty principle concerned with the uncertainty in simul-

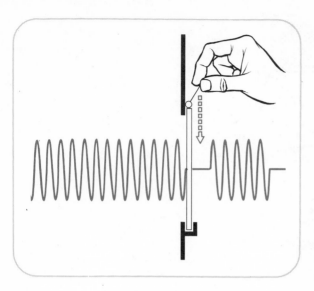

Figure 12-5 An attempt to fix the exact instant at which a photon passes through an opening.

taneously making measurements of time and energy. Imagine a beam of light with a single frequency passing through a hole that can be closed by a shutter (figure 12-5). We are trying to determine the exact instant at which a photon passes through the hole. We open the shutter and close it again a time τ seconds later. If a photon emerges on the other side, we then know that the instant it passed through the shutter lies somewhere within this time interval.

$$\text{Uncertainty in time} = \tau \tag{12-15}$$

However, the electromagnetic wave that passes through the hole is no longer an infinitely long wave, but a wave-packet that takes a finite time τ to pass by any point. It can be decomposed into Fourier components and theory reveals that these components are spread over a range of frequencies of width $1/\tau$.

$$\text{Uncertainty in frequency} \approx 1/\tau \tag{12-16}$$

A photon associated with the wave-packet does not have a definite energy $h\nu$. Its energy lies within a range of width h/τ because

$$\text{Uncertainty in energy} = h \text{ (Uncertainty in frequency)} \tag{12-17}$$

With the help of equation 12-16, this becomes

$$\text{Uncertainty in energy} \approx h/\tau \tag{12-18}$$

More Quantum Mechanics

Combining equations 12-15 and 12-18, we obtain the form of Heisenberg's uncertainty principle applicable to time and energy.

> *Heisenberg's uncertainty principle for time and energy*
>
> $$\text{(Uncertainty in time)} \times \text{(Uncertainty in energy)} \approx h \tag{12-19}$$

The apparatus was designed to fix as precisely as possible the instant of time at which the photon is in the hole. It does so only at the expense of producing a large uncertainty in the energy to be assigned to the photon after it emerges from the hole. If the shutter is left open for a shorter time, in order to fix the instant more precisely, the wave-packet emerging is shorter. A shorter wave-packet has Fourier components spreading over a wider range of frequencies. There is consequently a wider range of possibilities for the energy of the photon emerging.

12-4 Bohr's Theory of the Atom

Niels Bohr's famous theory of the atom was proposed in 1913. It came several years after Planck's discussion of the radiation emitted by a hot body and Einstein's explanation of the photoelectric effect. It preceded by more

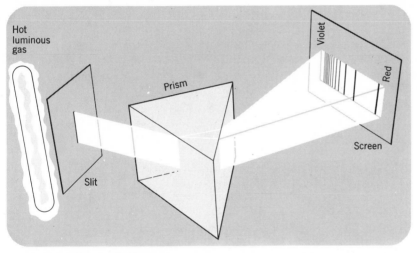

(a) Line spectrum of a luminous gas.

(b) The spectrum of monatomic hydrogen in the visible region.

Figure 12-6 An atom can emit only certain special frequencies of light.

than a decade the discovery of the wave nature of the electron. The rapid developments following this latter discovery revealed certain flaws in Bohr's theory and we do not now look on it as a completely adequate description of an atom. Nevertheless, it played a very important role in the development and understanding of quantum mechanics. It may also prove helpful to the reader as an intermediate link between the classical approach and the quantum approach.

The primary objective of the theory was to explain the line spectrum of a luminous gas of atoms. If the light emitted by the atoms is passed through a spectroscope, the spectrum is observed to consist of several narrow bright lines separated by dark regions (see figure 12-6). The atoms do not emit all frequencies of electromagnetic radiation, but only a few distinct frequencies. This is difficult to explain in terms of classical physics but can be given a simple explanation within the spirit of the quantum theory. The energy of an atom cannot have any conceivable value, but only one of certain particular values. Suppose that the atom initially has one of these permissible energies, ε, and then suddenly changes to another state in which it has a different permissible energy, ε'. Assuming that ε' is less than ε, an amount of energy $(\varepsilon - \varepsilon')$ is released, and this can be used up in the emission of a photon with energy $h\nu$, where

$$h\nu = (\varepsilon - \varepsilon') \tag{12-20}$$

Since there are only a limited number of permissible values for ε and ε', there are only a limited number of possible frequencies that can be emitted in this way. The problem, then, is to explain why the energy of an atom is restricted to these particular values and to calculate their magnitudes.

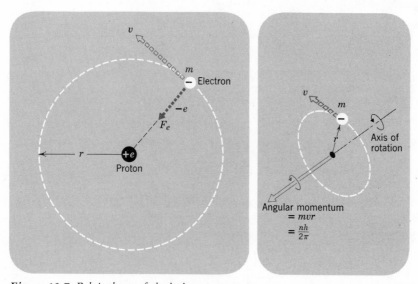

Figure 12-7 *Bohr's theory of the hydrogen atom.*

More Quantum Mechanics

We shall discuss the simplest atom, hydrogen, although, with a few refinements, the theory can readily be applied to more complicated atoms. Mainly as a result of the research of Ernest Rutherford, Bohr knew that the hydrogen atom consists of a small, positively charged nucleus (a proton) with a negatively charged electron moving around it. He imagined the electron to be moving in a circle with the proton at its center (figure 12-7), in much the same way that the earth moves around the sun. The classical treatment of the electron's motion is very similar to the treatment of the orbit of a satellite (section 3-7), except that the force on the electron is electrostatic rather than gravitational. The magnitude of this electrostatic force of attraction, F_e, exerted by the proton on the electron, is obtained from Coulomb's law (equation 6-3) by putting the two charges q_1 and q_2 both equal to the charge on the electron e and the distance between the charges equal to the radius of the circle, r.

$$F_e = \frac{e^2}{r^2}$$

(12-21)

If the velocity of the electron is v, its acceleration toward the center of the circle is

$$a = \frac{v^2}{r}$$

(12-22)

If m is the mass of the electron, the force may be equated to the product of mass and acceleration to give

$$\frac{e^2}{r^2} = m\frac{v^2}{r}$$

(12-23)

or $mv^2r = e^2$

(12-24)

There is nothing in this classical approach to impose any restriction on the value of the radius r provided that the velocity has the corresponding value to satisfy equation 12-23. The crucial assumption of Bohr's theory is that only certain values of r are permissible, and that an allowed value of r is such that the angular momentum of the electron is an integral multiple of a fundamental unit of angular momentum, $h/2\pi$. Angular momentum was defined in section 4-3 as mvr, so Bohr's assumption is

$$mvr = n\frac{h}{2\pi}$$

(12-25)

where n is an integer (1, 2, 3, 4 etc.)

Equation 12-25 yields the following expressions for v and v^2.

$$v = \frac{h}{2\pi mr}n$$

(12-26)

$$v^2 = \frac{h^2}{4\pi^2m^2r^2}n^2$$

(12-27)

When this value of v^2 is inserted into equation 12-24, we obtain

$$m\left(\frac{h^2}{4\pi^2m^2r^2}n^2\right)r = e^2$$

(12-28)

12-4 Bohr's
Theory of the
Atom

$$\text{or } \frac{h^2}{4\pi^2mr}n^2 = e^2$$

(12-29)

This can be rearranged to obtain an expression for the radius,

$$r_n = \frac{h^2}{4\pi^2e^2m}n^2$$

(12-30)

The radius has been written with a subscript n to indicate that it corresponds to a particular value of the integer n. If we insert known values of the fundamental constants h, e, and m,

$$r_n = 0.53 \times 10^{-8} n^2 \text{ centimeters} \tag{12-31}$$

In the normal state of the hydrogen atom $n = 1$, and the radius has its smallest value, which is 0.53×10^{-8} cm. This is in very satisfactory agreement with the known size of the atom. As n takes successive values of 1, 2, 3, 4, etc., the radii are in the ratio $1:4:9:16$ etc.

The *energy* of the atom has two parts, a negative electrostatic potential energy $-e^2/r_n$, and the positive kinetic energy of the electron, $\frac{1}{2}mv^2$. The rest mass energy need not be included because v is much less than c, and a classical, nonrelativistic treatment is adequate. The total energy of the atom in its nth orbit is therefore

$$\varepsilon_n = -\frac{e^2}{r_n} + \tfrac{1}{2}mv^2 \tag{12-32}$$

Notice, however, that equation 12-24 may be written in the form

$$mv^2 = \frac{e^2}{r_n} \tag{12-33}$$

whence $\varepsilon_n = -\dfrac{e^2}{r_n} + \dfrac{1}{2}\dfrac{e^2}{r_n}$

$$= -\tfrac{1}{2}\frac{e^2}{r_n} \tag{12-34}$$

Substituting the value of r_n from equation 12-30

$$\varepsilon_n = -\frac{2\pi^2 e^4 m}{h^2}\frac{1}{n^2} \tag{12-35}$$

This energy is negative because the negative potential energy predominates.

The next step is to calculate all the possible energies from equation 12-35 by inserting all possible values for the integer n. The lines in the spectrum should then correspond to all the possible transitions between pairs of these energy states, and their frequencies should be given by equation 12-20. The result is an almost perfect explanation of the spectrum of monatomic hydrogen, except for one thing. A line predicted by the theory often turns out to be a group of several lines close together, but with all their frequencies very close to the predicted value. The Bohr theory is clearly a first approximation to the truth, but there must be some further complications.

More Quantum Mechanics

12-5 *Clouds of Probability*

Bohr's circular orbits are clearly inconsistent with the uncertainty principle. It is not permissible to require that the electron shall always be exactly at a distance r from the proton and that its velocity shall always be exactly v. The correct way to describe a hydrogen atom is to state the value of the wave function, ψ, of the electron at each point P in the vicinity of the proton. ψ^2 is then a measure of the probability that the electron will be found near this point P.

Suppose that we devise a method of locating exactly the position of the electron in the atom at a particular instant of time. This does not violate the uncertainty principle as long as we do not demand any knowledge of the simultaneous velocity of the electron. It is not possible to predict where the electron will be found, although it is more likely to be found in those regions where ψ^2 is large. Having found the electron, let us mark its position with a dot. If we repeat the experiment a second time, the electron will be found in an entirely different position, which we again mark with a dot. After repeating the experiment a very large number of times, we obtain a cloud of dots occupying a region of space all around the proton. The density of this cloud in the vicinity of a point is proportional to the value of ψ^2 at that point.

The atom may therefore be visualized as a "cloud of negative electricity" surrounding a small positive nucleus. We must remember, though, that different parts of the cloud do not coexist simultaneously. A slightly better description of this picture of an atom is that it is a "cloud of probability"! Some typical clouds are shown in figure 12-8.

For an electron in an atom, the wave function ψ must not be thought to describe a wave traveling through space and transferring energy from place to place. It is more analogous to a **standing wave** on a plucked string. Figure 12-9 shows a few possible modes of vibration of a plucked string which is firmly held at each end. In this situation, one receives no impression that the wave is traveling to the right or to the left. Instead of the wave passing over each point of the string in succession, each point continually oscillates up and down with a fixed amplitude, which varies along the string. At points such as N, which are called *nodes*, there is never any oscillation. At points such as A, called *antinodes*, the amplitude always has its maximum value. This state of "unvarying oscillation" is the way we should visualize a hydrogen atom in a particular state with a particular energy. The amplitude of oscillation of the string is independent of time, but varies from point to point along the string. Similarly, the oscillating wave function ψ of the electron has an amplitude that is independent of time at a particular point, but varies from point to point. The variable density of a cloud in figure 12-8 is an indication of how ψ^2 varies from point to point.

There are many possible modes of vibration of a plucked string, only a few of which are shown in figure 12-9. To distinguish between them, each mode could be designated by the number of its antinodes. Similarly, there are many different **states** of a hydrogen atom, corresponding to different clouds of various sizes and shapes, a few of which are shown in figure 12-8. To designate a particular cloud, it is necessary to give the value of each of three integers, n, l, and m_l. These integers are called the **quantum numbers** of the electron in the atom.

The **principal quantum number** n is an integer similar to the integer n which was important in the Bohr theory. It determines the size of the cloud in the same way that it determined the radius of the Bohr orbit. It also determines the energy of the state represented by the cloud, except for a small

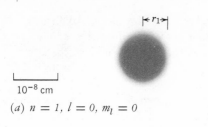

(a) $n = 1$, $l = 0$, $m_l = 0$

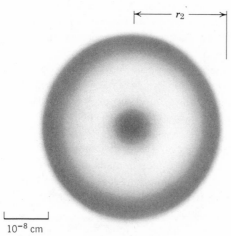

(b) $n = 2$, $l = 0$, $m_l = 0$

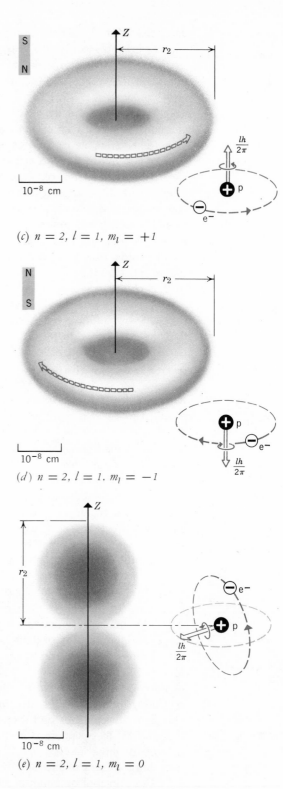

(c) $n = 2$, $l = 1$, $m_l = +1$

(d) $n = 2$, $l = 1$, $m_l = -1$

(e) $n = 2$, $l = 1$, $m_l = 0$

Figure 12-8 *Probability clouds for the first few states of the hydrogen atom. In parts c, d, and e the corresponding circular Bohr orbits are also shown.*

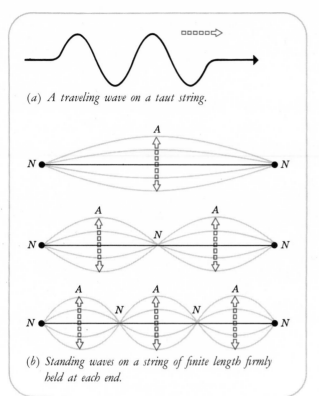

(a) A traveling wave on a taut string.

(b) Standing waves on a string of finite length firmly held at each end.

Figure 12-9 *The distinction between traveling waves and standing waves.*

contribution depending on the other quantum numbers. The formulas take exactly the same form that they did in the Bohr theory.

The radius of the cloud (the size of the atom)

$$r_n = \frac{h^2}{4\pi^2 e^2 m} n^2 \qquad (12\text{-}36)$$

n is the **principal quantum number** which may have any positive integral value, but may not be zero.

The energy associated with a cloud

$$\varepsilon = -\frac{2\pi^2 e^4 m}{h^2} \frac{1}{n^2} + \text{a small contribution depending}$$

on the other quantum numbers $\qquad (12\text{-}37)$

The angular momentum of the state is not $n\dfrac{h}{2\pi}$, as in the Bohr theory, but $l\dfrac{h}{2\pi}$. The **orbital quantum number** l is an integer which may have any value from 0 to $(n-1)$. The third quantum number m_l depends on the direction in which the angular momentum vector points, and it is not very important to us at the moment.

When n is fixed the overall size of the cloud is fixed, but it can have several different shapes, depending on the values of l and m_l. This is best understood by examining figure 12-8. Although the energy depends mainly on n, the different shapes correspond to slightly different energies. This partially explains the fact that, where the Bohr theory predicts one spectral line, we frequently find a group of lines close together. The reason why it does not provide a complete explanation will emerge in the next section.

Finally, let us form a picture of what happens when the atom emits a photon of light. It is initially in one state with particular values of n, l, and m_l, represented by a cloud of a certain size and shape. It suddenly collapses into a smaller cloud with a different shape. Since the smaller cloud has less energy, energy is released in the form of a photon. The energy $h\nu$, and hence the frequency ν, of this photon is determined mainly by the change in n. However, there are also small differences in energy depending on the shapes of the clouds. For fixed values of n and n' for the initial and final states, transitions between clouds of various shapes produce a group of spectral lines lying close together.

12-6 Spin

Our description of the atom is still not complete. The analysis of the previous section, involving the quantum numbers n, l, and m_l still does not explain all the lines in the spectrum of an atom! A line that would be expected to be single, according to this analysis, frequently turns out to be double. The additional concept needed was supplied by Uhlenbeck and Goudsmit in 1925. The electron has an intrinsic angular momentum.

The *orbital* angular momentum of the electron is due to its bodily motion around the nucleus. It is analogous to the angular momentum of the earth about an axis through the sun due to the annual orbital motion of the earth around the sun. It is well known that the earth also performs a daily rotation about an axis through its north and south poles, which results in an angular momentum about this axis. The electron is also rotating about an axis through itself and consequently has an *intrinsic* angular momentum, which is always present, even when the electron has zero translational velocity. In terms of the fundamental unit of angular momentum, $h/2\pi$, the intrinsic angular momentum of the electron is rather surprisingly found to be one half of a unit or $\frac{1}{2}h/2\pi$. This is frequently referred to as the spin of the electron.

The axis about which the electron is spinning is not allowed to point in any direction whatsoever. In any situation in which the electron finds itself there is always a preferred direction and the vector representing the intrinsic angular momentum can point either in this direction or in the exactly opposite direction. An observer looking in the preferred direction sees the electron spinning either clockwise or counterclockwise (figure 12-10). When the electron is in an atom the two directions of spin have slightly different energies and this explains the doubling of some spectral lines.

The information to be obtained from atomic spectra has still not been exhausted. If extremely refined techniques are used to examine a spectral line that would be expected to be single according to the above analysis, including n, l, m_l and electron spin, this line is often found to consist of a

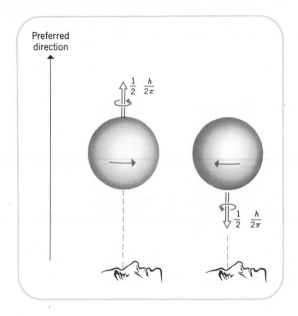

Figure 12-10 The spinning electron and its two possible orientations relative to a preferred direction.

group of lines, which are *very, very* close together. This is called **hyperfine structure.** It has several causes, but the one that concerns us at the moment is that the nucleus, like the electron, may have an intrinsic angular momentum. This angular momentum may have any one of several different orientations, each of which has a *slightly* different energy, thus giving rise to hyperfine structure in the spectrum.

A nucleus is believed to be made of protons and neutrons. A proton, or a neutron, is exactly like an electron in having an intrinsic spin with an angular momentum of one half a unit, or $\frac{1}{2}\frac{h}{2\pi}$. Again the angular momentum vector may point in one of two directions, corresponding to clockwise and counterclockwise spin.

Spin of an electron, a proton, or a neutron

The intrinsic angular momentum is

$$\text{one } half \text{ a unit} = \frac{1}{2}\frac{h}{2\pi} \tag{12-38}$$

The intrinsic angular momentum vector may point in a preferred direction or in exactly the opposite direction, corresponding to clockwise or counterclockwise spin.

12-6 Spin

It will now be realized that the total angular momentum of an atom is the vector sum of the orbital angular momenta of the electrons, plus the spin angular momenta of the electrons, plus the nuclear spin angular momentum. If we apply the law of conservation of angular momentum (section 4-4) to processes in which the atom emits a photon and changes its state and its total angular momentum, we are forced to the conclusion that the photon must have an intrinsic angular momentum of one whole unit.

Spin of a photon

Intrinsic angular momentum of a photon is

$$\text{one whole unit} = \frac{h}{2\pi} \tag{12-39}$$

We are not normally aware of this because ordinary light contains equal numbers of photons spinning in both directions. However, circularly polarized light, in which the electric vector does not oscillate up and down but swings around in a circle, consists of photons all spinning the same way. If these photons fall on a blackened disk and are absorbed by it, their angular momentum is transferred to the disk, which is observed to rotate (figure 12-11).

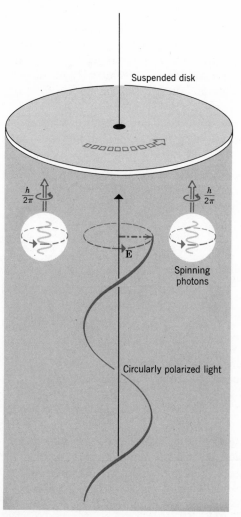

Suspended disk

$\frac{h}{2\pi}$ $\frac{h}{2\pi}$

E

Spinning photons

Circularly polarized light

Figure 12-11 *Circularly polarized light has all its photons spinning the same way. When they fall on a suspended disk, they transfer their angular momentum to it and cause it to rotate.*

We sometimes say, rather loosely, that an electron, a proton or a neutron has a spin of 1/2, whereas the photon has a spin of 1. There is an implicit assumption that the unit of angular momentum is $h/2\pi$.

12-7 The Pauli Exclusion Principle and Indistinguishability

In an atom containing several electrons, the electrons exert forces on one another and perturb one another's motion. Nevertheless, it is found to be a good approximation to treat the state of a single electron as though it were alone in a hydrogen atom and to describe it by the set of quantum numbers n, l, m_l and by the direction of its spin. From a consideration of atomic spectral lines and the set of energy values needed to explain them, Wolfgang Pauli discovered in 1925 that there is an important restriction on the way electrons may behave in atoms.

> *Pauli's exclusion principle*
>
> No two electrons in the same atom may be in exactly the same state.

This means, for example, that the two electrons may be in clouds of different sizes or at least different shapes. However, if they are in clouds of the same size and shape, so that they have identical values of the three integers n, l, and m_l, then they *must* be spinning in opposite directions.

Imagine a special type of piano with its mechanism designed in such a way that, when a key is depressed, the corresponding string is struck a blow of invariable strength and duration, quite independently of the force applied to the key or the length of time for which it is depressed. Adding electrons to an atom is somewhat similar to striking a chord of several simultaneous notes on such a piano. The number of different notes that can be struck simultaneously is limited only by the size of the keyboard, or by the number of players available and the number of fingers they possess. It is quite impossible, though, to strike any one note twice over at a single instant of time, to give it twice its normal loudness.

This analogy may also help to illustrate another important property of electrons, their **indistinguishability.** The sound emitted by the piano is independent of which fingers are used to strike the various notes. If two notes are struck, the first with the index finger of the right hand and the second with the index finger of the left hand, the sound is exactly the same as if the hands were crossed and the first note were struck by the index finger of the *left* hand. Similarly, all we can profitably say about an atom is that certain states are occupied by electrons, and it is not relevant to ask which electron is in which state. We cannot take two electrons, call them Tom and Bill, and put Tom into state 1 and Bill into state 2. Putting Tom into state 2 and Bill into state 1 would then clearly produce a different atom, but the fact is that no way has ever been found to distinguish between these two cases experimentally. An electron does not have a name and serial number, although it does have a rank in the sense that it can be distinguished from a proton.

The principle of indistinguishability of similar particles applies to all fundamental particles, for example protons, neutrons, photons, and even to atoms of the same element. A photon cannot be distinguished from any other photon with the same frequency, though it can be clearly distinguished from an electron. A helium atom in its normal state cannot be distinguished from any other helium atom in its normal state, though it can clearly be distinguished from a neon atom.

> *Indistinguishability of similar particles*
>
> It is sufficient to say that a state is, or is not, occupied by a particle of a particular kind. It is not permissible to try to establish the identity of the particle occupying the state.

Questions

A

1. I wish to determine the position of an electron with an uncertainty of about 10^{-4} cm, and at the same time determine its momentum as accurately as possible. Which of the following types of electromagnetic radiation would be most suitable for my purpose? (a) Radio waves, (b) microwaves, (c) visible light, (d) x-rays, (e) γ-rays.

B

2. Do the Bohr theory and the uncertainty principle both give the same order of magnitude for the velocity of an electron in an atom of given radius?

3. Since the velocity of a photon in a vacuum is always exactly c, is its position therefore infinitely uncertain?

C

4. Why is the hydrogen atom normally considered to be a stationary proton with an electron moving around it, rather than a stationary electron with a proton moving around it? Is either picture strictly true?

5. Very large molecules containing many atoms can be seen in a microscope. Imagine two such molecules with identical composition and structure in the field of view of a microscope. They can clearly be distinguished as "the left-hand molecule" and "the right-hand molecule." Is this consistent with the idea of indistinguishability?

Problems

A

1. The x component of the velocity of an electron is known to lie somewhere between 2.16×10^4 cm/sec and 2.35×10^4 cm/sec. Approximately what is the minimum uncertainty in its x coordinate?

2. The x coordinate of a proton is known to be somewhere between $+9.16534$ cm and $+9.16567$ cm. Approximately what is the minimum uncertainty in the x component of its momentum?

3. A radar pulse lasts for 2.5×10^{-7} sec. Approximately what is the uncertainty in the energy of its photons?

4. Show that the energy of the nth Bohr orbit is

$$\varepsilon_n = -\frac{2.18 \times 10^{-11}}{n^2} \text{erg}$$

$$= -\frac{13.6}{n^2} \text{eV}$$

B

5. The x coordinate of a 200 eV proton is known to lie somewhere between $+0.00123$ cm and $+0.00127$ cm. Approximately what is the minimum uncertainty in the x component of its velocity? How did you use the information given about the kinetic energy of the proton?

6. A beam of electrons with a velocity of 10^7 cm/sec passes through a slit 10^{-3} cm wide. What is the width of the central fringe of a diffraction pattern formed on a screen 2×10^2 cm away?

7. A beam of 10 eV electrons passes through a slit 10^{-4} cm wide. What is the width of the central fringe of a diffraction pattern formed on a screen 100 cm away?

8. The electron in a hydrogen atom is confined within a region of space with a diameter of about 10^{-8} cm. Approximately what is the minimum value of the uncertainty in its velocity? What then is the order of magnitude of its minimum permissible average velocity? Need you use relativistic formulas?

9. A hot gas has a prominent line in the blue region of its spectrum. The line is spread over a range of frequencies with a width equal to about 10^{-5} of the average frequency. What is the best accuracy with which one could pinpoint the instant of emission of a photon of this line by an atom of the gas?

10. What is the minimum frequency a photon must have to be able to lift the electron completely out of a hydrogen atom in its state of lowest energy? What type of electromagnetic radiation would it be?

11. What is the maximum frequency that appears in the line spectrum of hydrogen?

C

12. It was once thought that the nucleus of an atom contains protons and electrons. The diameter of a nucleus is about 10^{-12} cm. Set a lower limit on the order of magnitude of the average momentum of an electron inside a nucleus. Then calculate the corresponding average velocity and average energy (in ergs and electron volts). Need you use relativistic formulas?

13. A beam of 1 GeV electrons is passed through a shutter which is kept open for 10^{-20} sec. What is the uncertainty in the mass of the electrons that emerge? What is the uncertainty in their velocity?

14. An experiment is being designed to measure the mutual *gravitational* potential energy of two protons in a nucleus. The average distance between the two protons is 3×10^{-13} cm. If an accuracy of 1% is required, about how long will the experiment last?

15. What is the velocity of an electron in the first Bohr orbit of a hydrogen atom? What is the ratio of its mass to its rest mass?

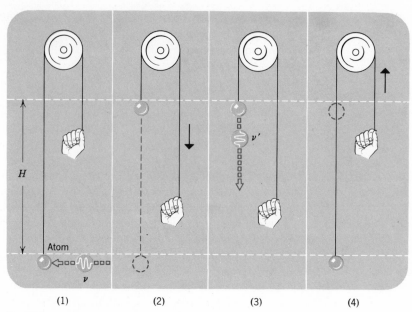

H

Atom

ν

ν'

(1)　　　　　(2)　　　　　(3)　　　　　(4)

Problem 12-18.

16. What is the minimum frequency of a photon that can be absorbed by a hydrogen atom in its ground state?

17. At what temperature would the average kinetic energy of the atoms of hydrogen gas be just sufficient to ionize a hydrogen atom during a collision?

18. With the help of the diagram, consider the following sequence of operations. (1) The atom absorbs a photon of frequency ν, which raises one of its electrons to a higher energy level. (2) The atom is then raised through a height H in the earth's gravitational field. (3) The electron falls back to the lower energy level, emitting a photon of frequency ν'. (4) The atom is lowered through a distance H back to its original position. Assuming that the law of conservation of energy is applicable, find the difference between ν and ν'. Compare your answer with equation 10-6.

13 *The Nucleus*

13-1 *Some Basic Properties of Nuclei*

Chemical properties of atoms depend almost entirely on the behavior of the electrons moving around the nucleus. Interatomic forces, which are the chief determining factor for the physical properties of solids and liquids, also depend almost entirely on the behavior of these electrons. The theory outlined in the previous chapter therefore leads to an almost complete understanding of the chemical and physical properties of macroscopic matter. In this connection, the principal importance of the nucleus is that it contains 99.9% of the mass of the atom. It also provides a positive charge that neutralizes the negative charge of the electrons and acts as an almost stationary center of attraction for the electrons. Study of the nucleus has served to elucidate only some of the finer details of the properties of macroscopic matter. On the other hand, it has opened up new, and still only partially explored, avenues in our search for the basic nature of matter and its ultimate constituents.

The nucleus is believed to be composed of protons and neutrons. A proton has a positive electric charge which is equal in magnitude to the negative charge on the electron. Its rest mass is 1836.1 times the rest mass of the electron. The neutron has no charge and its rest mass is 1838.6 times the rest mass of the electron, which means that it is slightly more massive than the proton, by about 0.14%. Both the proton and the neutron have an intrinsic angular momentum of $\frac{1}{2}(h/2\pi)$. It is common practice to refer to the protons and neutrons in the nucleus collectively as **nucleons**. The justification for this is that the properties of the proton and the neutron are almost identical except for their electric charges, and electromagnetic forces play only a subsidiary role inside the nucleus.

Since an atom is usually electrically neutral, the number of positively charged protons in its nucleus must equal the number of negatively charged electrons moving around the nucleus. This number is always represented by the symbol Z and is called the **atomic number**. If there are also N neutrons in the nucleus, the total number of protons and neutrons is

$$A = Z + N \qquad\qquad (13\text{-}1)$$

A is called the **mass number**. It is an exact integer and must not be confused with the **nuclear mass**. If the chemical symbol for an element is X, it is customary to denote its nucleus by $_Z X^A$. The subscript in front of the chemical symbol gives the number of protons and the superscript following it gives the total number of nucleons (protons and neutrons).

Atomic number, Z = number of protons in nucleus $\qquad (13\text{-}2)$

Mass number, A = total number of protons and neutrons
in nucleus $\qquad\qquad\qquad\qquad\qquad\qquad\qquad (13\text{-}3)$

Symbol denoting the nucleus of a chemical element X
is $_Z X^A$ $\qquad\qquad\qquad\qquad\qquad\qquad\qquad\qquad (13\text{-}4)$

Although, for a particular element, the number of electrons and hence the number of protons is fixed, the number of neutrons in the nucleus may be varied. The resulting nuclei, with various mass numbers and various atomic masses are called **isotopes** of the element. For example, ordinary uranium is $_{92}U^{238}$ and its nucleus contains 92 protons and $238 - 92 = 146$ neutrons. The famous isotope "uranium 235" is $_{92}U^{235}$, with 92 protons and $235 - 92 = 143$ neutrons. A few of the lighter nuclei are illustrated in figure 13-1, including the three isotopes of hydrogen (hydrogen, deuterium, and tritium) and the two isotopes of helium.

Although most of the mass of the atom resides in the nucleus, its volume is only about one trillionth of the total volume of the atom. If the atom were magnified to the size of Mount Everest, the nucleus would still be only about as large as a football. It follows that the density of matter inside the nucleus is very high. It is, in fact, about 2.3×10^{14} times greater than the density of water. This density is almost constant for all nuclei. Since protons and neutrons have approximately the same mass, this must mean that the average volume available per nucleon is approximately constant for all nuclei. Another way of expressing this is that the average distance between

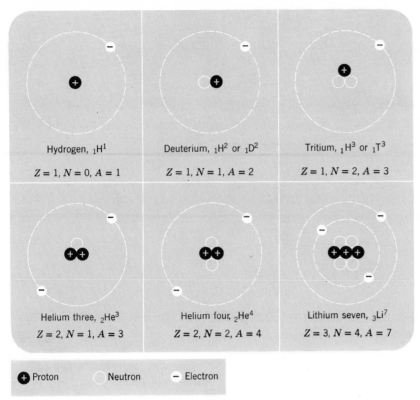

Hydrogen, $_1H^1$	Deuterium, $_1H^2$ or $_1D^2$	Tritium, $_1H^3$ or $_1T^3$
$Z = 1, N = 0, A = 1$	$Z = 1, N = 1, A = 2$	$Z = 1, N = 2, A = 3$
Helium three, $_2He^3$	Helium four, $_2He^4$	Lithium seven, $_3Li^7$
$Z = 2, N = 1, A = 3$	$Z = 2, N = 2, A = 4$	$Z = 3, N = 4, A = 7$

⊕ Proton ◯ Neutron **−** Electron

Figure 13-1 *Composition of some of the lighter atoms.*

adjacent nucleons is approximately constant. Its value is in fact very near to 1.9×10^{-13} cm for most nuclei. The radius of a proton or neutron is believed to be about 0.8×10^{-13} cm.

13-2 Nuclear Forces

The gravitational attraction between the nucleons is far too weak to hold the nucleus together and the electrostatic repulsion between the protons would only blow it apart. The forces that hold the nucleons together are called **nuclear forces** and they are very different in character from either gravitational or electromagnetic forces. Very little is known about them in detail and only a few qualitative remarks can be made with any certainty.

It is not even known exactly how the nuclear force varies with the distance between the nucleons, although it is quite certain that it does not vary inversely as the square of this distance, as in the case of gravitational forces and electrostatic forces. When the two nucleons are further apart than 10^{-12} cm, the nuclear force is negligible. As they are brought closer together, to distances less than 10^{-12} cm, the force is attractive and increases very rapidly (more rapidly than $1/R^2$). At a distance of about 2×10^{-13} cm, the force becomes much stronger than the electrostatic repulsion between two protons at the same distance apart. At distances below about 0.5×10^{-13} cm, the attraction probably changes into a strong repulsion. These qualitative aspects of the force are illustrated in figure 13-2.

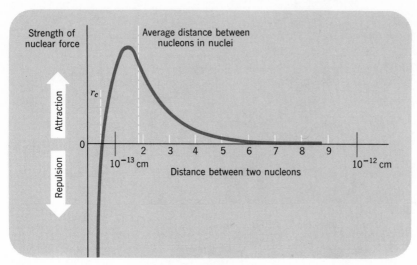

Figure 13-2 *Variation of nuclear force with distance (qualitative).*

The constant density of nuclear matter can be understood in terms of these characteristics of the nuclear force. It might be thought that two neighboring nucleons would settle down at that separation, r_c, where the force changes from an attraction to a repulsion. However, there is another consideration. If the average distance between nucleons is d, the uncertainty in the coordinate of a nucleon cannot be greater than d. There is then an uncertainty in the momentum of at least h/d. The nucleons are consequently in rapid motion and their average velocity increases as d decreases. If d were as small as r_c (about 0.5×10^{-13} cm) the motion would be so rapid that the nucleus would fly apart. As a compromise, the two neighboring nucleons settle down at a distance of 1.9×10^{-13} cm apart, where the attractive force is still large. The distance from a nucleon to its second nearest neighbors is then about 3.8×10^{-13} cm, but at this distance the attractive force is very much smaller. The force on a nucleon therefore depends primarily on its nearest neighbors and very little on the rest of the nucleus. The distance between nearest neighbors adjusts itself to its most favorable value (1.9×10^{-13} cm) independently of how many other nucleons are also in the nucleus.

The force between two nucleons seems to be almost independent of whether they are protons or neutrons. To an accuracy of about 1%, the force between two protons is the same as the force between two neutrons or the force between a proton and a neutron. This is called **charge independence** of the nuclear forces. It is one justification for the use of the word nucleon, which de-emphasizes the difference between protons and neutrons. One obvious difference, of course, is that in the case of two protons there is an additional electrostatic repulsion, but at a distance of 1.9×10^{-13} cm this is in fact less than 1% of the nuclear force.

The Nucleus

13-3 Mass Defects

Since the mass of a neutron is slightly greater than the mass of a proton, nuclear masses would not be expected to be integral multiples of the mass of a proton. It is significant, however, that the rest mass of a nucleus is always less than the sum of the rest masses of the protons and neutrons it contains. This is most easily discussed in the case of the deuteron, which is the nucleus of deuterium and contains one proton and one neutron (figure 13-1). The relevant masses are listed below.

$$\text{Rest mass of proton} = 1.67239 \times 10^{-24} \text{ gm} \tag{13-5}$$
$$\text{Rest mass of neutron} = 1.67470 \times 10^{-24} \text{ gm} \tag{13-6}$$

Rest mass of proton

+

$$\text{Rest mass of neutron} = 3.34709 \times 10^{-24} \text{ gm} \tag{13-7}$$
$$\text{Rest mass of deuteron} = 3.34313 \times 10^{-24} \text{ gm} \tag{13-8}$$

$$\text{Mass defect} = 0.00396 \times 10^{-24} \text{ gm} \tag{13-9}$$

The difference between the mass of a nucleus and the sum of the masses of its constituent nucleons is called its **mass defect.** For the deuteron, this is seen to be 3.96×10^{-27} gm.

Imagine a proton and a neutron initially at rest at an infinite distance apart. Then allow them to come together to form a deuteron. At distances less than 10^{-12} cm they have a strong attraction for one another, and this results in a negative potential energy, in exactly the same way that the attractive gravitational force produces a negative potential energy. When the deuteron has formed, this negative potential energy is only partially offset by the positive kinetic energy of the two nucleons as they move around inside the nucleus. The energy of the deuteron is consequently less than the energy of the proton and neutron at rest at infinity. The difference is called the **binding energy.**

Now let us make use of Einstein's principle of the interchangeability of mass and energy. If m_p is the rest mass of the proton and m_n the rest mass of the neutron,

Total energy of neutron and proton at rest at infinity
$$= m_p c^2 + m_n c^2 \tag{13-10}$$

If m_d is the rest mass of the deuteron,

$$\text{Total energy of deuteron} = m_d c^2 \tag{13-11}$$

Consequently,

$$\text{Binding energy of deuteron} = (m_p c^2 + m_n c^2) - m_d c^2 \tag{13-12}$$
$$= (m_p + m_n - m_d)c^2. \tag{13-13}$$

The binding energy is the mass defect multiplied by c^2.

These ideas are confirmed by observations of the **photodisintegration of the deuteron.** When deuterium is bombarded by γ-rays, a deuteron can absorb a photon and make use of its energy to break apart into a proton and a neutron (figure 13-3). A part of the photon's energy equal to the binding energy is required to separate the neutron and proton and place them at rest a large distance apart. Any energy in excess of this appears as kinetic

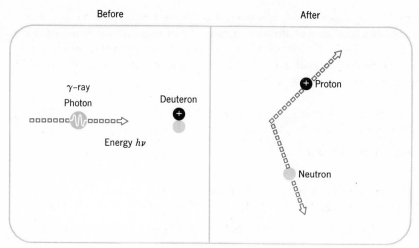

Before

γ-ray
Photon

Energy $h\nu$

Deuteron

After

Proton

Neutron

Figure 13-3 The photodisintegration of the deuteron.

energy of the nucleons after they have been separated. Clearly the process cannot occur at all unless the energy of the photon is at least as great as the binding energy. There is therefore a "threshold frequency," ν_c, given by

$$h\nu_c = (m_\mathrm{p} + m_\mathrm{n} - m_\mathrm{d})c^2 \tag{13-14}$$

Everything in this equation can be measured and its validity has been established with good accuracy. Inasmuch as it relates the energy of a photon to the rest masses of particles, it provides an excellent verification of the interchangeability of mass and energy.

In general, the binding energy of a nucleus is defined as the energy required to break up the nucleus into its constituent protons and neutrons and place them at rest at infinite distances from one another. The mass defect is defined as the difference between the rest mass of the nucleus and the sum of the rest masses of its constituent protons and neutrons. The binding energy is always equal to the mass defect multiplied by c^2. The mass defect divided by A, the number of nucleons in the nucleus, is a measure of the average binding energy per nucleon. This quantity indicates how strongly the nucleus is held together and how low is its energy as compared with separate neutrons and protons. In figure 13-4 it is plotted against the mass number A for naturally occurring nuclei. The curve has two features which will be important in our subsequent discussions. The helium nucleus, $_2\mathrm{He}^4$, stands out as being very strongly bound. The curve has a maximum at A = 56, and the most strongly bound nucleus is iron, $_{26}\mathrm{Fe}^{56}$.

13-4 *The Stability of Nuclei*

If the number of protons and the number of neutrons could be varied freely, the number of possible nuclei would be extremely large. Only a few of these possibilities have been found to exist and they are all shown in figure 13-5. Each nucleus is represented by a square symbol, ■, □, or ⊡. The center of this square has a horizontal coordinate equal to the number of

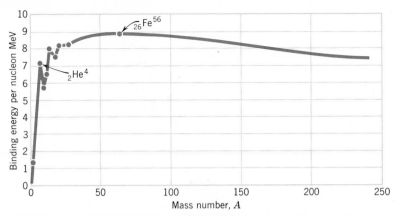

Figure 13-4 *The binding energy per nucleon and its variation with mass number for naturally occurring nuclei.*

protons, Z, in the nucleus and a vertical coordinate equal to the number of neutrons, N. All known nuclei lie within a narrow band, which should be compared with the straight line Z = N. For the lighter nuclei, the number of neutrons is never very different from the number of protons. For the heavier nuclei, there is a tendency for the number of neutrons to exceed the number of protons, but never by more than 60%.

The nuclei of figure 13-5 have been divided into three categories. Those denoted by the symbol ■ are stable nuclei which are found to occur naturally. They tend to lie near the center of the band. The nuclei denoted by the symbol □ are unstable and have to be produced artificially. They are very loosely bound and have an excessive amount of energy. They break up to form stable nuclei, which have less energy, and the excess energy is converted into kinetic energy of the fragments, enabling them to fly apart. The various ways in which this disintegration can occur will be discussed later. If these unstable nuclei ever existed in nature, they would long ago have disintegrated to form stable nuclei.

Some unstable nuclei do occur naturally. They are the **naturally radioactive nuclei** denoted by the symbol ⊡. There are two reasons for their natural occurrence. Some of them disintegrate very slowly, taking several billion years to disappear. Since there is reason to believe that the earth and its elements were formed only a few billion years ago, these long-lived unstable nuclei have not yet had time to disappear completely. Others of the naturally occurring radioactive nuclei are short-lived, but are produced during the disintegration of the long-lived radioactive nuclei. Thus, although they disappear soon after being formed, they are continually being replenished by disintegration of the long-lived nuclei. Notice that there are no stable nuclei beyond lead, which has Z = 82, and there are no naturally occurring nuclei at all beyond uranium, which has Z = 92.

It is possible to understand qualitatively why the stable nuclei have the compositions shown in figure 13-5. The first thing to appreciate is that a nucleus is stable when it has the lowest possible energy. Otherwise it would be able to undergo some change to a state of lower energy, giving off its excess energy in some way such as the emission of a photon or violent ejection of one of its particles. Suppose, then, that we try to build up nuclei by

13-4 The Stability of Nuclei

279

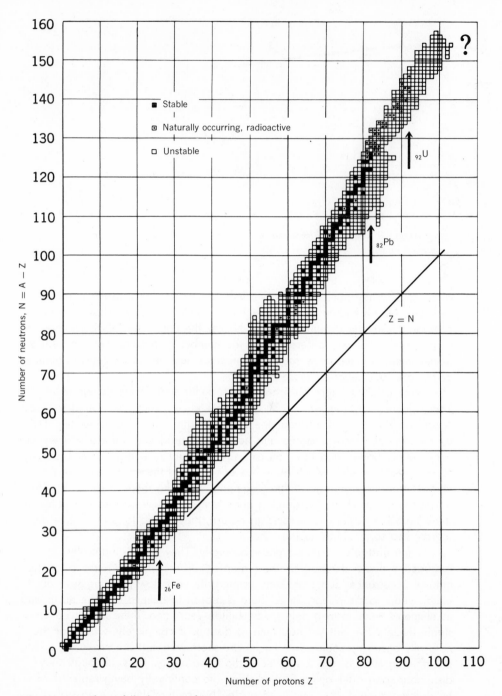

Figure 13-5 *Chart of the known nuclei.*

adding neutrons or protons one at a time in such a way that the energy of the nucleus is kept as small as possible.

As far as the nuclear forces and their potential energy are concerned, there is very little difference between a neutron and proton. However, a proton has the disadvantage that the electrostatic repulsion between the protons in the nucleus introduces a positive electrostatic potential energy. At

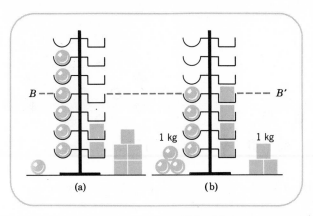

Figure 13-6 An analogy to illustrate the effect of the Pauli exclusion principle on the building up of a nucleus with the lowest possible energy. A fixed number of weights (8 in the case illustrated) must be placed on the rack with the expenditure of a minimum amount of work to lift the weights. The answer is to place an equal number of weights in the circular and square holders, as in (b). The advantage of (b) over (a) is that, instead of placing two spherical weights above the line BB', we place two cubical weights below this line.

first sight, it would therefore seem advantageous to construct a nucleus entirely from neutrons. In fact, there are no stable nuclei composed entirely of neutrons. Not even the di-neutron, composed of two neutrons bound together, is found to exist in a stable state. The explanation depends on the fact that protons and neutrons obey the Pauli exclusion principle, which plays as important a part here as it does in the case of electrons in atoms.

The neutrons in a nucleus must all be in different states of motion. The first neutron to go in may be put into the state of lowest energy, but succeeding neutrons must go into states of successively higher energy. However, a proton and a neutron, being dissimilar particles, may be in the same state of motion. The first proton that goes in may go into the state of lowest energy, even though there is a neutron already there. The second proton cannot go into the state of lowest energy because there is a proton already in it, but it can go into the next highest state, even though this is already occupied by a neutron. As far as this aspect of the situation is concerned, the lowest possible energy would be achieved by having equal numbers of protons and neutrons.

Consider the following analogy, which is illustrated in figure 13-6. A set of spherical 1 kilogram weights can be placed in the circular holders which are arranged in a vertical column at equal intervals of height above one another. A set of cubical 1 kilogram weights can be placed in square holders arranged in the same way at the same heights. The object of the game is to place a fixed number, A, of weights onto the rack with the minimum expenditure of work in lifting the weights, producing the arrangement with the minimum gravitational potential energy. The spherical weights represent neutrons and the cubical weights represent protons. There is a complete freedom of choice in selecting spherical or cubical weights, but not more than one weight may be placed in each holder. The answer is to choose an equal number of cubical and spherical weights, filling both sides of the rack up to the line *BB'*, as in *b*. In any other arrangement, such as *a*, one side of the rack must be filled above *BB'*, which requires a greater amount of work in lifting weights to a greater height than if the other side of the rack were first filled below *BB'*.

For the lighter nuclei the above consideration is the dominant one, and lighter nuclei do have approximately equal numbers of protons and neutrons (see figure 13-5). The other important feature of light nuclei is that the binding energy per nucleon increases (the energy per nucleon becomes *more*

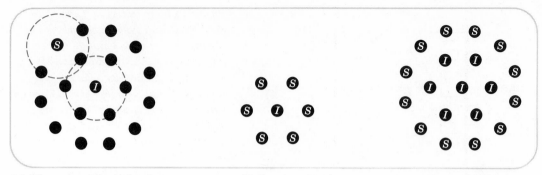

(a) *The nucleon I in the interior of the nucleus has 6 nearest neighbors. The nucleon S on the surface has only 3 nearest neighbors.*

(b) *Fraction of surface nucleons = $\frac{6}{7}$ = 0.857.*

(c) *Fraction of surface nucleons = $\frac{12}{19}$ = 0.632.*

Figure 13-7 *Surface energy of a nucleus. These patterns are 2-dimensional, of course, but a similar situation pertains in 3 dimensions.*

negative) as the nucleus becomes bigger (figure 13-4). The reason for this is that a nucleon in the interior of the nucleus is surrounded on all sides by other nucleons and has a negative mutual potential energy with each one. On the other hand, a nucleon at the surface of the nucleus has fewer nearest neighbors and a less negative total potential energy (figure 13-7a). As the nucleus becomes larger, the fraction of nucleons in the interior becomes larger and so the average potential energy per nucleon becomes more negative (figure 13-7b and c).

The positive electrostatic potential energy of the protons is not very important for lighter nuclei and has little effect on the arguments just advanced. However, it becomes progressively more important as the number of protons in the nucleus is increased. When we add another proton to a nucleus that already has Z protons, it acquires a *positive* electrostatic potential energy from each of the protons already present. For the heavier nuclei it becomes a real advantage to add an uncharged neutron rather than a proton, even though the exclusion principle requires the neutron to go into a state of higher energy. This is why the heavier nuclei contain appreciably more neutrons than protons. The existence of no stable nuclei above $Z = 82$ and no naturally occurring nuclei above $Z = 92$ is also related to the increasing importance of the electrostatic potential energy.

13-5 Nuclear Reactions

A chemical reaction is the reshuffling of the atoms of two reacting molecules to form different product molecules (section 5-1). A nuclear reaction is the reshuffling of the nucleons of two colliding nuclei to form different nuclei. In order to bring the two reacting nuclei close enough together, one of them must be accelerated to a high velocity, so that it will not be turned back too soon by the strong electrostatic repulsion between the two positively charged nuclei. The bombarding particle is usually a very light nucleus, such as a proton ($_1H^1$), a deuteron ($_1H^2$) or an α-particle (a helium nucleus,

Figure 13-8 *The first atom-smashing machine, built by Cockcroft and Walton in 1932. Protons were accelerated up to about 0.5 MeV. (Courtesy of J. Cockcroft.)*

$_2He^4$). Occasionally heavier nuclei are used. A simple method of achieving high impact velocities is to accelerate the particle through an electric potential difference of several million volts. One of the first "atom smashing" machines of this kind is shown in figure 13-8, and figure 13-9 shows a more modern version.

Let us consider the first nuclear reaction to be discovered (by Rutherford in 1919). A helium nucleus ($_2He^4$) strikes a nitrogen nucleus ($_7N^{14}$) and merges with it. A proton ($_1H^1$) is subsequently ejected, leaving behind an oxygen nucleus ($_8O^{17}$). This reaction is illustrated in figure 13-10 and is represented by the equation

$$_2He^4 + {_7N^{14}} \longrightarrow {_8O^{17}} + {_1H^1} \tag{13-15}$$

13-5 Nuclear Reactions

The 2 protons and 2 neutrons of $_2He^4$ mix with the 7 protons and 7 neutrons of $_7N^{14}$ to form a **compound nucleus** containing 9 protons and 9 neutrons. Since this compound nucleus has $Z = 9$ and $A = 18$, it must be an isotope of fluorine, $_9F^{18}$. This fluorine nucleus is not formed in its normal state, but in a state of much higher energy. This makes it very unstable and it gets rid of its excess energy by ejecting a proton. The compound nucleus exists for only a very short time (about 10^{-19} sec!) but, by nuclear time standards, that is long enough to give it a separate existence. The reaction

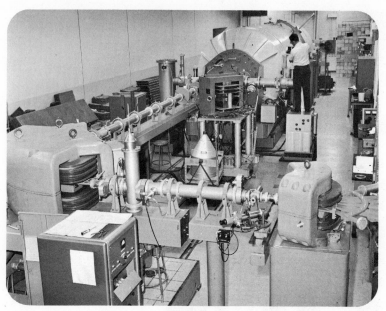

Figure 13-9 A modern atom-smashing machine. The tandem accelerator at the University of Pennsylvania. It accelerates light nuclei up to about 15 MeV.

should therefore be considered to take place in two separate stages

$$_2He^4 + {}_7N^{14} \longrightarrow [{}_9F^{18}] \tag{13-16}$$

$$[{}_9F^{18}] \longrightarrow {}_8O^{17} + {}_1H^1 \tag{13-17}$$

Nuclear reactions may proceed in many different ways. Except in the case of fission, which will be discussed in the next section, the compound nucleus usually ejects a proton or a neutron, or a small nucleus such as a deuteron, $_1H^2$, or an α-particle, $_2He^4$. Frequently the product nucleus is an

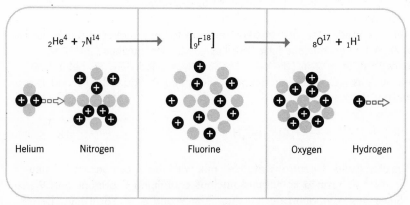

(a) *The helium nucleus strikes the nitrogen nucleus.*

(b) *A compound nucleus is formed and has a transitory existence.*

(c) *A proton is ejected and the nucleus left behind is an isotope of oxygen.*

Figure 13-10 A nuclear reaction.

unstable nucleus which does not occur in nature, and which is radioactive. It disintegrates over a period of time which is long compared with the typical life-time of a compound nucleus and it can therefore be separated out and studied at leisure. Such a nucleus is called a **radioisotope.** A few typical reactions are given below

$$_{1}H^{1} + {}_{13}Al^{27} \longrightarrow [{}_{14}Si^{28}] \longrightarrow {}_{12}Mg^{24} + {}_{2}He^{4} \qquad (13\text{-}18)$$

$$_{1}H^{2} + {}_{12}Mg^{26} \longrightarrow [{}_{13}Al^{28}] \longrightarrow {}_{11}Na^{24} + {}_{2}He^{4} \qquad (13\text{-}19)$$

$$_{0}n^{1} + {}_{13}Al^{27} \longrightarrow [{}_{13}Al^{28}] \longrightarrow {}_{12}Mg^{27} + {}_{1}H^{1} \qquad (13\text{-}20)$$

In the last equation the symbol $_{0}n^{1}$ represents a neutron. The use of a neutron as a bombarding particle has one particular advantage. Since the neutron is uncharged, it experiences no electrostatic repulsion as it approaches the target nucleus. A neutron with a very small velocity is therefore able to "trickle" into a nucleus and cause a reaction, whereas charged bombarding particles have to be accelerated to high velocities.

Since a nuclear reaction is a reshuffling process, the number of protons or neutrons remains unchanged. In the equation representing the reaction a subscript represents the number of protons in a nucleus and so the sum of the subscripts must be the same on both sides of the equation. In equation 13-18, for example, we start with 1 + 13 protons, form a compound nucleus with 14 protons, and end up with 12 + 2 protons. A superscript represents the total number of nucleons (protons and neutrons) in a nucleus and the sum of the superscripts must also be the same on both sides. In equation 13-18, we start with 1 + 27 nucleons, form a compound nucleus with 28 nucleons, and end up with 24 + 4 nucleons.

Nuclear reactions provide yet another means of checking the interchangeability of mass and energy. The sum of the rest masses of the reacting nuclei is never equal to the sum of the rest masses of the product nuclei. The difference is compensated by a difference in the kinetic energy of the reacting nuclei and the product nuclei. The sum of the rest mass energies of the reacting nuclei plus their kinetic energies *is* always equal to the sum of the rest mass energies of the product nuclei plus their kinetic energies.

13-6 Fission and Fusion

One of the important features of figure 13-4 is that the binding energy per nucleon of a heavy nucleus, with A in the vicinity of 200, is appreciably less than the binding energy per nucleon of a nucleus with A in the vicinity of 100. If the heavy nucleus were to split up into two approximately equal fragments, the average energy per nucleon would become more negative and kinetic energy would be made available for the fragments flying apart. Another way to look at this is that the combined rest mass of the two fragments would be less than the rest mass of the original nucleus and the difference would be converted into kinetic energy. This type of disintegration, in which a very heavy nucleus splits up into two parts of approximately equal size, is called **fission.** The nature of the fission fragments is not unique. A particular kind of heavy nucleus may split up in many different ways into a wide choice of product nuclei of intermediate masses.

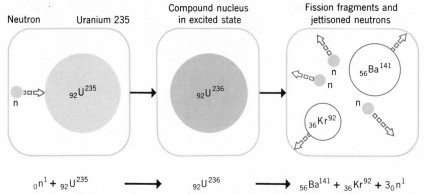

$$_0n^1 + _{92}U^{235} \longrightarrow _{92}U^{236} \longrightarrow _{56}Ba^{141} + _{36}Kr^{92} + 3_0n^1$$

Figure 13-11 Fission induced in uranium 235 by the addition of a neutron.

Although fission is energetically possible in naturally occurring heavy nuclei, it is usually a very rare event. For example, a nucleus of uranium 238, $_{92}U^{238}$, waits on the average for 8×10^{15} years before undergoing fission. It is more likely to emit an α-particle ($_2He^4$), which it does on the average after 4.5×10^9 years.

A heavy nucleus may be induced to undergo fission more rapidly by supplying it with extra energy so that it becomes more unstable. The most important way of doing this is to bombard the heavy nucleus with a neutron, which is able to enter the nucleus without having to be accelerated to a high velocity to overcome the electrostatic repulsion of the nuclear charge. In many cases the addition of an extra neutron produces a compound nucleus with an excess of energy which readily undergoes fission. The typical, but historically important, fission reaction of figure 13-11 is induced by adding a neutron to uranium 235.

Since very heavy nuclei contain a higher proportion of neutrons than stable nuclei of intermediate mass (figure 13-5), the fission fragments are initially formed with too many neutrons to be stable. They rapidly jettison these excess neutrons, in about 10^{-14} sec, and each fission usually results in the liberation of two or three free neutrons. Some of these neutrons may enter other fissionable nuclei and produce further fissions and still more neutrons. If, on the average, more than one neutron from each fission goes on to produce further fissions, the number of fissions taking place at each succeeding stage increases very rapidly (figure 13-12). This is called a **chain reaction.** Within a small fraction of a second, the number of nuclei having undergone fission corresponds to many grams of material and the energy released, initially in the form of kinetic energy of the fission fragments, is equivalent to the explosion of many thousands of tons of TNT. This is the principle underlying the atomic bomb.

If, on the average, exactly one neutron from each fission goes on to produce a further fission, the number of fissions occurring per second remains constant and the situation does not get out of hand (figure 13-13). This is the principle of a **nuclear reactor,** which is a controllable source of useful nuclear energy. A careful balance is achieved in its design, so that each fission produces more than one neutron which *could* produce a further fission, but a sufficient number of these neutrons are absorbed in nonfission processes to ensure that *exactly one* does go on to produce a further fission.

The Nucleus

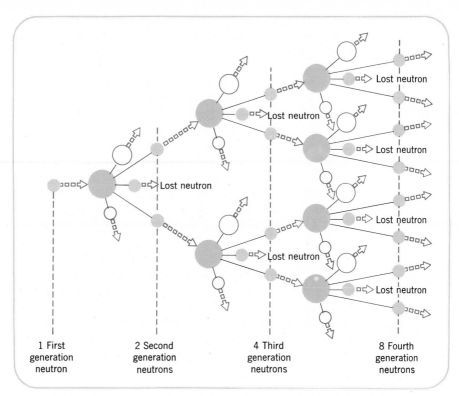

Figure 13-12 *An uncontrolled fission chain reaction. The principle of the atomic bomb.*

Another important feature of figure 13-4 is a strong tendency among the lighter nuclei for the binding energy per nucleon to increase with increasing mass number. This means that, if two light nuclei are fused together, the average energy per nucleon decreases and energy is released. This is called **nuclear fusion** and is the principle underlying the hydrogen bomb and thermonuclear power.

Before fusion can take place, the nuclei must be brought together against the opposition of their mutual electrostatic repulsion. This problem is the reverse of the escape velocity of a rocket (section 4-7). The rocket must be given sufficient initial kinetic energy to allow for the increase in its gravi-

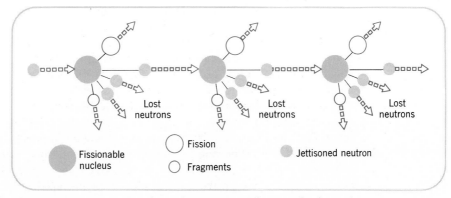

Figure 13-13 *A controlled fission chain reaction. The principle of a nuclear reactor.*

tational potential energy as it moves away from the earth. As the two nuclei move closer together, their mutual electrostatic potential energy becomes larger and more positive. Since energy is conserved, their kinetic energy decreases as their potential energy increases. They turn back when all their kinetic energy has been converted into potential energy. For fusion to take place, this must not happen before they are within the range of one another's *nuclear* forces.

Sufficiently high kinetic energies are readily achieved by accelerating one of the nuclei in an "atom smashing machine." In fact nuclear fusion is merely a special case of the nuclear reactions discussed in section 13-5. However, such experiments deal with a small number of nuclei and yield only small amounts of energy. To produce fusion in a large mass of material, the kinetic energies must be a result of the thermal motion of the nuclei, but sufficiently high velocities are realized only at temperatures of about 10^7 °K (ten million degrees absolute!) Nuclear fusion produced by a very high temperature is called a **thermonuclear reaction.**

Temperatures of 10^7 °K are not easy to achieve on earth, but they occur naturally near the center of a star, such as the sun. Thermonuclear reactions therefore proceed readily near the center of the sun and they are the immediate source of most of the energy poured out by the sun, or any similar star. The sun is the ultimate source of energy for all life on earth. Conventional fuels, such as coal, oil, and gasoline, are produced by decaying organic matter. The thermonuclear reactions at the center of the sun are therefore essential to our existence.

The thermonuclear reactions taking place inside the sun and the stars are not yet properly understood, but the net result is the fusing of four protons to produce a helium nucleus and some other small particles (positrons and neutrinos) which will be introduced to the reader later.

Source of stellar and solar energy

$$_1H^1 + _1H^1 + _1H^1 + _1H^1 \longrightarrow$$
$$_2He^4 + e^+ + e^+ + \nu + \nu \quad (13\text{-}21)$$

The binding energy of $_2He^4$ is particularly large (figure 13-4) and the decrease in mass in this fusion reaction is large. The reaction is therefore a very efficient way of converting mass into energy.

The explosion of an atomic bomb produces a temperature of about 5×10^7 °K, which is high enough to make fusion possible. A hydrogen bomb consists of a mixture of light nuclei, such as deuterium, tritium, and lithium, with an atomic bomb to act as a fuse and initiate the fusion of these light elements. The size of an *atomic* bomb is limited by the following design problem. Before firing there must be a situation in which most neutrons are lost and less than one per fission produces a further fission. The firing must produce a sudden change to a situation in which fewer neutrons are lost and more than one per fission produces a further fission. The size of a *hydrogen* bomb is not limited by these considerations and its destructive power depends only on the total quantity of light nuclei that can be incorporated into its design.

The Nucleus

288

Vigorous attempts are being made to produce controlled thermonuclear reactions in the laboratory, with a view to using them as a possible future source of commercial power. At the time of this writing the technical problems have not yet been solved.

13-7 Alpha-Rays and Gamma-Rays

The first nuclear transformations to be discovered were a consequence of natural radioactivity. In 1896 Henri Becquerel observed that uranium-bearing salts blackened a photographic plate in their vicinity, even if the plate was wrapped in several layers of paper opaque to visible light. During the next few years this was shown to be due to the emission of three kinds of rays. The α-rays were found to be positively charged and were identified as helium nuclei, $_2\text{He}^4$. The β-rays were found to be negatively charged and were identified as electrons. β-emission leads to some important new ideas and will not be discussed until the next chapter. The γ-rays were found to be photons, and they will be discussed shortly.

Alpha-emission is the ejection from the nucleus of an **α-particle,** which is a helium nucleus, $_2\text{He}^4$, containing two protons and two neutrons. The disintegrating nucleus therefore loses two protons and its atomic number Z decreases by 2. Altogether it loses 4 nucleons and its mass number is therefore decreased by 4. The general equation describing α-decay is

$$_Z\text{X}^A \longrightarrow \; _{Z-2}\text{Y}^{A-4} + \, _2\text{He}^4 \tag{13-22}$$

It is illustrated in figure 13-14. Although the possibility of one chemical element being transmuted into another is now a matter of common knowledge, its discovery had a dramatic impact on a scientific world that had come to regard the elements as distinct and immutable.

Alpha-emission occurs only for the heavier nuclei, in which the electrostatic repulsion between the protons has become inconveniently large. These are the nuclei which attempt to lower their energies by undergoing fission. Figure 13-4 reveals that the helium nucleus has a particularly large binding energy, which is not very much less than the binding energies of nuclei of intermediate mass. Suppose that a heavy nucleus with mass number A and mass M_A emits an α-particle of mass M_α and changes into a nucleus with mass number (A — 4) and mass M_{A-4}. This is energetically possible, with energy left over for kinetic energy, if M_A is greater than $(M_{A-4} + M_\alpha)$. This is found to be so for nuclei with Z greater than 82 and is the main reason why there are no permanently stable nuclei above lead, $_{82}\text{Pb}^{208}$.

Alpha-emission resembles fission inasmuch as it does not always occur readily, even when it is energetically possible. For example, a nucleus of bismuth, $_{83}\text{Bi}^{209}$, waits on the average 3×10^{17} years before emitting an α-particle. This is about one hundred million times the age of the earth!

After one of the many kinds of nuclear transformations we have been discussing, the product nucleus is sometimes formed in an excited state with excess energy. This excess energy is usually jettisoned in the form of one or more photons as the nucleus is de-excited to its normal state of lowest energy. Since nuclear energies are many times larger than the energies needed to excite electrons in atoms, the emitted photons have frequencies much

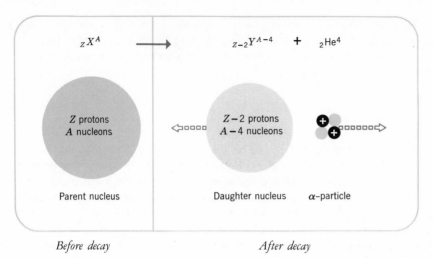

$$_ZX^A \longrightarrow \quad _{Z-2}Y^{A-4} \quad + \quad _2He^4$$

Z protons
A nucleons

Z − 2 protons
A − 4 nucleons

Parent nucleus

Daughter nucleus   α–particle

Before decay   *After decay*

Figure 13-14 *Alpha-emission.*

larger than visible light and wavelengths much shorter. Their frequencies are comparable with those of x-rays or even higher. They are the γ-rays and have energies up to about 20 MeV, as compared with about 5 eV for visible light.

The term γ-rays is customarily used to denote all high energy photons, whatever their origin. The highest energy γ-rays are now produced by accelerating charged particles to high energies and then decelerating them rapidly by allowing them to strike a target. This is exactly the same principle that is used in an x-ray tube. Photons with energies of several billion electron volts can be produced in this way. Their wavelengths are about 10^{-14} cm, which is less than the diameter of a proton or an electron.

Questions

1. Which of the following are possible reactions?

(a) $_1H^1 + _2He^3 \longrightarrow _2He^4$

(b) $_1H^1 + _3Li^7 \longrightarrow _2He^4 + _2He^4$

(c) $_{88}Ra^{224} \longrightarrow _{86}Rn^{219} + _2He^4$

(d) $_{94}Pu^{238} + _0n^1 \longrightarrow _{54}Xe^{141} + _{40}Zr^{97} + 2_0n^1$

(e) $_5B^{11} + _1H^1 \longrightarrow _4Be^8 + _2He^4$

(f) $_2He^4 + _{13}Al^{27} \longrightarrow _{15}P^{30} + _0n^1$

(g) $_1H^2 + _{15}P^{31} \longrightarrow _{13}Al^{29} + _2He^4$

The Nucleus

2. Insert the missing symbol.

(a) $_2He^4 + _6C^{12} \longrightarrow _7N^{15} + ?$

(b) $_{84}Po^{210} \longrightarrow _{82}Pb^{206} + ?$

(c) $? + _6C^{12} \longrightarrow _6C^{13} + _1H^1$

(d) $_1H^1 + ? \longrightarrow _2He^3 + _0n^1$

(e) $_{96}Cm^{240} + _0n^1 \longrightarrow ?$

3. What is wrong with the following sentence? "The half-life for α-decay of deuterium is 10.2 sec and so 25 percent of the original sample remains after 20.4 sec."

B

C

4. "In the photodisintegration of a nucleus, infinity means 10^{-12} cm or greater." Explain this statement.

5. Do you think a nucleus is more likely to be shaped like a sphere or a long thin cigar? Why?

6. Would you expect $_2He^3$ to have a greater or smaller binding energy than $_1T^3$? Why?

7. A 10 MeV proton strikes a stationary uranium nucleus, $_{92}U^{238}$. According to the theory of relativity, it is permissible to describe the collision in the frame of reference in which the incident proton is initially at rest. Find the kinetic energy of the approaching uranium nucleus in this new frame, and show that it is much greater than 10 MeV. This suggests the possibility of new reactions that cannot take place when the available kinetic energy is only 10 MeV, but that become possible when more initial kinetic energy is available. What is wrong with this argument?

Problems

A

1. If a nucleus emits a 20 MeV γ-ray, calculate the mass difference between its excited state and its ground state.

2. The α-particle emitted by $_{92}U^{235}$ has a kinetic energy of 4.58 MeV. Deduce the atomic mass of the $_{90}Th^{231}$ nucleus left behind after its emission. (Use the atomic masses given on the inside of the cover and express the answer in atomic mass units.)

3. $_{94}Pu^{239}$ emits a 5.2 MeV α-particle. Find its atomic mass as accurately as you can in atomic mass units. (Again use the atomic masses given on the inside of the cover).

B

4. Calculate the density of nuclear matter if the average distance between nucleons is 1.9×10^{-13} cm.

5. With the help of Heisenberg's uncertainty principle, make a rough estimate of the average velocity of a nucleon in a large nucleus in order to decide whether it is small compared with the velocity of light.

6. If photodisintegration of a deuteron is caused by absorption of a photon with a wavelength of 5×10^{-11} cm, how much kinetic energy is given to the escaping neutron and proton?

7. If the kinetic energy given to the proton and neutron after photodisintegration of a deuteron is 0.85 MeV, what was the frequency of the incident photon?

C

8. If a photon is capable of disintegrating a $_6C^{12}$ nucleus into three α-particles, set a lower limit on its frequency.

9. $_{92}U^{235}$ emits a 4.58 MeV α-particle. Find the velocity of the recoiling $_{90}Th^{231}$ nucleus.

10. Imagine that a 1 MeV proton strikes a stationary deuteron to produce the reaction

$$_1H^1 + {_1}H^2 \longrightarrow {_2}He^3$$

If the $_2He^3$ nucleus is produced in its state of lowest energy and no energy is lost by unknown processes, calculate the velocity of the $_2He^3$ nucleus. Consider the question of whether momentum can be conserved.

11. Approximately how much of the mass of 1 kg of iron is the mass equivalent of kinetic energy and potential energy, rather than rest mass of protons, neutrons, and electrons?

12. Calculate the classical escape velocity from the earth if it had its present radius but the density of nuclear matter. Comment.

13. Imagine a universe in which all the matter is in the form of an ideal monatomic gas of hydrogen at a temperature of about 10°K. Suppose that the hydrogen atoms are able to come together to form atoms of iron $_{26}Fe^{56}$ and that all the energy released is converted into thermal kinetic energy of these atoms. Calculate the final temperature of the ideal monatomic gas of iron vapor. At this temperature, would the iron remain in the form of neutral atoms?

14. Calculate the age of the sun from the following data. Assume that the solar radiation of 1.4×10^6 ergs per cm^2 per sec in the vicinity of the earth is a reliable guide to the total energy output of the sun, and that it has maintained this output at this level all its life. This energy is derived from the reaction of equation 13-21 and all other processes may be ignored. Five percent of the mass of the sun is now $_2He^4$, but it originally contained no $_2He^4$.

14 A Profusion of Particles and Processes

14-1 Beta-Decay

Beta-decay is the emission of an electron by a nucleus. All other properties of nuclei firmly indicate that they are composed of protons and neutrons, but contain no electrons. Where, then, does the electron come from? The only satisfactory explanation is that beta-decay occurs when a neutron inside the nucleus changes into a proton by emitting two small particles, an electron and an antineutrino. (The nature of neutrinos and antineutrinos will be explained in due course.) This differs fundamentally from the processes discussed in the previous chapter, which involved the reshuffling of neutrons and protons. We are now introducing a new and very important type of process in which a "fundamental" particle changes into other fundamental particles. This is often called a **fundamental process.**

Experiments on beams of free neutrons have shown that a neutron can indeed change into a proton with simultaneous emission of an electron. The half-life of the decay is about 11 minutes. The reader is reminded that this

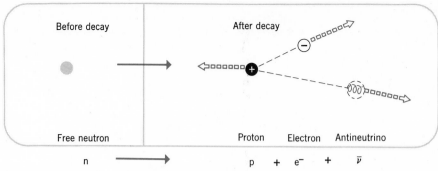

Figure 14-1 *Decay of a free neutron.*

means that, after 11 minutes half the number of neutrons originally present have decayed and the other half have not. A free neutron is therefore an unstable particle which does not live very long by human standards. The decay is depicted in figure 14-1 and is represented by the equation:

$$n \longrightarrow p + e^- + \bar{\nu} \qquad (14\text{-}1)$$
$$\text{neutron} \longrightarrow \text{proton} + \text{electron} + \text{antineutrino}$$

It is conventional to represent the neutron by the symbol n, the proton by p, the electron by e^-, and the neutrino by ν. The superscript associated with e indicates that the charge on the electron is negative. The proton might therefore be represented by p^+, but this is not normally done. The bar above the ν indicates that this particle is an *anti*neutrino rather than a neutrino. The distinction will be explained later.

When a neutron inside a nucleus changes into a proton, the electron and the antineutrino are violently ejected from the nucleus. The emitted electrons are the β-rays, which have long been known to constitute part of the radiation coming from radioactive materials. The nucleus loses one neutron and gains an extra proton. Its atomic number therefore increases from Z to Z + 1. The total number of neutrons plus protons remains the same and the mass number A is consequently unaltered. The process is depicted in figure 14-2 and is represented by the equation

$$_Z X^A \longrightarrow _{Z+1} Y^A + e^- + \bar{\nu} \qquad (14\text{-}2)$$

Why, then, do not all nuclei containing neutrons emit electrons? Why do not all neutrons decay into protons so that, after a time not very much longer than 11 minutes, very few neutrons are left and the universe is composed almost entirely of protons and electrons? The answer is that a neutron in a nucleus will spontaneously beta-decay into a proton if energy is released in the process, but not if additional energy has to be supplied to the nucleus before the decay can take place. The rest mass of a *free* neutron is greater than the sum of the rest masses of its decay products. Some mass is therfore available to be converted into the kinetic energies of the products, enabling them to fly apart. Inside a nucleus the situation is more complicated. The replacement of a neutron by a proton sometimes results in a siza-

A Profusion of Particles and Processes

294

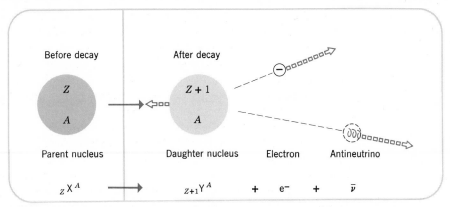

Figure 14-2 *Beta-decay. A neutron in the parent nucleus has become a proton in the daughter nucleus.*

ble *increase* in the energy of the nucleus. In a large nucleus this might be due to the positive electrostatic potential energy of the newly created proton and the other protons already in the nucleus. In addition, the Pauli exclusion principle does not allow the proton to be created in a state that is already occupied by an existing proton, and the available unoccupied states might all have energies in excess of the energy of the original neutron. In figure 13-6b, it is not possible to remove a spherical weight (neutron) from the left-hand rack and add a cubical weight (proton) to the right-hand rack without increasing the energy of the system. Thus, in many nuclei it is not possible for a neutron to change into a proton unless a large amount of energy is added from outside, which rarely happens in nature. In section 13-4 there was a discussion of the factors determining the most favorable ratio of neutrons to protons. Electron emission is most likely to occur in a nucleus with too many neutrons. The replacement of a neutron by a proton then brings the nucleus closer to the region of stable nuclei shown in figure 13-5.

14-2 The Neutrino

The neutrino is an elusive particle that does not make its presence felt in any obvious way. It was therefore originally thought that β-decay results in the emission of an electron only.

$$_{Z}X^A \longrightarrow _{Z+1}Y^A + e^- \tag{14-3}$$

If this process is to take place, the rest mass of the parent nucleus, M_X, must be greater than the rest mass of the daughter nucleus, M_Y, so that an amount of energy $(M_X - M_Y)c^2$ is released. Part of this is needed for the rest mass energy, m_0c^2, of the electron and the rest, $(M_X - M_Y - m_0)c^2$, is available for kinetic energy. Momentum must be conserved and the daughter nucleus $_{Z+1}Y^A$ must recoil in the opposite direction to the electron in much the same way that the earth recoils when a rocket is fired. The earth is much more massive than the rocket and the recoiling nucleus is also much more massive than the electron. Therefore, as explained in section 4-7, only a negligible fraction of the kinetic energy is given to the nucleus. We would therefore expect that the electron would always be emitted with the same velocity, v_m, corresponding to a kinetic energy $(M_X - M_Y - m_0)c^2$.

The electrons leaving the nucleus are actually observed to have all possible velocities from 0 up to v_m. A few of them do have a velocity v_m and the expected kinetic energy, but most of them are emitted with too small a velocity and some kinetic energy appears to be missing. Wolfgang Pauli suggested in 1930 that the missing kinetic energy is given to a new kind of particle emitted simultaneously with the electron. Enrico Fermi developed this idea and called the new particle a **neutrino,** which is Italian for "little neutral one." The particle accompanying an electron in β-decay is now known to be an *antineutrino,* but we shall ignore this distinction at the moment. The energy $(M_X - M_Y)c^2$ lost by the nucleus is shared between the electron and the neutrino in all possible proportions. Usually the electron is emitted with less than the total available energy. Occasionally, however, it receives all of the available energy and the neutrino receives none. This must mean that the rest mass of the neutrino is zero. Otherwise, a minimum amount of energy would be taken by the neutrino to provide its rest mass energy, even if it had zero velocity.

The only other known particle with zero rest mass is the photon. As explained in section 11-4, the photon, which has a velocity c, avoids having infinite energy by having zero rest mass. In the expression for its energy,

$$\epsilon = \frac{m_0 c^2}{\sqrt{1 - \dfrac{v^2}{c^2}}} , \qquad (14\text{-}4)$$

both the numerator and denominator are zero and the energy is not necessarily zero or infinite. The neutrino is similar. It has zero rest mass, but avoids having zero energy by always moving with the velocity of light.

The neutrino salvages not only the law of conservation of energy, but also the laws of conservation of linear momentum and angular momentum. If only an electron were emitted, the daughter nucleus would recoil in a direction exactly opposite to the direction of emission of the electron, with a linear momentum equal, but opposite, to that of the electron. It is found, however, that the electron and the nucleus do not usually fly apart in exactly opposite directions. Their two vector momenta therefore cannot possibly be added together vectorially to give zero, whereas the parent nucleus is usually moving so slowly that its momentum is very nearly zero. The assumption that a neutrino is simultaneously emitted remedies the situation. The vector sum of the three vectors representing the momenta of the daughter nucleus, the electron, and the neutrino is always found to be zero.

Similarly, we discover that angular momentum can be conserved during the decay only if the neutrino has a spin of $\frac{1}{2}(h/2\pi)$, like the electron. We can then see how a neutrino differs from a photon. Although they both have zero rest mass and the velocity of light, a neutrino has a spin of 1/2 a unit, whereas a photon has a spin of one whole unit, $h/2\pi$ (section 12-6).

Experimentally, it is very clear that a neutrino is not a photon. A γ-ray with the same energy as a neutrino would produce some easily observable effects such as the photoelectric effect (section 11-2) or the Compton effect (section 11-3). Interaction between a neutrino and any other form of matter is an extremely rare event. It was not until 1956, sixty years after the discovery of β-decay, that any such interaction was observed. The rarity of this interaction can be emphasized in the following way. A photon of appropriate energy passing through solid lead would travel on the average for about ten billionths of a second and cover a distance of about 300 cm before interacting with an atom. A neutrino with the same energy would travel for about fifty years before interacting! In order to stop the neutrino, we would have to use a lead shield with a thickness more than ten times the distance between the sun and the nearest star!

The weakness of its interaction with matter is conclusive proof that the neutrino carries no electric charge. (It is the "little neutral one"). Any charged particle moving rapidly through matter exerts strong electric fields on the atoms as it passes by and very soon pries loose an electron. The absence of charge on the neutrino may also be deduced from the **law of conservation of charge,** which says that there is no known process that creates or destroys electric charge (equal amounts of positive and negative charge being assumed to cancel one another to give zero net charge). The charge on a neutron is zero. When it decays, the positive charge of the proton is canceled by the negative charge of the electron. If the net charge is to remain zero, the neutrino must be uncharged.

Neutrinos are probably very abundant throughout the universe, although their presence is not very obvious because they interact so rarely with matter. The thermonuclear reactions taking place at the center of the sun produce neutrinos (equation 13-21). The sun is therefore continually bombarding the earth with a strong flux of neutrinos. As you read this, there are probably about one hundred thousand solar neutrinos inside your body passing through with the speed of light. This statement is independent of whether it is day or night because, even if the sun is on the other side of the earth, the neutrinos pass through the earth with negligible interference and then enter your body. As they pass through the earth, only about one in every trillion (10^{12}) is likely to interact with an atom of the earth. As far as the neutrinos passing through your body are concerned, you will probably have to wait several hours for the next interaction between one of these neutrinos and an atom of your body.

The neutrino, ν

Charge $= 0$	(14-5)
Rest mass $= 0$	(14-6)
Velocity $= c$ always	(14-7)
Intrinsic angular momentum $= \frac{1}{2}(h/2\pi)$	(14-8)

14-3 The Positron

The positron is identical with the electron in all respects, except that it is positively charged. It has exactly the same rest mass as the electron, its spin angular momentum is $\frac{1}{2}(h/2\pi)$, but its charge is $+e$ as compared with $-e$ for the electron. It is called the **antiparticle** of the electron and is represented by the symbol e^+, whereas the electron is represented by e^-.

Some nuclei decay by emitting a positron. This is called **positron-emission** and is very similar to electron-emission. The term β-decay is frequently used to denote both processes. Positron-emission may be looked on as the result of a proton in the nucleus changing into a neutron.

$$p \longrightarrow n + e^+ + \nu \qquad (14\text{-}9)$$

Notice that charge is conserved, since there is one unit of positive charge on the proton before the reaction, and the only charged product is the positron with one unit of positive charge. As in the case of electron-emission, a neutrino is needed, but this is now really a neutrino rather than its antiparticle, the antineutrino. (The reason for needing to distinguish between them and the difference between them will become clear later.)

A free proton does not decay into a neutron. This is energetically impossible, since the rest mass of a neutron is greater than the rest mass of a proton. However, in a nucleus that has too many protons, the substitution of a neutron for a proton often decreases the total energy. Positron-emission may be represented by the equation:

$$_Z X^A \longrightarrow {}_{Z-1} Y^A + e^+ + \nu \qquad (14\text{-}10)$$

The result is to lower the atomic number Z by one without changing the mass number A.

In some sense, not yet fully understood, a positron is the exact opposite

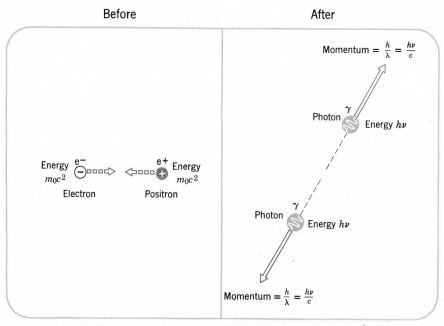

(a) *A positron and an electron approach one another. We are considering the important special case when their velocities are small compared with c.*

(b) *The electron and positron change into two photons with equal frequencies, moving in exactly opposite directions.*

Figure 14-3 *The annihilation of matter.*

of an electron. If a positron encounters an electron, they cancel one another out and both disappear. Their energy, momentum, and angular momentum reappear, usually in the form of two photons (although occasionally three or even more photons may be formed). This is called the **annihilation of matter** and is illustrated in figure 14-3. Using the conventional symbol γ to represent a photon, the equation describing the process is

$$e^+ + e^- \longrightarrow 2\gamma \tag{14-11}$$

This process is different from any other that we have so far discussed inasmuch as the rest mass is completely destroyed and completely converted into the energy of electromagnetic radiation. It therefore provides the final vindication of Einstein's principle of interchangeability of mass and energy. Consider the very common case when the electron and positron are moving so slowly that their kinetic energies are negligibly small compared with their rest mass energies. Then the total energy available before the collision is approximately $2m_0c^2$, where m_0 is the rest mass of either the electron or the positron. Since the total momentum before the collision is negligibly small, the two photons must be emitted in exactly opposite directions with equal, but opposite, momenta. The momentum of a photon is $h\nu/c$, and so the two photons must have equal frequencies and equal energies. The Einstein equation

$$E = mc^2 \tag{14-12}$$

takes the form

$$2h\nu = 2m_0c^2 \tag{14-13}$$

The energy of each photon is equal to the rest mass energy of an electron.

The reverse process of conversion of energy into rest mass occurs in **pair production,** which is sometimes called **the materialization of light.** If a photon strikes a nucleus, the strong electric field in the vicinity of the nucleus induces the photon to change into an electron-positron pair.

$$\gamma \longrightarrow e^+ + e^- \tag{14-14}$$

This is illustrated in figure 14-4. The initial energy of the photon is $h\nu$ and the energies of the electron and positron are m_ec^2 and m_pc^2. Here m_e and m_p are the "total" masses of the electron and positron, including both rest mass and kinetic energy. The law of conservation of mass plus energy is

$$h\nu = m_ec^2 + m_pc^2 \tag{14-15}$$

14-4 *Antiparticles*

Nearly all the fundamental particles have clearly distinguishable antiparticles. The antiparticle of the proton is the **antiproton,** which is represented by the symbol $\bar{p}$. A bar over the symbol for a particle always denotes its antiparticle. The antiproton is identical with the proton in all respects, except

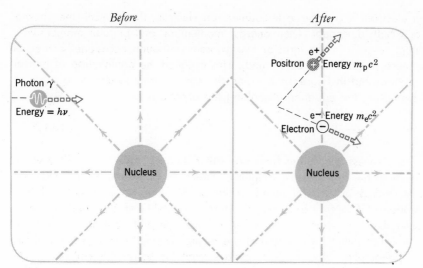

Figure 14-4 *The materialization of light (positron-electron pair production).*

that it is *negatively* charged. However, the difference between a particle and its antiparticle is more profound than a mere reversal of the sign of the charge. The uncharged neutron has an antiparticle, the **antineutron, $\bar{n}$,** which is clearly distinct from the neutron, as we shall see. We have already mentioned that there is an **antineutrino, $\bar{\nu}$.**

Since the rest mass of a proton is much larger than that of an electron, much more energy is needed to create a proton-antiproton pair than to create an electron-positron pair. The procedure is to accelerate a proton to an enormously high energy and then to allow it to strike a stationary proton. Both protons survive the collision, but part of the energy is used to create another proton and an antiproton.

$$(p + p) \longrightarrow (p + p) + (p + \bar{p}) \tag{14-16}$$

Sometimes a neutron and an antineutron are formed

$$(p + p) \longrightarrow (p + p) + (n + \bar{n}) \tag{14-17}$$

The first particle accelerator to achieve this feat was the Bevatron, which accelerates protons up to energies of 6.2 billion electron volts. It is shown in figure 14-5, which gives an impression of the enormous amount of effort and money devoted to research in this field.

The antiproton is a rare particle in nature. One that is produced in any way very soon encounters a proton and annihilates with it. The energy is usually converted into pions

$$p + \bar{p} \longrightarrow \pi^+ + \pi^0 + \pi^- \tag{14-18}$$

The pion, or π-meson, is yet another fundamental particle, and we shall say more about it later. In the particular reaction just quoted all three kinds of pion are produced with positive charge, negative charge, and no charge. A different number of pions may result, although conservation of charge re-

A Profusion of Particles and Processes

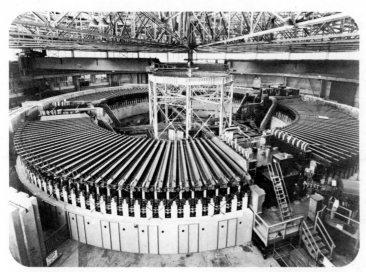

Figure 14-5 *This photograph shows the "Bevatron," a giant atom-smasher at the University of California Lawrence Radiation Laboratory, Berkeley. The machine accelerates protons to 6.2 billion electron volts. At the lower right is a linear accelerator that injects protons into a chamber in the giant circular 10,000 ton magnet (diameter, 120 feet). The protons circle this chamber 4 million times in 1.8 seconds, traveling over 300,000 miles, farther than from the earth to the moon. With this machine, the antiproton and numerous other particles have been discovered. Compare the size of the machine with the man in the bottom right-hand corner. (Courtesy of the Lawrence Radiation Laboratory, University of California, Berkeley.)*

quires that the number of positive and negative pions shall always be equal. Figure 14-6 is a bubble chamber photograph of an annihilation in which two positive and two negative pions are produced.

In a similar way, an antineutron can annihilate with a neutron. The reaction might be

$$n + \bar{n} \longrightarrow \pi^+ + \pi^+ + \pi^0 + \pi^- + \pi^- \qquad (14\text{-}19)$$

The existence of this process is a clear indication that an antineutron is different from a neutron, being in some way its opposite. Although they cannot be distinguished by the sign of their electric charge, there is at least one way in which they are clearly different. They are both spinning like tops and each has an angular momentum of $\frac{1}{2}(h/2\pi)$. Also, they both behave like small bar magnets. However, if a neutron and an antineutron are both spinning in the same direction, their bar magnets point in opposite directions (figure 14-7).

14-4

Antiparticles

The atoms of ordinary matter have positively charged nuclei composed of protons and neutrons, surrounded by negatively charged electrons. **Antimatter** has atoms with negatively charged nuclei, composed of antiprotons and antineutrons, surrounded by positively charged positrons (figure 14-8). Its properties are identical with those of ordinary matter in almost all respects, and it would be difficult to distinguish between them. Of course, if antimatter came into contact with ordinary matter, they would annihilate one

301

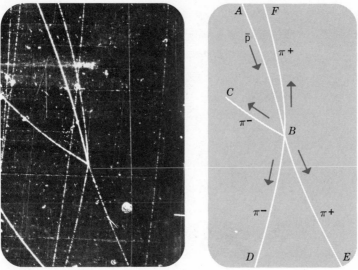

Figure 14-6 *A bubble chamber photograph of a proton annihilating with an antiproton. A white track is a chain of bubbles left behind by a charged particle as it passes through a superheated liquid. The incoming antiproton enters at A and collides with a proton at B. Since this proton is at rest it does not produce a track of bubbles. The four pions resulting from the annihilation produce the tracks BC, BD, BE, and BF. (Courtesy of the Lawrence Radiation Laboratory, University of California, Berkeley.)*

another and there would be an explosive release of large amounts of energy in the form of photons and pions. However, it is conceivable that some isolated regions of the universe are composed exclusively of antimatter. We may well be able to see some of these regions, but their visual appearance would be exactly the same as if they were composed of ordinary matter.

14-5 Sorting Out the Particles

The proliferation of the fundamental particles has proceeded to the point where, at the time this is being written, there are more than one hundred of them! It seems unreasonable to regard them all as fundamental, but there is no reason as yet to believe that any one of them is necessarily less funda-

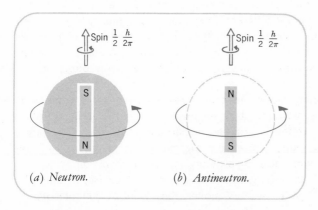

(a) *Neutron.* (b) *Antineutron.*

Figure 14-7 *If a neutron and an antineutron spin in the same direction, their magnetic effects are in opposite directions.*

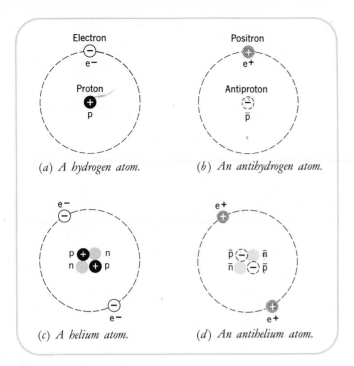

(a) A hydrogen atom.

(b) An antihydrogen atom.

(c) A helium atom.

(d) An antihelium atom.

Figure 14-8 *Matter and antimatter.*

mental than the original two, the electron and the proton. Although we have a very incomplete understanding of these particles, some progress has been made in classifying them and introducing order into their properties. Table 14-1 lists all the more important particles that live for a reasonably long time before decaying into something else. In this instance "reasonably long" means "much longer than 10^{-23} sec," which is characteristic of the half-life times of the very unstable particles, which have not been included in the table. The table also includes some of the properties of the particles and indicates some of the categories into which they can be classified. Notice that the rest mass increases as we go downward from the top of the table.

The first broad division is into **particles** and **antiparticles.** Every particle has a corresponding antiparticle. In the case of the three particles shown surrounded by a circle, the photon, the neutral pion, and the eta-meson, their properties are such that the particle and antiparticle are identical in all respects and cannot be distinguished from one another. (If -1, -2, -3, etc. are called the *antinumbers* of $+1$, $+2$, $+3$, etc., then 0 is indistinguishable from its antinumber.) The electric charge on a metastable fundamental particle is always $+e$, 0, or $-e$. If the particle is charged, the antiparticle has a charge of the opposite sign. Notice the interesting case of the sigma-particle, with three particles, Σ^+, Σ^0, and Σ^- and three distinguishably different antiparticles $\overline{\Sigma}^-$, $\overline{\Sigma}^0$, and $\overline{\Sigma}^+$.

The second broad division is into **fermions** and **bosons.** A fermion has a half-integral spin (1/2, 3/2, 5/2, etc.) and obeys the Pauli exclusion principle. Two fermions of the same kind cannot be in the same state. A boson has zero or integral spin (0, 1, 2, etc.), and is not restricted by the Pauli exclusion principle. Two bosons of the same kind are allowed to be in the same state.

Table 14-1 *The Metastable Particles*

	NAME	PARTICLES +e	PARTICLES 0	PARTICLES −e	ANTIPARTICLES +e	ANTIPARTICLES 0	ANTIPARTICLES −e	REST MASS (UNITS SUCH THAT REST MASS OF ELECTRON = 1)	SPIN. UNITS OF $\frac{h}{2\pi}$
Photon	photon		γ			γ		0	1 (Boson)
Leptons	neutrino, antineutrino		ν			$\bar{\nu}$		0	$\frac{1}{2}$ (Fermions)
	μ-neutrino anti-μ-neutrino		ν_μ			$\bar{\nu}_\mu$		0	
	electron, positron			e^-	e^+			1	
	muons			μ^-	μ^+			207	
Mesons	π-mesons (pions)	π^+	π^0			π^0	π^-	264 / 273	0 (Bosons)
	K-mesons (Kaons)	K^+	K^0			$\bar{K}^0$	K^-	966 / 974	
	η-meson (eta)		η^0			η^0		1074	
Baryons	proton, antiproton neutron, antineutron	p	n			$\bar{n}$	$\bar{p}$	1836.1 / 1838.6	$\frac{1}{2}$ (Fermions)
	Lambda		Λ^0			$\bar{\Lambda}^0$		2183	
	Sigma	Σ^+	Σ^0	Σ^-	$\bar{\Sigma}^+$	$\bar{\Sigma}^0$	$\bar{\Sigma}^-$	2328 / 2334 / 2343	
	Xi		Ξ^0	Ξ^-	$\bar{\Xi}^+$	$\bar{\Xi}^0$		2573 / 2586	
	Omega			Ω^-	$\bar{\Omega}^+$			3276	$\frac{3}{2}$

The fundamental particles are conveniently divided into four groups: (1) the Photon, (2) Leptons, (3) Mesons, and (4) Baryons. Mesons and baryons are sometimes grouped together and called **hadrons**. We are already familiar with the photon. It is the particle associated with the electromagnetic field. It has no charge and no rest mass and moves with the velocity of light. Its angular momentum is 1 unit of $h/2\pi$, so it is a boson and does not obey the Pauli exclusion principle.

Leptons all have a spin of 1/2 a unit and are therefore fermions, obeying the Pauli exclusion principle. The best-known members of the group are the electron, the positron, the neutrino, and the antineutrino. In addition there are two peculiar particles μ^- and μ^+, called **muons.** Associated with these muons are two particles, ν_μ and $\bar{\nu}_\mu$, which are very similar to neutrinos and which will be called **μ-neutrinos.**

The properties of the muons, μ^- and μ^+, are identical with the properties of the electron and positron, e^- and e^+, in all except two respects. A muon has a rest mass which is equal to 207 electron rest masses, and it is therefore sometimes described as a "heavy electron." In addition, a muon has a half-life of only 2.2×10^{-6} sec, for it very soon decays into an electron, a μ-neutrino, and an ordinary antineutrino.

$$\mu^- \longrightarrow e^- + \bar{\nu} + \nu_\mu \qquad (14\text{-}20)$$

One reason for grouping together electrons, muons, and neutrinos is the **law of conservation of leptons.** For the purpose of this law, each of the particles ν, ν_μ, e^-, and μ^- is to be counted as $+1$ lepton, whereas each of the antiparticles, $\bar{\nu}$, $\bar{\nu}_\mu$, e^+, and μ^+ is to be counted as -1 lepton. The law then states that a process in which fundamental particles are transformed into other fundamental particles can occur only if it leaves the total number of leptons unaltered. The number of leptons in the universe remains constant, provided that the sum of a particle and an antiparticle is counted as zero.

For example, in equation 14-20 the original negative muon, μ^-, counts as $+1$ lepton. Among its products, e^- counts as $+1$, $\bar{\nu}$ as -1, ν_μ as $+1$, and the law is satisfied because $+1 = +1 - 1 + 1$. In the fundamental equation of β-decay,

$$n \longrightarrow p + e^- + \bar{\nu} \qquad (14\text{-}21)$$

the original neutron, n, is not a lepton and counts as 0. The proton, p, also counts as 0, while the electron, e^-, counts as $+1$ and the antineutrino, $\bar{\nu}$, as -1. The law is then satisfied because $0 = 0 + 1 - 1$. This is the reason for establishing the convention that the neutrino accompanying electron-emission is the *anti*neutrino.

It seems probable that there is also a **law of conservation of μ-leptons,** in which μ^- and ν_μ are counted as $+1$ and μ^+ and $\bar{\nu}_\mu$ as -1. For example, in equation 14-20 we start with $+1$ μ-lepton, μ^-, and the only μ-lepton among the products is ν_μ, which also counts as $+1$. This law excludes, for example, the process

$$\mu^- \longrightarrow e^- + \bar{\nu} + \nu \qquad (14\text{-}22)$$

which conserves leptons of both types, but does not conserve μ-leptons by themselves. There is initially $+1$ μ-lepton, but there are no μ-leptons among the products.

Mesons have no spin, are bosons, and are not restricted by the Pauli exclusion principle. Any number of mesons may be in the same state. There is no conservation law for mesons analogous to the law of conservation of leptons. For example, a charged pion, which is a meson, decays into a muon and a μ-neutrino, which are both leptons

$$\pi^+ \longrightarrow \mu^+ + \nu_\mu \tag{14-23}$$

Pions interact strongly with nucleons and play an important role in the theory of nuclear forces, as we shall see later. They have been called "the glue that holds the nucleus together."

All **baryons** have half-integral spin, are fermions and obey the Pauli exclusion principle. The best-known baryons are the nucleons—the proton, the antiproton, the neutron, and the antineutron. The other members of the group are the lambdas, the sigmas, the xis, and the omegas, all of which have rest masses greater than the rest masses of the nucleons.

As in the case of leptons, there is a **law of conservation of baryons.** Particles count as $+ 1$ baryon, antiparticles as $- 1$ baryon, and particles outside the group as 0. In the now familiar equation for β-decay,

$$n \longrightarrow p + e^- + \overline{\nu} \tag{14-24}$$

the original neutron counts as $+ 1$ baryon. Among the products, the proton is $+ 1$ baryon and the electron and antineutrino are both leptons and count as 0 baryons. In the equation describing the method of producing antineutrons,

$$p + p \longrightarrow p + p + (n + \overline{n}) \tag{14-25}$$

p and n each count as $+ 1$, while $\overline{n}$ counts as $- 1$. So we start with $+ 2$ baryons and end up with $3 - 1$ baryons.

In recent years there has been considerable progress in sorting all these particles—stable, metastable, and unstable—into groups and subgroups. There is a partial understanding of the physical difference between groups, between subgroups within a group and between particles within a subgroup. We may be close to a much better appreciation of what is really going on here.

Questions

A

1. "Just before falling under the death-rays of the Martians' guns, the scientist has time to scribble down part of his secret formula
$$_{151}X^{362} \longrightarrow {}_{150}X^{362} + \cdots"$$
Can you guess what type of process he is describing and fill in the missing part?

2. Refer to figure 13-5 and decide whether the following nuclei are more likely to emit electrons or positrons. (a) $_{55}Cs^{145}$, (b) $_{80}Hg^{192}$, (c) $_{38}Sr^{81}$, (d) $_{36}Kr^{94}$, (e) $_{20}Ca^{49}$.

3. Rewrite equations 14-20 to 14-23 with all particles replaced by their antiparticles.

B

4. The particles e^+, e^-, p, and $\bar{p}$ are all equivalent to small bar magnets. In each case consider whether the angular momentum vector points from the south pole toward the north pole, or in the opposite direction.

5. A new particle, the subiota, and its antiparticle, the antisubiota, are produced in the following way. X-rays from a tube with 10,000 volts across it travel an average distance of 10^5 cm along a partially evacuated tube and then create subiota pairs. On the average the subiotas return to the x-ray tube 2×10^{-4} sec after the x-ray pulse is sent out. How much can you deduce about the properties of the subiota from this information?

6. For each of the following equations, state which conservation laws are violated.

(a) $\mu^+ \longrightarrow e^+ + \nu + \gamma$
(b) $\pi^+ \longrightarrow \mu^+ + \nu_\mu + \bar{\nu}$
(c) $n \longrightarrow p + \bar{p} + \pi^0$
(d) $\Sigma^+ \longrightarrow \Lambda^0 + \bar{p} + \pi^0$
(e) $p \longrightarrow \Sigma^+ + K^0$

7. For each of the following descriptions make a list of all the particles you know which satisfy this description:

(a) The antiparticle of the photon.
(b) A lepton moving with the speed of light.
(c) The meson of smallest rest mass.
(d) A "heavy electron"
(e) A baryon with a rest mass less than 1200 MeV.

C

8. From the atomic masses quoted on the inside of the cover, can you decide whether tritium, $_1T^3$, is able to emit an electron? It does in fact emit an electron with a maximum kinetic energy of 0.019 MeV. Is this consistent with the atomic masses quoted? What can you deduce from the maximum energy of the emitted electron? (Hint: The figures quoted are the *atomic* masses, not the *nuclear* masses.)

9. Occasionally an electron in an atom is captured by the nucleus which undergoes a kind of "inverse beta-decay." Can you guess exactly what is happening? If a negative muon were captured by a nucleus in a similar way, what would the process be?

10. Well up beyond the tropostrata
There is a region stark and stellar
Where, on a streak of anti-matter,
Lived Dr. Edward Anti-Teller.

Remote from Fusion's origin,
He lived unguessed and unawares
With all his anti-kith and kin
And kept macassars on his chairs.

One morning, idling by the sea,
He spied a tin of monstrous girth
That bore three letters: A.E.C.
Out stepped a visitor from Earth.

Questions

307

Then, shouting gladly o'er the sands,
Met two who in their alien ways
Were like as lentils. Their right hands
Clasped, and the rest was gamma rays.

H. P. F.

(Copyright 1956 The New Yorker Magazine, Inc.)

Do the last two words of this poem provide a complete description of the ultimate remnants of the encounter?

Problems

A

1. What is the wavelength of the lowest energy photon capable of pro- ducing a proton–antiproton pair? (Ignore the question of conservation of linear momentum, and assume that the proton and antiproton can be created with zero velocities.)
2. A positron and electron with negligible kinetic energies annihilate one another to produce two photons. What are their frequencies?
3. A 1 MeV electron encounters a 1 MeV positron traveling in exactly the opposite direction. What are the wavelengths of the two photons pro- duced?

B

4. An iota has a rest mass equal to 50 electron rest masses. What is the frequency of a photon which is just able to produce an iota anti-iota pair? (Ignore the correction caused by the need to conserve momen- tum.)
5. Calculate the energy in ergs and electron volts that is released when a helium atom encounters an antihelium atom. Include the energy of all the products.
6. If an η-meson at rest decays into two photons, calculate their wave- lengths.
7. Is it energetically possible for an η-meson to decay into (a) four pions (which four?), (b) five pions?
8. What is the momentum of a photon that has just sufficient energy to create an electron-positron pair? Calculate the velocity of an electron with an equal momentum. What is the kinetic energy of this electron in electron volts?
9. A newly discovered particle and its antiparticle come slowly together and produce two photons. One of these photons creates an electron- positron pair. Both the electron and the positron have velocities of 0.99c. Calculate the rest mass of the new particle.

C

10. 10^{-12} gm of antihelium is introduced into 4 gm of helium gas. Put an upper limit on the resulting rise in temperature. Why will the actual rise in temperature be less than this?
11. What is the maximum kinetic energy in ergs and in electron volts of an electron produced by the β-decay of a free neutron at rest?

12. Show that, in the absence of a nearby nucleus, the reaction

$$\gamma \longrightarrow e^+ + e^-$$

cannot possibly conserve both energy and momentum. How does the nearby nucleus save the situation?

13. Calculate as precisely as you can the energy in electron volts released in the fusion process of equation 13-21 when four protons fuse to form a helium nucleus. Use the atomic masses given on the inside of the cover and remember that they are not nuclear masses. Include any energy that might result from the subsequent fate of the positrons. Explain why all of the energy released is not necessarily available to convert into useful power. Calculate the maximum amount that might not be available. What further information would you need in order to calculate more precisely the actual available energy in a macroscopic process.

14. Light takes about 10^{10} years to travel across the observable universe. The average density of matter in the whole of the universe probably lies somewhere between 10^{-30} gm/cm^3 and 10^{-28} gm/cm^3. Does a neutrino have a good chance of traveling right across the observable universe without being absorbed? How could you have guessed this without doing a calculation?

15 *Inklings*

15-1 *Forces, Interactions, and Fundamental Processes*

The long list of "fundamental" particles in the previous chapter is a clear indication that physics has not yet arrived at a simple, elegant description of
the universe. Nevertheless, some orderliness is beginning to emerge, and
there are inklings of a new conception of the basic nature of the universe.

The universe of Newtonian mechanics consisted of isolated bodies moving through empty space and exerting instantaneous forces on one another.
The classical theory of electromagnetism, achieving its final expression in
Maxwell's equations, shifted the emphasis to a field existing at all points in
space. Interactions between charged bodies took place through the interme-
diary of this field and traveled from one body to another with the speed of
light. The picture of energy, in the form of an electromagnetic wave, traveling through the field from one body to a distant body created the impression that the field was a very "real" entity. This impression was strengthened
by the quantum mechanical description of an electromagnetic wave as a

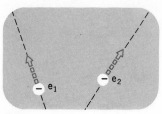

(a) Two electrons, e_1 and e_2.

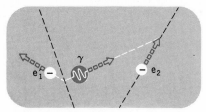

(b) e_1 emits a photon.

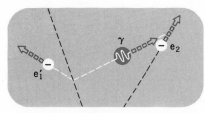

(c) The photon in flight.

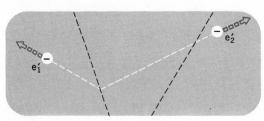

(d) The photon is absorbed by e_2.

Figure **15-1** *The electromagnetic interaction between two electrons can be described in terms of the interchange of photons between them.*

stream of photons. In the list of fundamental particles, the photon has a valid claim to inclusion on an equal footing with the charged particles that exert electromagnetic forces on one another. In fact, the electromagnetic interaction between two charged particles, instead of being described in terms of these forces, may now be visualized somewhat differently in the following way.

In figure 15-1 we consider two interacting electrons, but the argument is applicable to any two charged fundamental particles. The electron e_1 first emits a photon by means of the fundamental process

$$e_1 \longrightarrow e_1' + \gamma \qquad (15\text{-}1)$$

The prime on the symbol e_1' is intended to indicate that this electron has transferred to a different state of motion, giving up energy and momentum to the photon. The photon then travels through space with a velocity c until it reaches the other electron e_2, which absorbs it by means of the fundamental process

$$\gamma + e_2 \longrightarrow e_2' \tag{15-2}$$

The net result is that the two electrons have exchanged energy and momentum and each has changed its state of motion. In classical terms, the two electrons have "accelerated" one another. As the electrons proceed on their way, a continual stream of photons passes backward and forward between them, maintaining a continuous interaction between them.

The electromagnetic interaction is consequently seen to depend on the two fundamental processes of equations 15-1 and 15-2. However, there are many different fundamental processes and they are all presumably capable of producing interactions in a similar way. We shall now turn to the fundamental processes responsible for nuclear forces.

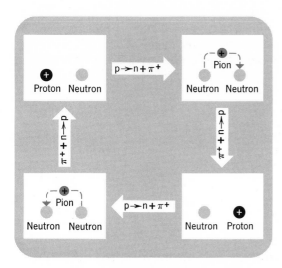

Figure 15-2 *The nuclear force between a proton and a neutron is partly due to the passage of a positive pion, π^+, backward and forward between them. Other contributions to the force arise from a similar interchange of negative pions, π^-, or neutral pions, π^0.*

15-2 Nuclear Forces

Electromagnetic forces are caused by the passing of photons backward and forward between charged particles. Nuclear forces are caused by the passing of *pions* backward and forward between nucleons. Consider a proton and a neutron inside a nucleus (figure 15-2). The proton can emit a positive pion and change into a neutron, the fundamental process being

$$p \longrightarrow n + \pi^+ \tag{15-3}$$

This pion then travels to the original neutron, which absorbs it and changes into a proton. This second fundamental process is the reverse of the previous one.

$$\pi^+ + n \longrightarrow p \tag{15-4}$$

As a result, the two nucleons have not only exchanged energy and momentum, but the proton and neutron have exchanged places. The neutron which caught the pion, and changed into a proton, then throws it back again and restores the situation to its initial state. The pion is continually thrown backward and forward, producing a continuously acting nuclear force between the two nucleons.

There are similar processes involving the exchange of a negative pion, π^-, or a neutral pion, π^0. They also make their contributions to the nuclear force.

Since the rest mass of a pion is about 140 MeV, a nucleon must have an excess energy greater than this to be able to emit a pion. Energies this large are not normally available inside a nucleus, so we are faced with the question of how the processes we have been discussing can possibly occur. The answer is provided by the uncertainty principle, which allows us to violate the law of conservation of energy by an amount ε for a time τ, if

$$\tau \approx \frac{h}{\varepsilon} \tag{15-5}$$

If the energy is uncertain by an amount ε, we obviously cannot require that it should remain constant to an accuracy better than ε.

If m_π is the rest mass of a pion, the extra energy needed by a nucleon to emit a pion is

$$\varepsilon \approx m_\pi c^2 \tag{15-6}$$

This extra energy is available only for the very short time

$$\tau \approx \frac{h}{m_\pi c^2} \tag{15-7}$$

$$\approx 3 \times 10^{-23} \text{ sec} \tag{15-8}$$

The velocity with which the pion is emitted is certainly less than c, and so the maximum distance it can hope to escape from the nucleon during the time τ is

$$R_n = c\tau \tag{15-9}$$

$$\text{or, } R_n \approx \frac{h}{m_\pi c} \tag{15-10}$$

$$\text{that is, } R_n \approx 0.89 \times 10^{-12} \text{ cm} \tag{15-11}$$

A particle that is produced in this way, without enough energy being available to ensure it a permanent existence, is called a **virtual particle.** The process that produces it is called a **virtual process.** The virtual pion exists only by the grace of the uncertainty principle, and it must return to its parent nucleon before its time runs out. The situation can be described more precisely in terms of probability. There is a good chance that the pion will be emitted, travel a distance less than R_n, and then return. There is even a small chance that it will travel much further than R_n, but this is a rare occurrence and becomes less and less likely the further the pion travels.

During its journey the virtual pion may encounter another nucleon and be absorbed by it. In this case an exchange has taken place and we are discussing the process responsible for nuclear forces. The exchange is likely to occur only when the distance between the two nucleons is not much greater than $R_n = 0.89 \times 10^{-12}$ cm. This explains the fact that nuclear forces are short-range forces and become very weak when the distance between the two nucleons exceeds a range of the order of 10^{-12} cm.

This theory provides us with a very interesting picture of an isolated nucleon. It must be imagined as continually emitting pions and snatching them back again. A proton, or neutron, is surrounded by a cloud of virtual pions, most of which are within 10^{-13} cm of the center, but a few stray out to 10^{-12} cm or further. For an appreciable fraction of the time a proton is split up into a neutron and a positive pion,

$$p \rightleftharpoons n + \pi^+ \tag{15-12}$$

Similarly, for an appreciable fraction of its existence a neutron consists of a proton and a negative pion,

$$n \rightleftharpoons p + \pi^- \tag{15-13}$$

This picture can explain the puzzling observation that the neutron has magnetic properties, like a small bar magnet. Since magnetic effects are caused by rotating electric charges, it is difficult to see how they can be produced by an uncharged neutron. The explanation is that, for part of the time, the neutron consists of a proton with a doughnut-shaped cloud of negative pions rotating around it (figure 15-3).

15-3 Agitated Chaos

In place of the old concept of force, mysteriously acting at a distance, we now have a very different picture of two bodies exerting forces on one another by throwing particles backward and forward between themselves. In the case of an electromagnetic force, the particle thrown backward and forward is the photon. In the case of a nuclear force, it is the pion. The force on a particle is defined classically as the momentum transferred to it per sec-

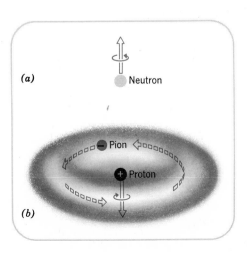

(a)

Neutron

(a) *Only part of the time is the neutron really a single uncharged particle.*

Pion

Proton

(b)

(b) *For part of the time it is a proton surrounded by a doughnut-shaped rotating cloud of negative pions (compare figure 12-8c and the associated discussion of clouds of probability).*

Figure 15-3 *The explanation of the neutron's magnetic properties.*

ond, and this momentum is now seen to come from the photons or pions that the particle absorbs or emits.

There is a similar change in our concept of the nature of a field. Previously we have considered a field to be a peculiar property of a point in space which enables it to exert a force on a particle placed there. For example, the electric field **E** at a point is such that an electron placed at that point experiences a force $-e\mathbf{E}$. We can now form a much more vivid picture of what is happening. The electron placed at the point experiences an electromagnetic force because it absorbs some of the real and virtual photons already present in the vicinity of the point.

The field **E** at a point can be considered to be caused by the presence of charged particles at distant points. Each of these charged particles is continually emitting real and virtual photons. A real photon corresponds to an electromagnetic wave radiated outward, never to return. A virtual photon, on the other hand, acquires its energy because the uncertainty principle allows a temporary violation of the law of conservation of energy, but it must eventually return to the charge that emitted it. The electromagnetic field surrounding a charged particle may therefore be visualized as a cloud of photons increasing in density as the particle is approached. The reader should compare section 11-5, where the oscillating electric vector of an electromagnetic wave was related to the density of photons in space.

We thus obtain a very peculiar picture of apparently empty space. A small volume of space contains virtual photons coming from nearby charges and virtual pions coming from nearby nucleons. It also contains other virtual particles related to other types of interaction and their corresponding fundamental processes. However, there is even more. If we look at the small volume for a short time τ, the uncertainty principle may allow it to have an extra energy $\varepsilon \sim h/\tau$. If τ is very small, this energy may be large enough to create an electron-positron pair or even a proton-antiproton pair or some other allowed combination of particles. The pair must, of course, annihilate again before its time runs out. Space is therefore in a state of agitated chaos, full of virtual particles and virtual pairs, which are continually being created and almost immediately annihilated.

Potential energy can also be viewed in a more revealing light. Consider an isolated charged particle at rest surrounded by its cloud of virtual photons and concentrate on a "snapshot" taken at a fixed instant of time. Each virtual photon has an energy $h\nu$, and the sum of the quantities $h\nu$ for all the photons present may be looked on as the potential energy residing in the electric field. An identical argument can be applied to a second charged particle far away from the first. Now bring the two particles together until they are a distance R apart. The electric field is then different, and the dis-

Table 15-1   The Four Types of Interaction

TYPE OF INTERACTION	RELATIVE STRENGTH (COUPLING CONSTANTS)
Strong interactions	1
Electromagnetic interactions	10^{-2}
Weak interactions	10^{-14}
Gravitational interactions	10^{-40}

tribution of virtual photons is different. The sum of the quantities $h\nu$ is not the same as when the two charged particles were far apart, and the difference is nothing more than the quantity q_1q_2/R, which was previously called mutual electrostatic potential energy. Thus, even the rather nebulous concept of potential energy finds a ready explanation in terms of virtual particles and fundamental processes.

15-4 Types of Interaction

There are very many fundamental processes and consequently many different ways in which fundamental particles can interact. Fortunately, all interactions can conveniently be divided into the four types shown in table 15-1. The classification is based partly on the nature of the fundamental processes and partly on the strength of the interaction.

The strength of an interaction depends on the probability that the associated fundamental process will occur. Consider two nucleons interacting by exchanging pions (figure 15-2). The strength of their interaction is proportional to the frequency with which the pion is thrown backward and forward. In this case the pion is exchanged about 10^{23} times every second! If the emission of a pion by a nucleon were a very improbable event, occurring say once every second, the nuclear forces would be very much weaker, by a factor of about 10^{23}.

The probability that a fundamental process will occur depends on many factors, such as the excess energy made available for kinetic energy of the products. Also, some interactions such as nuclear forces are confined to short distances, whereas others, such as electromagnetic forces, can extend over great distances. Nevertheless, when the theory has been formulated in a suitable mathematical form, it is found that, in addition to these factors, the strength of the interaction is proportional to a quantity called the **coupling constant,** which depends only on the type of interaction. The relative values of the coupling constants are shown in table 15-1.

The strongest interactions are the ones similar to nuclear forces which involve exchange of mesons. They are called **strong interactions.** Next come the **electromagnetic interactions,** which involve the exchange of photons. They are weaker than the strong interactions by a factor of about 100. There is an interesting difference between strong and electromagnetic interactions. Strong interactions have a range somewhat smaller than 10^{-12} cm, beyond which they become extremely weak. This range is related to the rest mass of a pion according to equation 15-10. Since the rest mass of a photon is zero, the corresponding range for electromagnetic forces is infinite. Thus, although electromagnetic forces fall off as $1/R^2$, they have no tendency to decrease very rapidly after a certain distance and are still effective at large distances.

The third category contains the **weak interactions,** with a coupling constant one hundred trillion (10^{14}) times smaller than that of the strong interactions. A fundamental process associated with a *strong* interaction has a typical half-life of 10^{-23} sec. The half-life of a process associated with a *weak* interaction is typically of the order of 10^{-9} sec, longer by the enormous factor 10^{14}. Although the obvious initial reaction to these lifetimes is that they are both unimaginably short, their ratio is the same as the ratio of one hour to the age of the earth.

Two typical weak interactions are β-decay

$$n \longrightarrow p + e^- + \bar{\nu} \qquad\qquad (15\text{-}14)$$

and the decay of a pion

$$\pi^+ \longrightarrow \mu^+ + \nu_\mu \qquad\qquad (15\text{-}15)$$

The fourth and final category of interactions contains the very, very weak **gravitational interactions,** weaker than the strong interactions by the stupendous factor of 10^{-40}. It is by no means certain that gravitational forces can be reconciled with quantum mechanics in its present form. In the general theory of relativity gravitational forces are caused by the curvature of space-time (section 10-4). If gravitational interactions are to be included in the present scheme, they must be assumed to be due to the passing backward and forward between two gravitating masses of an undiscovered fundamental particle, which has been called a **graviton.** Since gravitational forces are not short range, but obey the inverse square law like electromagnetic forces, the graviton must have zero rest mass and travel with the speed of light. It is probably a boson, like the pion that is responsible for nuclear forces, or the photon that is responsible for electromagnetic forces. Whereas the pion has a spin of zero and the photon has a spin of 1 unit, the graviton is expected to have a spin of 2 units. In section 14-2 we saw how difficult it is to detect a neutrino. Detection of a graviton is likely to be even worse by a factor of about 10^{26}, which is the ratio of the coupling constants for weak interactions and gravitational interactions.

15-5 Symmetry and Conservation Laws

When the diversity of the fundamental particles is properly understood, it is possible that our description of the universe will be reduced to a few basic concepts. Meanwhile, to act as a guide to this future revelation, we have at our disposal certain sweeping generalizations, which seem to transcend the details of any particular theory. Some of these are conservation laws, like the law of conservation of energy. Others are called **symmetry principles,** like the postulate of special relativity that the basic laws of physics are independent of the velocity of the observer. In some cases a conservation law can be seen to be a mathematical consequence of a symmetry principle.

In common usage the word symmetry implies a "balance" between two sides of an object. Referring to figure 15-4, the shape shown in part a is immediately recognized as symmetrical, and the shape of part b as unsymmetrical. In physics the word has a more precise, but more general meaning. Something has a particular type of symmetry if a particular operation can be performed on it and still leave it unchanged in some essential respect. This is best illustrated by examples. The symmetry of the shape in figure 15-4a is associated with the rather complicated operation illustrated in part c of the same figure. This is the operation of reflection about the central line AB. If P is any point on the figure, draw a line from P perpendicular to AB

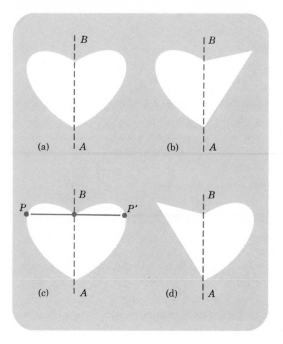

(a)   B   A

(b)   B   A

P •—————•—————• P'

(c)   B   A

(d)   B   A

Figure 15-4 *Illustrating the common meaning of symmetry.*

and move P to the point P' an equal distance on the other side of AB. Since the figure is symmetrical, P' coincides with a point that was originally on the figure. The point that was originally at P' is moved by this operation over to P. The net result is merely to interchange the two sides of the figure and, since the two sides are identical, the figure remains unaltered. However, in the case of the unsymmetrical shape of part b, the reflection operation produces the shape of part d, which is clearly different since it is the other way round.

There are some simpler symmetry operations, which nevertheless have a profound significance in physics. One of them is the operation of moving an object bodily through space. In figure 15-5 the triangle has been translated from position 1 to position 2, without rotation, but its size and shape remain unaltered. We say that the triangle is symmetrical under the operation of translation in space. In the realm of geometry this sounds rather trivial, but when applied to real physical situations it can be far from trivial. Suppose that a triangle is made by looping a rubber band round two nails and then hanging a weight from the midpoint of its lower section (figure 15-6). If this device is moved to the moon it does not preserve its shape. On the moon

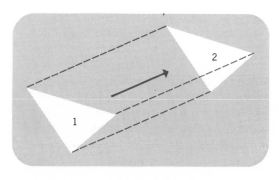

Figure 15-5 *Translation in space has not changed the shape of the triangle.*

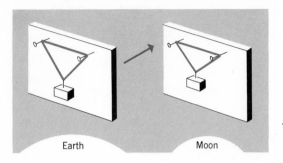

Figure 15-6 *The triangle is made by wrapping a rubber band around two nails and hanging a weight from the midpoint of its lower section. If this device is taken to the moon, the triangle does not preserve its shape.*

Earth   Moon

the weight is not so heavy and therefore it does not pull the band down so far.

This last device seems to have been specially designed to cause trouble, but the effect is present to a lesser degree with all real physical objects. Suppose that the triangle of figure 15-5 is cut out of cardboard. Like the rubber band, this cardboard triangle sags a little under the strain of its own weight and the bottom vertex is a little lower than it would be in the absence of gravity. Since the earth's gravitational field is not constant but varies from place to place, when the triangle is translated to a new position in this field the amount of sag changes a little and so the triangle changes its shape, just like the rubber band.

After a thorough searching analysis of this kind, is there anything that does remain unaltered when a physical system is moved from one place to another? We believe that the basic laws of physics, which explain the behavior of the system, are independent of its location in space. The gravitational force acting on the weight which distorts the rubber band is given by Newton's law of universal gravitation,

$$F = \frac{GM_1M_2}{R^2}$$ (15-16)

M_1 is either the mass of the earth or the mass of the moon and R is either the radius of the earth or the radius of the moon. Some of the quantities in the equation do have values that vary from place to place. The important thing is that the *form* of the equation is always the same. The force always obeys an inverse square law and is not, say, inverse square on the earth but inverse cube on the moon. Moreover, the fundamental constant G is the same everywhere and does not vary in an arbitrary fashion from place to place with no apparent reason. Similarly, the extent to which the rubber band sags depends on its elasticity. This depends on the fact that, when the rubber is stretched, its atoms move further apart and then exert forces on one another in an attempt to restore the initial situation. Interatomic forces are electromagnetic and are explained by Maxwell's equations and the laws of quantum mechanics. Maxwell's equations and the laws of quantum mechanics have the same form on the moon as on the earth.

We therefore say that "*The form of the basic laws of physics is symmetrical under the operation of translation in space.*" If we then look at the present form of the basic laws and consider the question of whether momentum is conserved, we discover an interesting fact. The law of conservation of momentum is a mathematical consequence of the fact that the basic laws have this property of assuming the same form at all points in space. The conservation law is a consequence of the symmetry principle.

There is reason to believe that the symmetry principle is more fundamental than the detailed form of the conservation law. The law of conservation of momentum first turned up in Newtonian mechanics, but it had to be modified in the special theory of relativity by allowing the mass to vary with velocity. In the general theory of relativity the basic equation of gravitation is not as simple as equation 15-16. Moreover, it is not true in the general theory that Maxwell's equations are the same on the moon as on the earth. Electromagnetic phenomena are influenced slightly by the strength of the gravitational field. Nevertheless, the general theory is able to formulate basic equations which have the property of assuming the same form at all points in space. These equations almost certainly lead to a law similar to the law of conservation of momentum, although the exact form of this law is not yet quite certain.

It is clear that, however hard pressed, the physicist will always try to trim his basic laws to make them symmetrical. The point at issue is whether he will always be successful, or whether he will eventually have to give up in despair. So far he has been quite successful.

15-6 *A List of Symmetry Principles and Conservation Laws*

In this section and the next, we shall collect together all the known symmetry principles and conservation laws. The subject is still controversial and it is not always certain how the conservation law follows from the symmetry principle and whether the connection is logically rigorous. It is therefore convenient in some cases to place the emphasis on the symmetry principle and in other cases on the conservation law. Also, some symmetry principles and conservation laws seem to be applicable only under restricted conditions, but nevertheless are very interesting and important.

A. Symmetry Under Translation in Space. The form of the basic laws of physics is the same at all points in space. This leads to the Law of Conservation of Linear Momentum.

B. Symmetry Under Rotation in Space. The basic laws of physics describing a system apply in the same form if the system is rotated through a fixed angle. The laws of physics have the same form in all directions. This leads to the Law of Conservation of Angular Momentum.

C. Symmetry Under Translation in Time. The form of the basic laws of physics does not change with time. Once the really basic laws have been found, they apply equally well to events that occurred a billion years ago and events that will occur a billion years in the future. This leads to the Law of Conservation of Energy.

D. Symmetry Under Reversal of Time. Imagine any physical process that can take place and therefore, of course, obeys the basic laws. Imagine this process running backward in time, like a movie that is passing through the projector in the wrong direction. Then this reversed process also obeys the basic laws of physics and is therefore a possible process. In section 5-5 we quoted the example of a body dropped from rest at a height h and striking the ground with a velocity v ($= \sqrt{2gh}$). Corresponding to this there is a possible time-reversed motion in which the body is thrown upward with a velocity v and comes to rest at a height h.

This particular symmetry principle is not an easy one to accept. Suppose that I drop a plate on the floor and it shatters to smithereens. Can I maintain that there is another possible motion in which the fragments fly together and reform an unblemished plate? Yes! There is nothing in the laws of physics to prevent this from happening. However, it would be necessary to throw the fragments together with *exactly* the right velocities so that they hit one another in *exactly* the right places. This is quite impossible to achieve in practice and would certainly never occur of its own accord. The breaking of the plate is a possible occurrence and a reasonably probable occurrence. The reconstitution of the plate by the time-reversed process is possible, but highly improbable. Symmetry under time reversal says that a time-reversed process *can* occur, but it gives no guarantee that it *will* occur. This has been discussed at greater length in Chapter 5, sections 5, 6, and 7.

Applied to fundamental processes, time reversal says that any process that can go in one direction can also go in the opposite direction. For example, a neutron with sufficient energy can disintegrate into a proton and a negative pion.

$$n \longrightarrow p + \pi^- \tag{15-17}$$

Conversely, a negative pion can combine with a proton to form a neutron

$$\pi^- + p \longrightarrow n \tag{15-18}$$

β-decay is the disintegration of a neutron into a proton, an electron, and an antineutrino.

$$n \longrightarrow p + e^- + \bar{\nu} \tag{15-19}$$

Presumably an antineutrino, an electron, and a proton can come together and form a neutron

$$\bar{\nu} + e^- + p \longrightarrow n \tag{15-20}$$

The reason this reaction has not been observed is associated with the very low probability that all three particles will be in the same place at the same time.

In recent years, certain subtle experiments on the decay of the neutral kaon, $K^°$, have suggested that this particular fundamental process may not be time reversible. However, the evidence is not yet as overwhelming as it should be to persuade us to abandon so basic a principle as time reversal.

E. Relativistic Symmetry. The basic laws of physics have the same form for all observers, independently of their motion. In the special theory of relativity the basic laws have the same form for all unaccelerated observers. In the general theory, which is less well established, the basic laws are assumed to have the same form for all observers, however complicated their motion.

F. Symmetry Under Interchange of Similar Particles. Fundamental particles do not have individual identities and the interchange of two similar particles makes no difference to a physical process. This was explained in more detail in section 12-7, where it was called the Indistinguishability of Similar Particles.

G. The Law of Conservation of Electric Charge. If negative charge is considered to cancel an equal amount of positive charge, there is no known physical process that changes the net amount of electric charge. In a fundamental process, the sum of the charges is the same on both sides of the equation. This conservation law is believed to be related to certain symmetry properties of the quantum mechanical wave function ψ, but the situation is not understood well enough to be easily explicable in simple terms.

H. The Law of Conservation of Leptons. If an antiparticle is considered to cancel its corresponding particle, there is no known physical process which changes the net number of leptons (section 14-5). There may be a similar Law of Conservation of μ-like Leptons. It is not yet known if there is a symmetry principle underlying these laws.

I. The Law of Conservation of Baryons. If an antiparticle is considered to cancel its corresponding particle, there is no known physical process that changes the net number of baryons (section 14-5). No symmetry principle underlying this law has yet been discovered, either. It is interesting that there are conservation laws for fermions but not for bosons. The net number of photons, pions, kaons, etas, or gravitons is not conserved.

The symmetries that remain to be discussed are sometimes called "broken symmetries" for the following reason. While they can be usefully applied to a wide range of physical phenomena, they break down when applied to other phenomena. This may be because nature is constructed according to a scheme of partial or imperfect symmetry. It is more likely, however, that we do not yet understand the situation properly and that we have not yet formulated the symmetry principles in such a way that they are universally applicable.

J. Charge Independence. The nuclear force between two nucleons is independent of whether they are neutrons or protons. This suggests the symmetry principle that the strength of a strong interaction is unaltered by changing the electric charge on the particles. That is, for example, by changing neutral n into positive p or Σ^+ into Σ^-. In the formal theory there is a quantity called **isotopic spin,** an integral or half-integral number which is different for n and p and for Σ^+ and Σ^-, thereby helping to distinguish between them. There is a Law of Conservation of Isotopic Spin which says that the sum of the isotopic spins on one side of the equation for a fundamental process is equal to the sum on the other side.

However, although n and p are equivalent as far as nuclear forces are concerned, their electromagnetic effects are very different because one is charged and the other is not. Consequently, it is found that isotopic spin is conserved in fundamental processes related to strong interactions, but not in processes related to electromagnetic and weak interactions.

K. The Law of Conservation of Strangeness. There is another integral quantum number called **strangeness,** S, which can be associated with some of the fundamental particles. This number is 0 for "well-behaved" particles like n, p, and π but is not zero for "strange" particles like the kaon, K, or the sigma, Σ. Strangeness is conserved in those fundamental processes that proceed rapidly, in about 10^{-23} sec, and are associated with strong interactions. If strangeness is not conserved, the fundamental process proceeds more slowly and is associated with weak interactions.

L. Particle–Antiparticle Symmetry. For every particle there is a corresponding antiparticle, which seems to be its exact opposite. For every fundamental process that is known to occur, there is another possible process obtained by changing every particle into its antiparticle. We have already given several examples of this. As a further example, corresponding to ordinary β-decay,

$$n \longrightarrow p + e^- + \bar{\nu} \tag{15-21}$$

there is a process of "anti-β-decay," in which an antineutron changes into an antiproton,

$$\bar{n} \longrightarrow \bar{p} + e^+ + \nu \tag{15-22}$$

Changing a particle into its antiparticle is sometimes called **charge conjugation.** As we shall see shortly, the indiscriminate change of particles into antiparticles sometimes leads to trouble, so this is classified as a broken symmetry.

M. Mirror Symmetry. This symmetry principle states that, for every known physical process there is another possible process which is identical with the mirror image of the first. This idea has led to some interesting developments in recent years, and so we shall discuss it more fully in the next section. As we shall see, it corresponds to the Law of Conservation of Parity. Mirror symmetry and parity conservation are valid for strong interactions and electromagnetic interactions, but not for weak interactions.

15-7 *Holding a Mirror Up to Nature*

Consider a clock, which is a complicated mechanism containing toothed wheels, spindles, and levers, obeying the laws of mechanics. Suppose that it is powered by a mainspring, the springiness of which can be ultimately traced to electromagnetic forces between its atoms, so that the laws of electromagnetism are also involved. The image of this clock in a mirror has every appearance of being a possible mechanism that would work and obey the laws of physics (figure 15-7). It is true that the numbers on its dial are Ɩ, ᒣ, Ɛ, and so on, and its hands go round counterclockwise. Nevertheless, there seems to be no reason why we should not construct in the real world a real copy of the image. It would look rather odd, but we would expect it to work, to obey the laws of physics and keep the same time as an ordinary clock.

To see that this is not all trivial and obvious, let us consider carefully

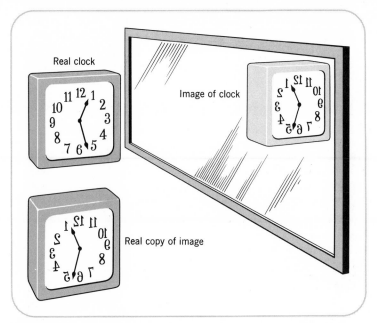

Figure **15-7** *A real clock, its image in a mirror, and a real copy of the*
image.

the difference between an object and its mirror image. The symmetry operation that produces a mirror image is very similar to the one associated with the popular notion of symmetry. From any point *P* on the object draw a line perpendicular to the surface of the mirror and move *P* to *P'* an equal distance on the other side of the mirror (figure 15-8). This procedure changes a right-hand glove into a left-hand glove. Try shaking hands with yourself in a mirror and you will discover that you can shake hands with the image of your left hand (almost!), but not with the image of your right hand. Now a left-hand glove is an essentially different thing from a right-hand glove. You

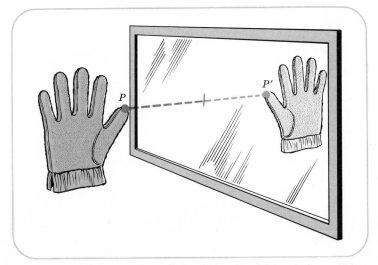

Figure **15-8** *The image of a right-hand glove is a left-hand glove, which*
is an essentially different thing.

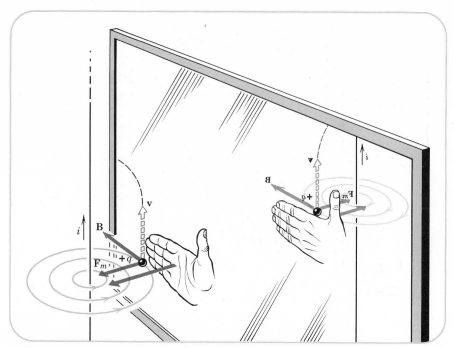

Figure 15-9 *In a mirror, electromagnetism has two left-hand rules instead of two right-hand rules.*

cannot put a left-hand glove on your right hand. There is no way of turning a left-hand glove around in space so that it falls exactly on top of a right-hand glove. You cannot step into a mirror and coincide with your image, because you would have to turn around, and your right hand would then be on top of the image of your left hand. Moreover, your real heart would be on your left side, while the image heart would be on the right side of your image. The image of an object is an essentially different thing from the object itself.

The mainspring of a clock relies on the laws of electromagnetism. In formulating these laws we made use of the right hand (sections 7-2 and 7-3). Since reflection changes a right hand into a left hand, we must consider carefully what happens to the laws of electromagnetism for a mirror image. Consider a current in a long straight wire exerting a magnetic force on a positive charge $+q$ moving in the same direction as the current (figure 15-9). Right-hand rule number 1 tells us that the magnetic field lines go around the current in the manner shown. Right-hand rule number 2 then tells us that this magnetic field exerts a magnetic force on the moving charge which deflects it toward the current. Now look at the mirror image. The magnetic field lines do not go the right way round the current and the image therefore violates right-hand rule number 1. If there is a real physicist using his right hand to figure out the direction of the magnetic field, then the image physicist is using his left hand, and this is why he gets it the wrong way round.

However, the magnetic field is just a mathematical convenience which we use at an intermediate stage in the calculation. What matters from a physical point of view is the behavior of the moving charge and we can see that the image charge is behaving itself and deflecting *toward* the current.

The reason is that the image physicist must use his left hand *twice* in order to discover the direction of the force on the charge. You can easily convince yourself that this gives the same result as using the right hand twice. If we concentrate on the things that really matter, we find that mirror images do obey the laws of electromagnetism. In fact there is nothing in classical physics to make us doubt mirror symmetry.

In modern physics mirror symmetry would imply that the mirror image of any known fundamental process is also a possible process consistent with the basic laws. This seems to be true for strong interactions, electromagnetic interactions, and gravitational interactions. However, in 1956 C. N. Yang and T. D. Lee suggested that some puzzling features of the behavior of kaons might be resolved if weak interactions violated mirror symmetry. Within two years several experiments had been performed to confirm this suggestion. Mirror symmetry can be expressed mathematically in terms of a quantity called **parity, P,** and there is a corresponding **Law of Conservation of Parity.** Weak interactions do *not* conserve parity, although all other types of interaction do.

One of the weak interactions is β-decay and the first experiment that established a violation of mirror symmetry involved the β-decay of cobalt 60 nuclei. With the cobalt nucleus spinning in the direction shown in figure 15-10a, it was found that the electron is *always* emitted downward and the antineutrino is *always* emitted upward. (More precisely, the electron is emitted in all directions with varying degrees of probability, but it is most likely to be emitted directly downward and it is never emitted directly upward. The antineutrino is most likely to be emitted directly upward and it is never emitted directly downward.) The angular momentum of the cobalt nucleus

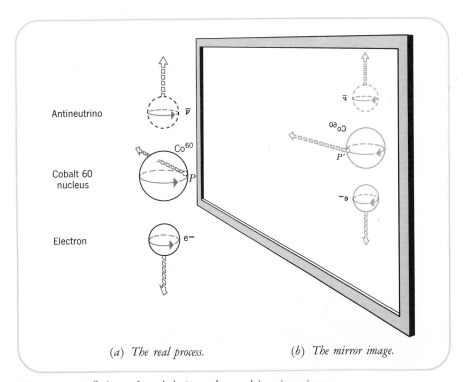

(a) *The real process.*   (b) *The mirror image.*

Figure 15-10 *β-decay of a cobalt 60 nucleus and its mirror image.*

decreases from 5 to 4 units of $h/2\pi$. The electron and antineutrino each have a spin of $\frac{1}{2}$ a unit and so, in order to conserve angular momentum, they must both be spinning in the same direction as the cobalt nucleus, as shown. The mirror image of this process is shown in part *b* of the same figure. Notice how reflection changes the direction of spin. The point *P* reflects into the point *P'* and both are moving into the paper.

The important difference between the real process and its mirror image is this. Let us always look at the cobalt nucleus in the direction in which it appears to spin clockwise, that is, upward for the real process and downward for the image process. Then, in the real process the electron is emitted toward us and this is what *always* occurs. In the image process the electron is emitted away from us, and this *never* occurs. So clearly the mirror image of a known process is not a possible process and mirror symmetry is violated.

The explanation is probably to be found in the nature of the antineutrino. There is good reason to believe that, if you look at an antineutrino in such a direction that it is moving away from you, then it is always rotating clockwise and never counterclockwise. Notice that this is so for the real β-decay process, but not for the image process, and so the object produced in the image process could not possibly be an antineutrino. In this respect an antineutrino is like a right-handed screw, the ordinary kind of screw. To make the screw move away from you as you screw it in, you must turn it clockwise. The other kind of screw, a left-handed screw, is an essentially different object which cannot be made to coincide with a right-handed screw however much it is turned around in space. Figure 15-11 shows that a left-handed screw is the mirror image of a right-handed screw. A right hand is shown turning the right-handed screw clockwise and driving it forward.

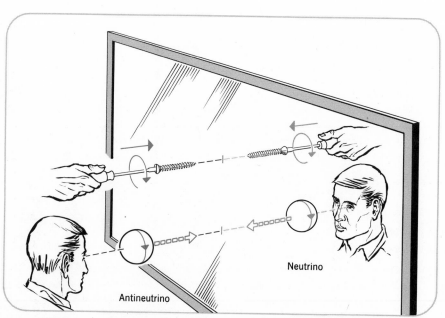

Figure 15-11 *The analogy between an antineutrino and a right-handed screw and between a neutrino and a left-handed screw.*

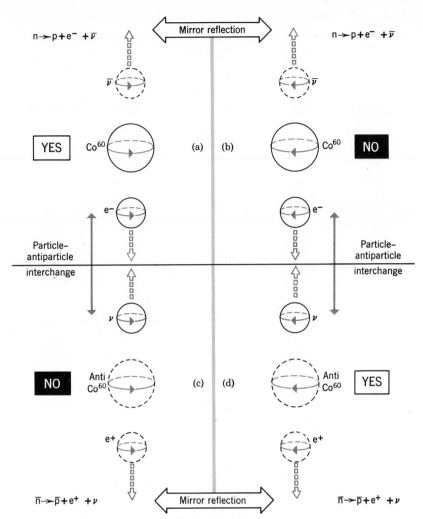

Figure 15-12 *The β-decay of cobalt 60 reflected in a mirror and subjected to particle-antiparticle interchange.*

The mirror image of this is a left hand turning a left-handed screw *counter-clockwise* to drive it forward.

The same figure shows an antineutrino moving toward the mirror and rotating clockwise from the point of view of an observer looking toward the mirror. The mirror image of this also moves toward the mirror, but rotates *counterclockwise* from the point of view of an observer behind the mirror looking toward the mirror. Such an entity is believed to be a neutrino. *The neutrino and antineutrino are mirror images of one another. The neutrino is the analogue of a left-handed screw and the antineutrino is the analogue of a right-handed screw.*

This brings us back to the question of particle–antiparticle symmetry. Figure 15-12a is the β-decay of cobalt 60 again, and part c of the same figure is the result of changing every particle into its antiparticle. The anti-nucleus is composed of antiprotons and antineutrons and it emits a positron and a neutrino. If everything else remains unaltered, including the directions of the spins, the process in c is not possible. The particle emitted upward is

15-7 Holding a Mirror Up to Nature

a right-handed screw and cannot possibly be a neutrino. By this and many similar examples, we are forced to conclude that all weak interactions violate the principle of particle–antiparticle symmetry.

Now consider the process of figure 15-12d, which is obtained from the observed process a by changing particles into antiparticles and also reflecting in a mirror. This is a possible process, because the particle emitted upward is a left-handed screw, as a neutrino should be. Therefore, although each symmetry principle is violated separately, the process is symmetrical under the double operation of mirror reflection plus particle–antiparticle interchange.

The end result of this discussion restores our faith in the elegance of nature. Figure 15-12 has a very satisfying symmetry. The expression "broken symmetry" seems unnecessarily harsh when applied to this situation. We have no reason to doubt that the more we come to understand the working of this universe, the more we shall be impressed by the beauty of its design.

Questions

A

1. Now that you have come to the end of the book, how would you answer Question 4 of Chapter 6?
2. Which of the following quantities is *always* conserved? (a) Momentum (b) angular momentum, (c) total rest mass, (d) charge, (e) number of baryons, (f) parity, (g) force, (h) isotopic spin.
3. A newly discovered particle, the iota, decays into an electron, e^-, and an antineutrino, $\bar{\nu}$. Which of the following possibilities can be excluded and why?
   (a) It is a lepton. (b) It is a baryon. (c) It is a meson. (d) It is a photon. (e) It is a boson.
4. Which conservation law (or laws) does each of the following violate?
   (a) $\mu^- \longrightarrow e^- + e^+ + \nu$
   (b) $\eta^0 \longrightarrow e^+ + \bar{\nu}$
   (c) $p \longrightarrow e^+ + \nu$
   (d) $p + \bar{n} \longrightarrow n + \bar{p}$

B

5. A cosmologist suggests that gravitational forces have a "range" of about 10^{10} light years, which is the size of the observable universe. What is the order of magnitude of the corresponding rest mass of the graviton?
6. Discuss the β-decay of the neutron from the point of view of each of the conservation laws and each of the symmetry principles.
7. Discuss the following statement. "Of course the basic laws of physics obey the symmetry principles, because if a law did not we would deny that it was truly basic."
8. Sometimes the decay of a particle A into two other particles B and C, ($A \longrightarrow B + C$), is allowed by all the known laws of physics, except that the rest mass of A is smaller than the sum of the rest masses of B and C by an amount m. Is the process possible if A is moving alone through otherwise empty space with a kinetic energy greater than mc^2? (Hint:

Inklings

The answer is obvious once you have related the question to the relevant symmetry principle.)

9. Write an essay on "Energy." Include an enumeration of the various forms of energy and detailed discussions of processes that transform one kind of energy into another kind. Say which are the basic kinds of energy and explain how some forms of energy, such as heat and chemical energy, can be analyzed in terms of these basic forms.

10. Section 15-3 contains the sentence, "Consider an isolated charged particle at rest surrounded by its cloud of virtual photons, and concentrate on a snapshot taken at a fixed instant of time." Although this conveys the right idea, it is really nonsense because the requirement of "a fixed instant of time" makes the energy infinitely uncertain. How would you reformulate the discussion to avoid this difficulty?

11. You have established radio contact with a friendly alien spaceship approaching the solar system. During the course of an exchange of knowledge, you find it necessary to explain to them which is your right hand. *Precisely* what would you say under the following circumstances? (a) You can modify your radio transmission by swinging it through an angle in space or by changing the direction of the oscillating electric vector. (b) You cannot modify your radio transmission but you know that the visitors are able to see the constellations. (c) Neither of the above is possible. The only possibility is to talk about the nature of physical phenomena. (d) The only possibility is to talk about physical phenomena, but you do not know whether the visitors are made of matter or antimatter.

12. A current from a storage battery passes through a coil of wire and the resulting magnetic field deflects a magnetic compass needle. The poles of the battery are marked $+$ and $-$, but there is no indication which is the north pole of the compass needle. Can you decide from the direction in which the compass needle deflects whether you are looking at the real apparatus or its reflection in a mirror? Suppose that all the protons, neutrons, and electrons in the apparatus were suddenly changed into antiprotons, antineutrons, and positrons. Would the compass needle still deflect in the same direction? Do not answer this question by applying known symmetry principles. Convince yourself that the symmetry principles are valid in this case.

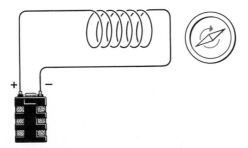

Question 15-12.

13. Would the process of figure 15-10b be a permissible physical process if we conceded that the particle emitted upward is a neutrino and not an antineutrino?

14. Without actually going there, is there any way of deciding whether a distant star is made of matter or antimatter? Our knowledge of the star may be obtained from the light emitted by its surface and also from the neutrinos emitted as a result of processes such as the ones summarized in equation 13-21.
15. Discuss the effect on physical processes of (a) reflection twice in two mirrors at right angles to one another, (b) reflection three times in three mirrors mutually at right angles (like two walls and the floor of a room all meeting in a corner).
16. In various places throughout this book we have had to admit our ignorance on some important issues. Make a list of what you consider to be the outstanding unsolved problems in physics. Can you guess how some of them might be resolved? If you had been a nineteenth century physicist, do you think you could have guessed the nature of quantum mechanics?

Mathematical Appendix

The student with little previous experience of mathematics should study this appendix carefully before proceeding beyond Chapter 1.

A Algebra
A1 Fundamentals

Algebra is arithmetic in which letters are used to represent numbers. In arithmetic, we would say

$$3 \times 5 = 15 \tag{1}$$

A similar statement in algebra is

$$ab = c \tag{2}$$

which means that, when the number a is multiplied by the number b, the result is the number c. If a represents the number 3, and b represents the number 5, then c must represent the number 15, and equations 1 and 2 are different ways of saying the same thing. However, it is possible for a and b to represent other numbers, for example,

$$a = 2\frac{1}{2} \quad \text{and} \quad b = 4\frac{1}{3} \tag{3}$$

Once a and b have been chosen, there is no freedom of choice for c because

$$c = ab \tag{4}$$

that is, $c = (2\frac{1}{2}) \times (4\frac{1}{3})$ \hfill (5)

or $\quad c = 10\frac{5}{6}$ \hfill (6)

Equation 2 is therefore a very general equation in which a and b may represent any two numbers whatsoever, but c is restricted to a particular value which is the product of a and b.

The **symbols** such as a, b, and c which represent numbers may be combined in various ways to produce various **algebraic expressions.** For example, the expression $a(b + c)$ means that the number represented by b is added to the number represented by c and the resulting sum is multiplied by the number represented by a. The expression $[a + (b + c)d](a + c)$ means that (1) b is added to c; (2) the sum is then multiplied by d; (3) a is then added; (4) the result of (3) is then multiplied by the sum of a and c. The expression $(ab)/(b + c)$ means that a is multiplied by b and the result is divided by the sum of b and c.

EXAMPLE

Evaluate the expression

$$p = \frac{(r + s)[rs + t(3 + s)]}{5st(r + t)}$$

when $r = 2$, $s = 5$, and $t = \frac{1}{2}$.

Mathematical
Appendix

Inserting the given values of r, s, and t,

$$p = \frac{(2 + 5)[(2 \times 5) + \frac{1}{2}(3 + 5)]}{5 \times 5 \times \frac{1}{2}[2 + \frac{1}{2}]}$$

$$= \frac{7[10 + (\frac{1}{2} \times 8)]}{\frac{25}{2} \times [2\frac{1}{2}]}$$

334

$$= \frac{7[10 + 4]}{\frac{25}{2} \times \frac{5}{2}}$$

$$= \frac{7 \times 14}{\frac{125}{4}}$$

Multiplying numerator and denominator by 4,

$$p = \frac{7 \times 14 \times 4}{125}$$

$$= \frac{98 \times 4}{125}$$

$$= \frac{392}{125}$$

$$= 3\frac{17}{125}$$

It is preferable to express this in decimal form

$$p = 3.136$$

Problems

Express the meaning of the following expressions in words.

1. $(a + b)(c + d)$

2. $\left(a + \dfrac{b}{c}\right)(b + c)$

3. $\dfrac{a}{c}\left[2a + bc\left(3 + \dfrac{1}{a}\right)\right]$

Evaluate

4. $a(b + c) + \frac{1}{2}b(c + 2a)$ when $a = 2, b = 3, c = 4$

5. $(r + s)\left(\dfrac{t}{r} + \dfrac{s}{t}\right)$ when $r = \frac{1}{2}, s = 5, t = 3\frac{1}{2}$

A2 *Negative Numbers*

The expression $a - b$ means that the number b is *subtracted* from the number a. It is clearly different from the expression $a + b$, which is the result of *adding* b to a. However, the idea that b can represent any number may be extended to allow b to represent both positive and negative numbers. If b is a negative number, the addition of b to a is really a subtraction. For example, if

$$a = +7 \text{ and } b = -3 \qquad (7)$$

then

$$
\begin{aligned}
a + b &= +7 + (-3) & (8) \\
&= +7 - 3 & (9) \\
&= +4 & (10)
\end{aligned}
$$

The subtraction of a negative number is equivalent to addition. In the above case

$$
\begin{aligned}
a - b &= +7 - (-3) & (11) \\
&= +7 + 3 & (12)
\end{aligned}
$$

To ensure that the normal procedures of arithmetic shall give consistent results when applied to negative numbers, it is necessary to adopt the following rules for multiplication. **The multiplication of two positive numbers produces a positive number.**

For example, $(+9) \times (+7) = +63$ $\qquad (13)$

The multiplication of a positive number and a negative number produces a negative number.

For example, $(+1.2) \times (-2.5) = -3$ $\qquad (14)$

and $\qquad (-\tfrac{1}{2}) \times (+3.6) = -1.8$ $\qquad (15)$

The multiplication of a negative number by another negative number produces a positive number.

For example, $(-5) \times (-0.32) = +1.6$ $\qquad (16)$

The value of a positive or negative number, regardless of its sign, is called its **numerical value** or sometimes its **absolute magnitude.** The numbers $+3$ and -3 both have a numerical value of 3. It is important to distinguish clearly between a negative number and its numerical value. The statement "The electric charge at P has a negative value of $-q$" implies that the symbol q represents the numerical value of the charge and q is therefore a positive number. If, at a later stage, one is told that the charge is actually -2.5 statcoulombs, the number which must be inserted for q in any formula is $+2.5$ and not -2.5.

EXAMPLE 1

Evaluate $p = \dfrac{(a - b)(ab + c)}{(cb - 2a)}$

when $a = +2$, $b = -\tfrac{1}{2}$, $c = -3$

Inserting the given values of a, b, and c

$$p = \frac{[+2 - (-\frac{1}{2})][(+2)(-\frac{1}{2}) + (-3)]}{[(-3)(-\frac{1}{2}) - 2(+2)]}$$

$$= \frac{[+2 + \frac{1}{2}][-1-3]}{[+1\frac{1}{2} - 4]}$$

$$= \frac{(2\frac{1}{2})(-4)}{-(2\frac{1}{2})}$$

The $2\frac{1}{2}$ in the numerator cancels the $2\frac{1}{2}$ in the denominator.

$$p = \frac{-4}{-1}$$

Multiplying numerator and denominator by -1

$$p = \frac{+4}{+1}$$

$$p = +4$$

EXAMPLE 2

When an electric charge of $+q_1$ statcoulomb is at a distance of R cm from another electric charge of $+q_2$ statcoulomb, their mutual electrostatic potential energy in ergs is

$$\Phi_e = \frac{q_1 q_2}{R}$$

Evaluate this energy when $q_1 = +1.5$ statcoulomb, $q_2 = -0.2$ statcoulomb, and $R = 0.1$ cm.

The inexperienced reader need not be disconcerted by an algebraic expression like this. The Greek letter capital phi, Φ, is merely a symbol representing a number, just like a or x or r. The subscript e is used merely to remind us that we are talking about electrostatics and to avoid confusion with a similar quantity, Φ_G, which is the mutual *gravitational* potential energy of two bodies. The symbol q is commonly used for an electric charge, but in this problem there are two charges with different numerical values and it is necessary to distinguish between them. This is done by the subscripts 1 and 2. Therefore, we merely insert the numerical values into the formula in the usual way.

Algebra

$$\Phi_e = \frac{(+1.5) \times (-0.2)}{0.1}$$

$$= \frac{-0.3}{0.1}$$

$$\Phi_e = -3.0 \text{ erg}$$

Notice that, since one charge is positive and the other is negative, the mutual electrostatic potential energy comes out negative.

Problems

6. Express in words the meaning of

$$\frac{(a - b)(ab + c)}{(cb - 2a)}$$

Evaluate

7. $a - b + ab$
when $a = -0.1$, $b = +20$

8. $\dfrac{(xy + y + z)(x - 1)}{xyz}$

when $x = +2$, $y = -3$, $z = -\frac{1}{2}$

9. $\dfrac{(\alpha\beta + \beta + \gamma)(\alpha - 1)}{\alpha\beta\gamma}$

when $\alpha = +2$, $\beta = -3$, $\gamma = -\frac{1}{2}$

10. What is the mutual electrostatic potential energy of two charges of value -2.5 statcoulomb and -15.0 statcoulomb, when the distance between them is 5 cm?

A3 *Exponents*

The expression a^n means that the number a is multiplied by itself n times. n is called the **exponent** of a. For example, "a squared" is

$$a^2 = a \times a \tag{17}$$

"a cubed" is

$$a^3 = a \times a \times a \tag{18}$$

"a to the fifth power" is

$$a^5 = a \times a \times a \times a \times a \tag{19}$$

The expression

$$a^n \times a^m$$

clearly means that a is multiplied by itself n times and then by a number which is the result of multiplying a by itself m times. This is the same as multiplying a by itself $n + m$ times, and so

$$a^n \times a^m = a^{n+m} \qquad (20)$$

For example

$$2^2 \times 2^3 = 2^{2+3} = 2^5 \qquad (21)$$
$$\text{because } (2 \times 2) \times (2 \times 2 \times 2) = (2 \times 2 \times 2 \times 2 \times 2) \qquad (22)$$

Similarly,

$$\frac{a^n}{a^m} = a^{n-m} \qquad (23)$$

because it means that a is multiplied by itself n times and then divided by a m times, which is equivalent to multiplying a by itself $n - m$ times.
For example,

$$\frac{2^6}{2^2} = 2^{6-2} = 2^4 \qquad (24)$$

$$\text{because } \frac{(2 \times 2 \times 2 \times 2 \times 2 \times 2)}{(2 \times 2)} = (2 \times 2 \times 2 \times 2) \qquad (25)$$

We can therefore establish the convention that

$$a^{-m} = \frac{1}{a^m} \qquad (26)$$

So that
$$\frac{a^n}{a^m} = a^n \times a^{-m} \qquad (27)$$
$$= a^{n+(-m)} \qquad (28)$$
$$= a^{n-m} \qquad (29)$$

If n is equal to m, then

$$\frac{a^n}{a^n} = a^{n-n} = a^0 \qquad (30)$$

Algebra

but it is also obviously equal to 1, so

$$a^0 = 1 \qquad (31)$$

whatever the value of a.

Since a^n means a multiplied by itself n times, the expression $(a^n)^m$ means a^n multiplied by itself m times. This is equivalent to multiplying a by itself nm times.

$$(a^n)^m = a^{nm} \tag{32}$$

For example,

$$(5^2)^3 = (5 \times 5)^3 \tag{33}$$
$$= (5 \times 5) \times (5 \times 5) \times (5 \times 5) \tag{34}$$
$$= (5 \times 5 \times 5 \times 5 \times 5 \times 5) \tag{35}$$
$$= 5^6 \tag{36}$$
$$\text{or} \quad (5^2)^3 = 5^{2 \times 3} \tag{37}$$

Fractional exponents can be introduced by assuming the general validity of equation 32, even when n and m are not integers. Then

$$(a^{1/n})^n = a^{n/n} \tag{38}$$
$$= a^1 \tag{39}$$

or

$$(a^{1/n})^n = a \tag{40}$$

$a^{1/n}$ is the number which, when multiplied by itself n times, produces a. It is therefore the "nth root" of a. For example, the square root of a is $a^{1/2}$ and

$$(a^{1/2})^2 = a \tag{41}$$

More generally,

$$(a^{1/n})^m = a^{m/n} \tag{42}$$

For example,

$$4^{3/2} = (4^{1/2})^3 \tag{43}$$
$$= 2^3 \tag{44}$$
$$= 8 \tag{45}$$

Mathematical
Appendix

EXAMPLE 1

Evaluate $\dfrac{3^5 3^2}{3^3}$

$$\frac{3^5 3^2}{3^3} = \frac{3^{5+2}}{3^3}$$

$$= \frac{3^7}{3^3}$$

$$= 3^{7-3}$$
$$= 3^4$$
$$= 3 \times 3 \times 3 \times 3$$
$$= 9 \times 9$$
$$= 81$$

EXAMPLE 2

Evaluate $(x^2 + y^2)^{3/2}$ when $x = 3$, $y = 4$.

$$(3^2 + 4^2)^{3/2} = (9 + 16)^{3/2}$$
$$= (25)^{3/2}$$
$$= (25^{1/2})^3$$
$$= 5^3$$
$$= 125$$

EXAMPLE 3

Evaluate $9^{-3/2}$

$$9^{-3/2} = \frac{1}{9^{3/2}}$$

$$= \frac{1}{(9^{1/2})^3}$$

$$= \frac{1}{3^3}$$

$$= \frac{1}{3 \times 3 \times 3}$$

$$= \frac{1}{27}$$

Problems

Algebra

Evaluate

11. 2^5

12. $3^2 \times 3^3$

13. $3^2 \times 2^3$

14. $\frac{3^5}{3^7}$

15. $\left(\frac{1}{5}\right)^{-2}$

16. $(\frac{1}{4})^{-1/2}$

17. $(2^3)^2$

18. $(4^3)^{1/2}$

19. $4^{-3/2}$

20. $\dfrac{a^3 + b^3}{a^2 + b^2}$ when $a = 2$, $b = 3$

21. $(a^2 + b^2)^{1/2}$ when $a = 3$, $b = 4$

A4 *Powers of Ten*

Numbers such as 25,310,000 and 0.00000537 are cumbersome, and calculations are made much easier if such numbers are expressed in terms of powers of ten.
For example,

$$25{,}310{,}000 = 2.531 \times 10{,}000{,}000 \tag{46}$$
$$= 2.531 \times 10^7 \tag{47}$$

$$\text{Also } 0.00000537 = \frac{5.37}{1{,}000{,}000} \tag{48}$$
$$= 5.37 \times 10^{-6} \tag{49}$$

In each case the result is a number between 1 and 10 multiplied by the appropriate power of ten.

A simple way to find the exponent of ten is the method of shifting the decimal point. If the number is greater than ten, place the point of your pen or pencil on the decimal point and shift it to the left one figure at a time. Each shift is equivalent to dividing by ten, and must therefore be compensated by multiplying by one power of ten. Continue until there is only one figure to the left of the decimal point, meanwhile counting the total number of shifts, which is then the required exponent of ten. If the number is less than 1, again place the point of your pen or pencil on the decimal point and shift it to the *right*, one figure at a time. Each shift is equivalent to multiplying by ten, and must be compensated by dividing by ten, or multiplying by 10^{-1}. Continue until the decimal point is to the right of the first figure which is not zero. The required exponent of ten is then minus the number of shifts.

EXAMPLE 1

Express 25,310,000 in powers of ten.
Introducing a decimal point, the number is

	25310000.0
After 1 shift	2531000.00×10^1
After 2 shifts	253100.000×10^2
After 3 shifts	25310.0000×10^3
After 4 shifts	2531.00000×10^4
After 5 shifts	253.100000×10^5
After 6 shifts	25.3100000×10^6
After 7 shifts	2.53100000×10^7

Since the decimal point then has only one figure to the left of it, this is the desired answer. Unless the five zeros following the 1 have special significance, the answer would normally be written $25{,}310{,}000 = 2.531 \times 10^7$.

EXAMPLE 2

Express 0.00000537 in powers of ten.

After 1 shift	$00.0000537 \times 10^{-1}$
After 2 shifts	$000.000537 \times 10^{-2}$
After 3 shifts	$0000.00537 \times 10^{-3}$
After 4 shifts	$00000.0537 \times 10^{-4}$
After 5 shifts	$000000.537 \times 10^{-5}$
After 6 shifts	$0000005.37 \times 10^{-6}$

The six initial zeros are clearly redundant and so $0.00000537 = 5.37 \times 10^{-6}$.

EXAMPLE 3

Evaluate $\sqrt{2 \times 10^9}$

$$\sqrt{2 \times 10^9} = (2 \times 10^9)^{1/2}$$
$$= 2^{1/2} \times 10^{9/2}$$
$$= 1.414 \times 10^{4\frac{1}{2}}$$

However, this is *not* what we want, because fractional powers of ten do not have the same simple significance as integral powers of ten. The correct procedure is:

$$(2 \times 10^9)^{1/2} = (20 \times 10^8)^{1/2}$$
$$= 20^{1/2} \times 10^{8/2}$$
$$= \sqrt{20} \times 10^4$$
$$= 4.47 \times 10^4$$

EXAMPLE 4

Evaluate $v = \sqrt{\dfrac{GM_e}{R}}$ when $G = 6.67 \times 10^{-8}$, $M_e = 6.0 \times 10^{27}$, and $R = 7 \times 10^8$.

Algebra

$$v = \left(\frac{6.67 \times 10^{-8} \times 6.0 \times 10^{27}}{7 \times 10^8}\right)^{1/2}$$

$$= \left(\frac{6.67 \times 6.0 \times 10^{27-8}}{7 \times 10^8}\right)^{1/2}$$

$$= \left(\frac{40.02 \times 10^{19}}{7 \times 10^8}\right)^{1/2}$$

$$= \left(\frac{40.02}{7} \times 10^{19-8} \right)^{1/2}$$
$$= (5.72 \times 10^{11})^{1/2}$$
$$= (57.2 \times 10^{10})^{1/2}$$
$$= \sqrt{57.2} \times 10^{10/2}$$
$$= 7.56 \times 10^{5}$$

Problems

22. If you really understand powers of ten, you should be able to tell immediately, just by glancing at the figures, which is the greater of each pair of numbers
    (i) 10^9 or 10^3 (ii) 10^3 or 10^{-9} (iii) 10^{-4} or 10^{-6} (iv) 4×10^6 or 2×10^6 (v) 4×10^{-6} or 2×10^{-6} (vi) 2×10^{-3} or 4×10^{-4}

23. Express in powers of ten
    (i) 53,200 (ii) 216,532,800 (iii) 232
    (iv) 0.0013 (v) 0.00000027 (vi) 0.000956

24. Evaluate
    (i) $(9 \times 10^8)^{1/2}$ (ii) $\sqrt{1.6 \times 10^7}$ (iii) $\sqrt{0.000064}$
    (iv) $(2.7 \times 10^4)^{1/3}$ (v) $(0.000125)^{1/3}$

25. Evaluate $\dfrac{(3.4 \times 10^5) \times (5 \times 10^{-11})}{(2 \times 10^3)^2}$

26. Evaluate $\dfrac{(2 \times 10^5)^2 (3 \times 10^2)^3}{(5 \times 10^{-6})^2}$

27. Evaluate $T = 2\pi \sqrt{\dfrac{R^3}{GM_e}}$

    when $R = 1.6 \times 10^9$ $G = 6.67 \times 10^{-8}$
    and $M_e = 8.5 \times 10^{31}$

28. Evaluate $\lambda = \sqrt{\dfrac{\eta T}{\pi \rho}}$

    when $\eta = 2 \times 10^{-5}$, $T = 30$
    and $\rho = 0.145$

A5 Algebraic Procedures

The procedure for multiplying brackets is to take the terms in the first bracket one by one and multiply each into all the terms in the second bracket one by one. This is best understood by a careful study of the following examples:

$$(a + b)(c + d) = a(c + d) + b(c + d) \tag{50}$$
$$= ac + ad + bc + bd \tag{51}$$

$$(a + b + c)(d + e + f) = a(d + e + f) + b(d + e + f)$$
$$+ c(d + e + f) \tag{52}$$
$$= ad + ae + af + bd + be + bf$$
$$+ cd + ce + cf \tag{53}$$

The following special cases are particularly important.

$$\text{First, } (a + b)(a - b) = a(a - b) + b(a - b) \tag{54}$$
$$= a^2 - ab + ba - b^2 \tag{55}$$

Since ab and ba mean the same thing, they cancel one another, and so

$$(a + b)(a - b) = a^2 - b^2 \tag{56}$$

$$\text{Second, } (a + b)^2 = (a + b)(a + b) \tag{57}$$
$$= a(a + b) + b(a + b) \tag{58}$$
$$= a^2 + ab + ba + b^2 \tag{59}$$

$$(a + b)^2 = a^2 + 2ab + b^2 \tag{60}$$

$$\text{Third, } (a - b)^2 = (a - b)(a - b) \tag{61}$$
$$= a(a - b) - b(a - b) \tag{62}$$
$$= a^2 - ab - ba + b^2 \tag{63}$$

$$(a - b)^2 = a^2 - 2ab + b^2 \tag{64}$$

An equation remains valid if the same quantity is added to both sides or subtracted from both sides. To illustrate this, suppose that we are told that

$$2x - 3 = x + 7 \tag{65}$$

and we wish to find the value of x. Subtract x from both sides of the equation, which then becomes

$$2x - x - 3 = x - x + 7 \tag{66}$$
$$\text{or} \quad x - 3 = 7 \tag{67}$$

Algebra

Now add 3 to both sides of the equation

$$x - 3 + 3 = 7 + 3 \tag{68}$$
$$\text{or} \quad x = 10 \tag{69}$$

which is the desired result.

An equation remains valid if both sides are multiplied by the same quantity or divided by the same quantity. To simplify the equation

$$\frac{x + y}{x - y} = \frac{x - y}{x + y} \tag{70}$$

let us multiply both sides by $(x - y)$. Then

$$\frac{(x + y)(x - y)}{(x - y)} = \frac{(x - y)(x - y)}{(x + y)} \tag{71}$$

On the left-hand side, multiplying by $(x - y)$ and then dividing by $(x - y)$ leaves us where we started, and so the $(x - y)$ in the numerator cancels the $(x - y)$ in the denominator. Therefore,

$$(x + y) = \frac{(x - y)^2}{(x + y)} \tag{72}$$

Similarly, multiplying both sides by $(x + y)$,

$$(x + y)^2 = (x - y)^2 \tag{73}$$

Using equations 60 and 64,

$$x^2 + 2xy + y^2 = x^2 - 2xy + y^2 \tag{74}$$

Subtracting $x^2 + y^2$ from both sides,

$$2xy = -2xy \tag{75}$$

Adding $2xy$ to both sides,

$$4xy = 0 \tag{76}$$

Dividing both sides by 4,

$$xy = \frac{0}{4} \tag{77}$$

or $xy = 0$ \qquad (78)

A little thought will convince the reader that this is possible only if

either $x = 0$ \qquad (79)

or \quad $y = 0$ \qquad (80)

Fractions are added by arranging for them all to have a common denominator. Consider the expression

$$p = \frac{1}{a - b} - \frac{1}{(a + b)} \tag{81}$$

Since a quantity is unchanged by multiplying it by a second quantity and simultaneously dividing by this second quantity,

$$p = \frac{1}{(a-b)} \frac{(a+b)}{(a+b)} - \frac{1}{(a+b)} \frac{(a-b)}{(a-b)} \qquad (82)$$

Both fractions now have the same denominator $(a+b)(a-b)$, so

$$p = \frac{(a+b) - (a-b)}{(a+b)(a-b)} \qquad (83)$$

$$= \frac{a+b-a+b}{(a^2 - b^2)} \qquad (84)$$

$$\text{or } p = \frac{2b}{(a^2 - b^2)} \qquad (85)$$

EXAMPLE 1

Find the value of α which satisfies the equation

$$(\alpha + 2)(2\alpha + 5) = 2(\alpha - 1)^2$$

The equation may be written

$$\alpha(2\alpha + 5) + 2(2\alpha + 5) = 2(\alpha^2 - 2\alpha + 1)$$
$$\text{or } 2\alpha^2 + 5\alpha + 4\alpha + 10 = 2\alpha^2 - 4\alpha + 2$$

Subtracting $2\alpha^2$ from each side and adding the terms 5α and 4α on the left side,

$$9\alpha + 10 = -4\alpha + 2$$

Adding 4α to each side,

$$13\alpha + 10 = +2$$

Subtracting 10 from each side,

$$13\alpha = -8$$

Algebra

Dividing both sides by 13,

$$\alpha = -\frac{8}{13}$$

$$\text{or } \alpha = -0.615$$

29. Evaluate $(3 + 5)(2 + 8)$ by first performing the additions and then the multiplication. Then evaluate it by the rule for multiplying brackets and check that you obtain the same answer.

30. Multiply the brackets in the following expressions
    (i) $(\alpha - \beta)(\alpha^2 + \beta^2)$
    (ii) $(a + b)(a^2 - ab + b^2)$
    (iii) $(a - b)(a^2 + ab + b^2)$
    (iv) $(x + y + z)(xy + yz + zx)$

31. Find x, given that
    (i) $3x + 4 = 2x - 1$
    (ii) $2(x + 5) = 3(x - 2)$
    (iii) $(x - 1)^2 + 4 = (x + 1)^2$
    (iv) $\dfrac{x - 1}{x + 5} = \dfrac{x + 3}{x - 2}$

32. Express the following with a common denominator
    (i) $\dfrac{1}{a} + \dfrac{1}{b} + \dfrac{1}{c}$
    (ii) $\dfrac{x}{yz} + \dfrac{y}{zx} + \dfrac{z}{xy}$
    (iii) $\dfrac{x}{(x + a)} - \dfrac{y}{(y + b)}$

A6 *Infinity*

"When R becomes infinite, $1/R$ becomes zero." What does this statement mean? The concept of infinity in mathematics is a subtle one and we shall not attempt to discuss it with meticulous detail. From a practical point of view the statement may be taken to mean "If R is made large enough, $1/R$ can be made so small that it can be neglected." It is clear that the larger we make R, the smaller $1/R$ becomes. When R is 100, $1/R$ is 0.01. When R is 1,000,000, $1/R$ is 0.000001. When R is 1,000,000,000, $1/R$ is 0.000000001, and so on. How small $1/R$ has to be before it can be neglected depends on the particular problem that is being considered. Nevertheless, however small it has to be, it can obviously be made this small by making R large enough.

Another simple way to express this is as follows. The statement "R becomes infinite" means that R becomes larger than any number you can think of, however large you try to make it. The statement "$1/R$ becomes zero" means that $1/R$ becomes smaller than any number you can think of, however small you try to make it. Infinity is sometimes treated as though it were an ordinary number and is represented by the symbol ∞. The preceding discussion can then be summarized by the equation

$$\frac{1}{\infty} = 0$$

(86)

Now consider another statement of a similar kind. "When α becomes zero, $1/\alpha$ becomes infinite." This means that, by making α small enough, we can make $1/\alpha$ as large as we please. For example, if α is 0.01, then $1/\alpha$ is 100. If α is 0.000001, then $1/\alpha$ is 1,000,000. If α is 0.000000001, then $1/\alpha$ is 1,000,000,000, and so on. In simplified symbolic form,

$$\frac{1}{0} = \infty \qquad (87)$$

The quantity 0/0 has no well-defined meaning. It might seem obvious that the zero in the numerator cancels the zero in the denominator, giving 1, since any number divided by itself gives 1. However, consider a quantity such as

$$\frac{2(1 - x)}{(1 - x)} = 2 \qquad (88)$$

If we put $x = 1$, then both numerator and denominator become 0 but it would now seem reasonable to continue to take the answer to be 2. On the other hand, $\dfrac{3(1 - x)}{(1 - x)}$ would be most reasonably taken to be 3. In fact it seems that 0/0 might be any number, depending on the circumstances.

B Euclidean Geometry

B1 Triangles

The sum of the three angles of a triangle is equal to 180°. For example, in figure 1 the angle ABC (which means the angle at the point B) is $110°$ and the angle BCA is $25°$. The third angle CAB must therefore be $180° - 110° - 25° = 45°$.

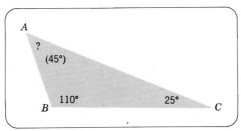

Figure 1 *The sum of the three angles of a triangle is 180°.*

A **right angled triangle** (often called a **right triangle**) has one of its angles equal to 90°. The sum of the other two angles must then be 90°. The **hypotenuse** is the side opposite the right angle. In figure 2 it is the angle BAC which is equal to 90°, and the hypotenuse is the side BC. Right angled triangles obey Pythagoras' theorem, which states that: "**The square of the length of the hypotenuse is equal to the sum of the squares of the**

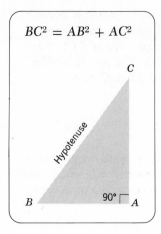

$$BC^2 = AB^2 + AC^2$$

C

Hypotenuse

B   90° ⌐ *A*

Figure 2 *A right triangle.*

lengths of the other two sides.'' For triangle ABC of figure 2, this would imply that

$$BC^2 = AB^2 + AC^2 \tag{89}$$

B2 *Circles*

Figure 3 illustrates the meaning of various concepts associated with a circle. It should be self-explanatory.

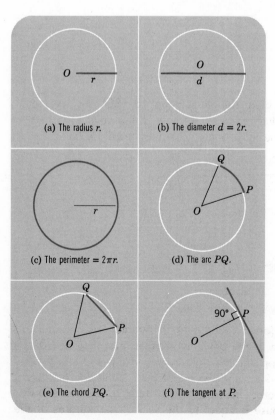

Figure 3 *Some concepts associated with a circle.*

Length of the perimeter of a circle

$$= 2\pi \times (\text{radius}) \qquad (90)$$

$$= 2\pi r \qquad (91)$$

Area enclosed by a circle

$$= \pi \times (\text{radius})^2 \qquad (92)$$

$$= \pi r^2 \qquad (93)$$

$$\pi = 3.14159 \ldots \qquad (94)$$

Unless greater accuracy is obviously needed, π can be taken to three significant figures when working problems in this book.

$$\pi \approx 3.14 \qquad (95)$$

B3 *Solid Geometry*

Surface area of a sphere

$$= 4\pi \times (\text{radius})^2 \qquad (96)$$

Volume of a sphere

$$= \frac{4\pi}{3} \times (\text{radius})^3 \qquad (97)$$

Volume of a cylinder

$$= (\text{Area of base}) \times (\text{height}) \qquad (98)$$

C *Answers to Problems in Mathematical Appendix*

4. 26

5. $\dfrac{649}{14} = 46\dfrac{5}{14} = 46.357$

7. -22.1

8. $-\dfrac{19}{6} = -3\dfrac{1}{6} = 3.167$

9. -3.167

10. $+7.5$ erg

11. 32

12. 243

13. 72

14. $\dfrac{1}{9} = 0.111$

15. 25

16. 2

17. 64

18. 8

19. $\dfrac{1}{8} = 0.125$

20. $\dfrac{35}{13} = 2\dfrac{9}{13} = 2.692$

22. (i) 10^9 (ii) 10^3 (iii) 10^{-4} (iv) 4×10^6
    (v) 4×10^{-6} (vi) 2×10^{-3}

23. (i) 5.32×10^4 (ii) 2.165328×10^8 (iii) 2.32×10^2
    (iv) 1.3×10^{-3} (v) 2.7×10^{-7} (vi) 9.56×10^{-4}

24. (i) 3×10^4 (ii) 4×10^3 (iii) 8×10^{-3}
    (iv) 30   (v) 5×10^{-2}

25. 4.25×10^{-12}

26. 4.32×10^{28}

27. 1.69×10^2

28. 3.63×10^{-2}

29. 80

30. (i) $\alpha^3 + \alpha\beta^2 - \alpha^2\beta - \beta^3$
    (ii) $a^3 + b^3$
    (iii) $a^3 - b^3$
    (iv) $x^2y + zx^2 + xy^2 + y^2z + yz^2 + z^2x + 3xyz$

31. (i) $x = -5$
    (ii) $x = +16$
    (iii) $x = +1$
    (iv) $x = -\dfrac{13}{11} = -1.182$

32. (i) $\dfrac{bc + ca + ab}{abc}$

    (ii) $\dfrac{x^2 + y^2 + z^2}{xyz}$

    (iii) $\dfrac{(bx - ay)}{(x + a)(y + b)}$

Tables

THE GREEK ALPHABET					
Alpha	A	α	Nu	N	ν
Beta	B	β	Xi	Ξ	ξ
Gamma	Γ	γ	Omicron	O	o
Delta	Δ	δ	Pi	Π	π
Epsilon	E	ε	Rho	P	ρ
Zeta	Z	ζ	Sigma	Σ	σ
Eta	H	η	Tau	T	τ
Theta	Θ	θ, ϑ	Upsilon	Υ	υ
Iota	I	ι	Phi	Φ	ϕ, φ
Kappa	K	κ	Chi	X	χ
Lambda	Λ	λ	Psi	Ψ	ψ
Mu	M	μ	Omega	Ω	ω

The Elements

ELEMENT	SYM-BOL	ATOMIC NUMBER	MASS NUMBER[a]	ELEMENT	SYM-BOL	ATOMIC NUMBER	MASS NUMBER[a]
Actinium	Ac	89	227	Lead	Pb	82	208
Aluminum	Al	13	27	Lithium	Li	3	7
Americium	Am	95	243	Lutetium	Lu	71	175
Antimony	Sb	51	121	Magnesium	Mg	12	24
Argon	Ar	18	40	Manganese	Mn	25	55
Arsenic	As	33	75	Mendeleevium	Md	101	256
Astatine	At	85	210	Mercury	Hg	80	202
Barium	Ba	56	138	Molybdenum	Mo	42	98
Berkelium	Bk	97	247	Neodymium	Nd	60	142
Beryllium	Be	4	9	Neon	Ne	10	20
Bismuth	Bi	83	209	Neptunium	Np	93	237
Boron	B	5	11	Nickel	Ni	28	58
Bromine	Br	35	79	Niobium	Nb	41	93
Cadmium	Cd	48	114	Nitrogen	N	7	14
Calcium	Ca	20	40	Nobelium	No	102	254
Californium	Cf	98	251	Osmium	Os	76	192
Carbon	C	6	12	Oxygen	O	8	16
Cerium	Ce	58	140	Palladium	Pd	46	106
Cesium	Cs	55	133	Phosphorus	P	15	31
Chlorine	Cl	17	35	Platinum	Pt	78	195
Chromium	Cr	24	52	Plutonium	Pu	94	244
Cobalt	Co	27	59	Polonium	Po	84	209
Copper	Cu	29	63	Potassium	K	19	39
Curium	Cm	96	248	Praseodymium	Pr	59	141
Dysprosium	Dy	66	164	Promethium	Pm	61	147
Einsteinium	Es	99	254	Protactinium	Pa	91	231
Erbium	Er	68	166	Radium	Ra	88	226
Europium	Eu	63	153	Radon	Rn	86	222
Fermium	Fm	100	253	Rhenium	Re	75	187
Fluorine	F	9	19	Rhodium	Rh	45	103
Francium	Fr	87	223	Rubidium	Rb	37	85
Gadolinium	Gd	64	158	Ruthenium	Ru	44	102
Gallium	Ga	31	69	Samarium	Sm	62	152
Germanium	Ge	32	74	Scandium	Sc	21	45
Gold	Au	79	197	Selenium	Se	34	80
Hafnium	Hf	72	180	Silicon	Si	14	28
Helium	He	2	4	Silver	Ag	47	107
Holmium	Ho	67	165	Sodium	Na	11	23
Hydrogen	H	1	1	Strontium	Sr	38	88
Indium	In	49	115	Sulfur	S	16	32
Iodine	I	53	127	Tantalum	Ta	73	181
Iridium	Ir	77	193	Technetium	Tc	43	97
Iron	Fe	26	56	Tellurium	Te	52	130
Krypton	Kr	36	84	Terbium	Tb	65	159
Lanthanum	La	57	139	Thallium	Tl	81	205
Lawrencium	Lw	103	257	Thorium	Th	90	232

The Elements (continued)

ELEMENT	SYM-BOL	ATOMIC NUMBER	MASS NUMBER[a]	ELEMENT	SYM-BOL	ATOMIC NUMBER	MASS NUMBER[a]
Thulium	Tm	69	169	Xenon	Xe	54	132
Tin	Sn	50	120	Ytterbium	Yb	70	174
Titanium	Ti	22	48	Yttrium	Y	39	89
Tungsten	W	74	184	Zinc	Zn	30	64
Uranium	U	92	238	Zirconium	Zr	40	90
Vanadium	V	23	51				

[a] For stable elements the mass number refers to the most abundant isotope. For unstable elements it refers to the isotope with the longest half-life.

Vector Code

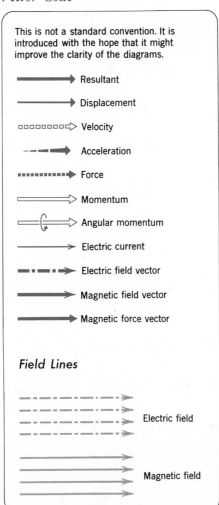

This is not a standard convention. It is introduced with the hope that it might improve the clarity of the diagrams.

Resultant

Displacement

Velocity

Acceleration

Force

Momentum

Angular momentum

Electric current

Electric field vector

Magnetic field vector

Magnetic force vector

Field Lines

Electric field

Magnetic field

Tables

Periodic Table of the Elements

	GROUP I	GROUP II					TRANSITION ELEMENTS						GROUP III	GROUP IV	GROUP V	GROUP VI	GROUP VII	GROUP O
Period 1	H 1																	He 2
Period 2	Li 3	Be 4											B 5	C 6	N 7	O 8	F 9	Ne 10
Period 3	Na 11	Mg 12											Al 13	Si 14	P 15	S 16	Cl 17	Ar 18
Period 4	K 19	Ca 20	Sc 21	Ti 22	V 23	Cr 24	Mn 25	Fe 26	Co 27	Ni 28	Cu 29	Zn 30	Ga 31	Ge 32	As 33	Se 34	Br 35	Kr 36
Period 5	Rb 37	Sr 38	Y 39	Zr 40	Nb 41	Mo 42	Tc 43	Ru 44	Rh 45	Pd 46	Ag 47	Cd 48	In 49	Sn 50	Sb 51	Te 52	I 53	Xe 54
Period 6	Cs 55	Ba 56	* 57–71	Hf 72	Ta 73	W 74	Re 75	Os 76	Ir 77	Pt 78	Au 79	Hg 80	Tl 81	Pb 82	Bi 83	Po 84	At 85	Rn 86
Period 7	Fr 87	Ra 88	† 89–?															

*Rare earth metals	La 57	Ce 58	Pr 59	Nd 60	Pm 61	Sm 62	Eu 63	Gd 64	Tb 65	Dy 66	Ho 67	Er 68	Tm 69	Yb 70	Lu 71
†Actinide metals	Ac 89	Th 90	Pa 91	U 92	Np 93	Pu 94	Am 95	Cm 96	Bk 97	Cf 98	Es 99	Fm 100	Md 101	No 102	Lw 103

GROUP VII: ?

The Metastable Fundamental Particles

NAME		PARTICLES CHARGE			ANTIPARTICLES CHARGE			REST MASS (UNITS SUCH THAT REST MASS OF ELECTRON = 1)	SPIN. UNITS OF $\frac{h}{2\pi}$
		+e	0	−e	+e	0	−e		
Photon	photon		γ			γ		0	1 (Boson)
Leptons	neutrino, antineutrino		ν			$\bar{\nu}$		0	1/2 (Fermions)
	μ-neutrino anti-μ-neutrino		ν_μ			$\bar{\nu}_\mu$		0	
	electron, positron			e^-	e^+			1	
	muons			μ^-	μ^+			207	
Mesons	π-mesons (pions)	π^+	π^0			π^0	π^-	264 273	0 (Bosons)
	K-mesons (Kaons)	K^+	K^0			$\bar{K}^0$	K^-	966 974	
	η-meson (eta)		η^0			η^0		1074	
Baryons	proton, antiproton neutron, antineutron	p	n			$\bar{n}$	$\bar{p}$	1836.1 1838.6	1/2 (Fermions)
	Lambda		Λ^0			$\bar{\Lambda}^0$		2183	
	Sigma	Σ^+	Σ^0	Σ^-	$\bar{\Sigma}^+$	$\bar{\Sigma}^0$	$\bar{\Sigma}^-$	2328 2334 2343	
	Xi		Ξ^0	Ξ^-	$\bar{\Xi}^+$	$\bar{\Xi}^0$		2573 2586	
	Omega			Ω^-	$\bar{\Omega}^+$			3276	3/2

Answers to Odd-Numbered Problems

Chapter 2

1. 140 new mph.   **3.** (a) 8.05×10^3 m. (b) 5×10^{-2} m. (c) 30 m. (d) 5 kg.
(e) 5×10^{-3} kg. (f) 4.74×10^7 sec. (g) 3.6×10^2 sec. (h) 4.47×10^2 m/sec.
(i) 1.39 m/sec. (j) 2×10^{-4} m/sec. (k) 15 m/sec.   **7.** 3.36×10^3 cm/sec^2.
9. 72 mph.   **11.** (a) 760 days. (b) 820 years. (c) 330 centuries. (d) 220 million
years.   **13.** 2.96×10^3 cm/sec.   **15.** 0.85 miles, 30.2° south of west.
17. 9.2×10^{24} cm/sec$^2 \approx 10^{22}$ g.   **19.** Greater than 2×10^4 cm.   **21.** 40 mph,
0 mph.   **23.** (a) 90 m. (b) 80.9 m. (c) 0 m.   **25.** 5.54 sec, 54.3 m/sec.
27. 63 m/sec.

Chapter 3

1. 16 dynes.   **3.** 19.6 dynes.   **5.** 1.67×10^{-10} cm/sec^2.   **7.** 1.6×10^2 cm/sec^2.
9. 9.0×10^8 cm.   **11.** 6.7×10^{35} gm.   **13.** 3.7×10^{27} dynes.
15. 9.6×10^4 dynes.   **17.** 1.83×10^2 cm/sec.   **19.** 1.41 days.
21. 8.2×10^2 cm/sec^2.

Chapter 4

1. 5.01×10^{-18} gm cm/sec.   3. 0.29 cm/sec.   5. 20 cm/sec.
7. -5.84×10^{44} ergs.   9. 21 cm/sec.   11. 96 cm/sec.   13. 44.6 gm cm²/sec.
15. 5.88×10^8 ergs.   17. $v = \sqrt{2gh}$ 19. 1.88×10^5 gm cm/sec.
21. 4.9×10^{15} ergs/sec, 4.9 million bulbs, a large city.
23. 2.4×10^5 cm/sec = 5,300 mph.   25. 6.7×10^8 cm.

Chapter 5

3. 1.87×10^3 dyne/cm².   5. 2.4×10^5 dyne/cm².   7. $T_2 = \dfrac{p_2 V_2}{p_1 V_1} T_1$

9. 1.38×10^{-15} dyne/cm².   13. 59.4 miles.   15. 3.77×10^{-16} ergs/deg N,
136.61°N.   17. 7.4×10^{79} light years.

Chapter 6

1. 8.75 dynes, $+75.5$ ergs.   3. 100 dynes.   5. 2.87×10^{17} cm/sec².
7. -4.8×10^{-7} ergs.   9. (a) 5.5×10^{21} cm/sec². (b) 1.01×10^{25} cm/sec².
11. 5.6×10^{-3} dynes/statcoulomb.   13. 2.08×10^4 electrons/sec.
15. 2.56×10^5 volts.   17. 6.95×10^8 cm/sec, 1.55×10^7 mph.
19. 3 dynes/statcoulomb.   21. 87 cm.   23. 1.7×10^{23} statcoulombs.
25. 5.3×10^{25} cm/sec.   27. 3.12×10^{15} eV.

Chapter 7

1. 9.2×10^{-8} gauss.   3. 2.4 gauss, horizontal, toward the east.
5. 1.2×10^{-5} dyne, horizontal, toward the east.   7. 8.33 dyne, horizontal,
toward the west.   9. 3.47×10^2 gauss, horizontal, toward the south-east.
11. 1.04 cm.   13. 33.3 gauss.   15. (a) 5.3×10^8 cm/sec.

(b) 4.8×10^{-19} gm cm/sec. (c) 79 eV.   17. $T = \dfrac{2\pi mc}{eB}$.

Chapter 8

1. 4×10^{-4} sec.   3. 4.8×10^{-5} cm.   5. (a) 10^{-9} sec. (b) 5×10^{-7} sec.
(c) 5×10^{-5} sec. (d) 1.5×10^{-2} sec. (e) 0.13 sec. (f) 497 sec = 8 min 17 sec.
7. 2.12×10^3 cm—a house; 1.7 cm—a thumb nail.   9. 3.3×10^{-5} dyne/cm².
11. 2.46×10^{-2} cm.   13. 0.183 cm.   15. 7.9×10^2 atmospheres.

Chapter 9

1. 61.3.   3. 2.09×10^{-24} gm.   5. 1.11×10^{-21} gm.   7. (a) 1.72 sec.
(b) 1.35 sec. (c) 1.53 sec.   9. 5.2×10^{-6} sec.   11. 6.7 gm.
13. 2.62×10^{10} cm/sec.   15. \$450,000.   17. 0.9999998 c.   19. $c/\sqrt{2} =$
2.12×10^{10} cm/sec.   21. $1/\sqrt{1 - V_r^2/c^2}$ meter. B maintains that the distance
between the two brushes is $\sqrt{1 - V_r^2/c^2}$ meter, but that they are not parallel to
one another.   23. 2.05×10^{-17} gm cm/sec.   25. 1.4×10^{-13} cm.
27. 4×10^{13} gm, 0.11 mile.

Chapter 10

1. 1.2×10^5 cycles per sec. **3.** The rates of the two clocks differ by about 1.55 parts in 10^{39}. **5.** (b) is twice (a). **7.** The energy requirement is equivalent to 450 years of the present power output.

Chapter 11

1. 3.02×10^{12} cycles per sec. **3.** 1.99×10^{-16} cm. **5.** 4.15×10^5 volts. **7.** 2.21×10^{-6} cm. **9.** 6.63×10^{-10} erg $= 4.15 \times 10^2$ eV. **11.** 1.24 eV. **13.** 3.28×10^{-33} gm. **15.** 5.11×10^5 volts. **17.** 1.9×10^{-32} cm. **19.** 5.5×10^{-8} cm. **21.** 1.6×10^{14} cycles per sec. **23.** 3.2×10^{-10} cm. **25.** 1.006×10^{20} cycles per sec. **27.** About 10^{-32} cm.

Chapter 12

1. 3.8×10^{-3} cm. **3.** 2.7×10^{-20} erg. **5.** 100 cm/sec. **7.** 7.8×10^{-2} cm. **9.** Within about 10^{-10} sec. **11.** 3.29×10^{15} cycles per sec. **13.** About 7×10^{-28} gm. The velocity is very nearly equal to c and its uncertainty has a very small value of about 10^{-10} c. **15.** 2.18×10^8 cm/sec, 1.000026. **17.** $105,000°K$.

Chapter 13

1. 0.0215 atomic mass unit $= 3.6 \times 10^{-26}$ gm. **3.** 239.0521 atomic mass units. **5.** About 10^{10} cm/sec. **7.** 7.44×10^{20} cycles per sec. **9.** 2.6×10^7 cm/sec. **11.** About 9 gm. **13.** 3.7×10^{12} °K, No.

Chapter 14

1. 6.6×10^{-14} cm. **3.** 8.2×10^{-11} cm. **5.** 1.19×10^{-2} erg $= 7.46 \times 10^9$ eV. **7.** (a) $\eta^0 \longrightarrow 4\pi^0$ is possible, $\eta^0 \longrightarrow \pi^+ + \pi^0 + \pi^0 + \pi^-$ is borderline. (b) No. **9.** 7.07 electron rest masses $= 6.45 \times 10^{-27}$ gm. **11.** 1.26×10^{-6} erg $= 0.79$ MeV. **13.** 26.7 MeV.

Answers to
Odd-Numbered
Problems

Index

Index

366

Index

369

Atomic Masses of Some Important Isotopes

1 atomic mass unit $= 1.66 \times 10^{-24}$ gm
$= 931$ MeV

Except in the case of the neutron and the proton, the mass given refers to the neutral atom.

NAME	SYMBOL	ATOMIC MASS
Carbon 12	$_6C^{12}$	12.000000
neutron	n	1.008665
proton	p	1.007272
Hydrogen	$_1H^1$	1.007825
Deuterium	$_1H^2$ or $_1D^2$	2.014102
Tritium	$_1H^3$ or $_1T^3$	3.016049
Helium 3	$_2He^3$	3.016029
Helium 4	$_2He^4$	4.002603
Nitrogen 14	$_7N^{14}$	14.003074
Oxygen 16	$_8O^{16}$	15.994915
Uranium 235	$_{92}U^{235}$	235.0439
Uranium 238	$_{92}U^{238}$	238.0508

IMPORTANT NOTE:

For practical reasons it is convenient to take the unit of atomic mass to be $\frac{1}{12}$ of the mass of a $_6C^{12}$ atom. This is approximately, but not exactly, equal to the mass of a hydrogen atom, $_1H^1$.